Foundations of Rational Agency

APPLIED LOGIC SERIES

VOLUME 14

SCOPE OF THE SERIES

Logic is applied in an increasingly wide variety of disciplines, from the traditional subjects of philosophy and mathematics to the more recent disciplines of cognitive science, computer science, artificial intelligence, and linguistics, leading to new vigor in this ancient subject. Kluwer, through its Applied Logic Series, seeks to provide a home for outstanding books and research monographs in applied logic, and in doing so demonstrates the underlying unity and applicability of logic.

The titles published in this series are listed at the end of this volume.

Foundations of Rational Agency

edited by

MICHAEL WOOLDRIDGE

Queen Mary and Westfield College,
London, United Kingdom

and

ANAND RAO

Mitchell Madison Group,
Melbourne, Australia

KLUWER ACADEMIC PUBLISHERS

DORDRECHT / BOSTON / LONDON

A C.I.P. Catalogue record for this book is available from the Library of Congress.

ISBN 0-7923-5601-2

Published by Kluwer Academic Publishers,
P.O. Box 17, 3300 AA Dordrecht, The Netherlands.

Sold and distributed in North, Central and South America
by Kluwer Academic Publishers,
101 Philip Drive, Norwell, MA 02061, U.S.A.

In all other countries, sold and distributed
by Kluwer Academic Publishers,
P.O. Box 322, 3300 AH Dordrecht, The Netherlands.

Logo design by L. Rivlin

Printed on acid-free paper

Printed in the Netherlands.

EDITORIAL PREFACE

The Editors are very pleased to have the present collection on agents. Agent theory is an important field at the crossroads of philosophy, artificial intelligence, computer science, practical reasoning and logic.

All of the above disciplies are trying to model and understand human agents, each for their own historical reasons, some dating back to Plato and Aristotle.

This area is, therefore, a meeting ground for future cooperation between all these communities. In fact, the state of the research and application is such that cooperation is necessary, in my opinion, for the successful future development of each strand.

Let us hope that the publication of this book, through the agency of our series, will serve to advance cooperation between all agent resarch communities.

The Editors

CONTENTS

A. RAO AND M. WOOLDRIDGE

FOUNDATIONS OF RATIONAL AGENCY

1 AGENTS AND RATIONAL AGENTS

The word *agent* is widely used in computer science these days. It can mean anything from a few lines of code that gets executed automatically when a certain condition is satisfied (e.g., a daemon that gets scheduled by the Operating System every day at midnight), to a more sophisticated program that can reason about its own behaviour and achieve long-term goals (e.g., an autonomous intelligent vehicle on Mars) [13]. In addition to the simplicity or sophistication of these software entities, agents are also characterised into different types based on their functionality. Interface agents, mobile agents, believable agents, reactive agents, learning agents, and emergent agents are just a few of these types. The level of autonomy (from being completely autonomous to being intelligent assistants to humans) and the level of their interaction with physical entities (from being totally in the software environment — softbots — to being intimately connected with their physical environment — robots) are also used to characterise agents.

The aim of this book is not to cover this entire spectrum of agents, but rather to examine one particular type of agents — the so called *rational agents* [19]. Rationality as a concept has been around for a long time even before the advent of computers. Rational agents are software entities that perceive their physical or software environment through appropriate sensors; have a model and can reason about the environment that they inhabit; and based on their own mental state take actions that change their environment. The key aspects of rationality include:

- balancing reactive and pro-active behaviour;
- balancing perception, deliberation, and action, especially when there are limited resources;
- balancing self-interest and community-interest.

It is these types of agents that will be of primary interest to us. Such rational agents have been investigated in a number of application domains, including intelligent information retrieval, modeling the tactics of groups of pilots in combat scenarios, aiding air-traffic controllers in the optimal sequencing and monitoring of aircraft, and diagnosis of faults in the Space Shuttle [13, 11]. Although, we will not be looking at these applications in detail, the foundations and theories discussed in this book form an integral part of developing such systems.

M. Wooldridge and A. Rao (eds.), Foundations of Rational Agency, 1–10.

2 HISTORY

Like most sub-fields of Artificial Intelligence (AI), the study of rational agents is a truly inter-disciplinary area. One can trace the roots of rational agents to philosophy, computer science, economics, sociology, and psychology. We explore some of these fields and their contributions to the understanding of rationality and the science of building rational agents below.

2.1 Practical Reasoning

Practical reasoning is a branch of philosophy that deals with how a rational agent decides to act based on its internal *mental state*. Some of the major concerns of this branch of philosophy include: What are the primitive mental attitudes? How do these mental attitudes translate into actions by an agent? What are the processes involved in the cycle of perception-thinking-action? How do these attitudes and processes change under limited resources?

Arguably, the work from this branch of philosophy that has most influenced the notion of building rational agents is the seminal work of Bratman [3]. In his 1987 book *Intentions, Plans, and Practical Reason*, he argues for the primacy of *intentions* in human practical reasoning, and also argues convincingly that intentions are irreducible to other mental states — in particular, desires. He thereby lays the groundwork for the attitudes of beliefs, desires, and intentions, as three separate and irreducible mental states. Beliefs capture the informational state of an agent. Unlike desires and intentions, beliefs do not play any motivational role within an agent. Desires and intentions both play a motivational role and are referred to as *pro-attitudes*. However, while desires merely influence the actions of an agent, intentions control the actions that are taken by an agent. Hence, desires are referred to as *conduct-influencing* pro-attitudes and intentions as *conduct-controlling* pro-attitudes.

Bratman's notion of intention is closely tied with the notion of *plans*. Intentions are the building blocks of larger plans. Plans are used in two contexts — as abstract structures or recipes, and as mental states that are intertwined with one's beliefs and desires. It is in the latter sense that intentions are part of the adopted or committed plans of an agent. Bratman's contribution to rational agency extends from the static analysis of these mental attitudes and their interrelationships to the processes of means-end-reasoning, deliberation, and reconsideration that continuously shape and modify these attitudes [4].

While the three mental attitudes of beliefs, desires, and intentions may be a good starting point, and arguably necessary mental attitudes for the study of rational agents, it is by no means a sufficient or complete analysis of mental attitudes of rational agents. A number of researchers have examined modeling other related attitudes, such as obligation, commitment, capability, wishes, wants, power, influence, and countless others [19]. Mental attitudes are not restricted only to single agents, but can also be ascribed to groups or "societies" of agents. Mutual belief, common knowledge, joint intention, joint goal, and social commitment are just some of the multi-agent mental attitudes that have been studied (see e.g., [14]).

Rational agents are not solitary entities living in an uninhabited static environment; they are embedded in a continuously changing environment and have to constantly interact with other agents. Hence, communication with other agents and interaction with the environment are key concerns within this field. The theory of *speech acts* and its extensions, as well as philosophical theories of action, time, and change have greatly influenced the study of rational agents (see, e.g., [6]).

2.2 *Philosophical Logic*

While philosophical theories are based on arguments, refutations, and informal claims, philosophical logic — a synthesis of philosophy and mathematical logic — takes these claims to the next level in formalizing the assumptions, intuitions, and claims as discussed in the philosophical theories. A variety of logics based on variants of classical mathematical logics and modal logics have been used as the basis of formalizing the mental attitudes of agents and their interaction with other agents and the environment.

Modal and temporal logics [7] have been extensively used to capture the mental attitudes discussed earlier and analyze their properties. The *possible-worlds semantics* of such logics provide an useful abstraction with which to analyze the behaviour of rational agents. Action and program logics [12] have been used to capture the connection between mental attitudes and action have also influenced the operational semantics of rational agents.

2.3 *Decision Theory and Game Theory*

While philosophical logic is concerned with capturing the relationships between different attitudes, decision theory and game theory — branches of mathematics and economics — provide a quantitative account of rationality. While decision theory considered individual decision-making under uncertainty and individual preference, game theory parameterised decision-making with respect to the decisions of other agents in the environment [1].

2.4 *Artificial Intelligence*

Philosophical logic and classical decision theory contribute a great deal to the understanding of the general principles of rational agency. However, they invariably consider *idealized* conditions in which *computational* considerations are not an issue. Artificial Intelligence is, as much as anything, about *building* rational agents [18], and as such it is often the test bed for experimenting with new philosophical, psychological, and sociological theories.

Bratman's philosophical theory of intentions and the role of plans in practical reasoning came at a time when AI researchers in planning were getting disillusioned with the state-of-the art classical planning techniques (see, e.g., [9]). Search based classical planning techniques, which worked in toy problems, did not scale well to real-world problems. In particular, the lack of explicit modeling of other agents and their actions, the failure to account for a continuously changing environment, and the limited resources available to real computational agents, were some of the more serious drawbacks. As a result, Bratman's theory of intentions, which explicitly addressed at least two of these three drawbacks, was embraced warmly by a group of planning researchers within AI. Planning that used pre-compiled plans or recipes rather than doing first-principles classical planning came to be known as *reactive planning* [10].

These developments led to one of the early implementations of rational agents — the Procedural Reasoning System (PRS) [10]. This system explicitly modeled the attitudes of beliefs, goals, and intentions. The system was programmed by specifying the plans that would achieve certain goals, under certain belief conditions. The agent dynamically adopted these hierarchical plans based on its mental state (i.e., its current beliefs, goals, and intentions).

Around the same time, a number of researchers in AI planning also started addressing the drawbacks of classical planning using decision theory. A number of decision-theoretic and utilitarian reasoning mechanisms were developed for planning with limited resources in uncertain environments [2].

Both these streams influenced the architectures that emerged for rational agents. The above architectures and techniques were also enhanced to account for communication, coordination, synchronization, and joint decision-making with other rational agents.

2.5 Software Engineering

As ideas mature from philosophy to implemented systems it gains widespread currency amongst a larger group of practitioners requiring easy-to-use tools, efficient languages, and well-defined analysis and design methodologies. This is when the field makes a transition from AI to a software engineering discipline. Over the past couple of years the field of rational agents is slowly making this transition. This can be seen by the increasing emphasis on architectures, language constructs, reliability, and usability concerns within the rational agent community.

3 RATIONAL AGENTS: CURRENT RESEARCH PARADIGMS

3.1 Foundations

Rationality and Intelligence by Stuart Russell sets the foundation for this collection by examining the very definition of *rational agency*. He proposes four formal definitions of rationality and discusses the merits of each definition. The last of these four formal definitions — *bounded optimality* — grounds the notion of rationality with respect to agent programs that can run on abstract machines. The model takes an external designer's perspective on rationality. Russell discusses the research directions for this emerging area of *bounded optimality* and how it can bridge the gap between theory and practice in agent systems.

While Russell was primarily concerned with the very notion of rational agency, the paper by Aaron Sloman examines how to realize such rational agents. Sloman addresses the concept of *architecture* — different types of architectures, ways of evaluating such architectures, and how architectures can be made to exhibit specific behaviours. In particular, Sloman introduces the notions of *niche space* or the space of possible sets of requirements and *design space* or the space of possible designs, and mappings between them. Within this framework he examines a particular architecture consisting of *reactive*, *deliberative*, and *meta-management* subsystems.

Sloman's meta-management subsystem is similar in spirit to Russell's meta-level rationality. The reactive and deliberative subsystems capture the object-level rationality with respect to the observations, actions and computations performed by agents. In summary, the two papers by Russell and Sloman lay the foundation for examining theories of rational agency as well as architectures (and languages) to realize rationality amongst agents.

The paper by Pollock on *Planning Agents* addresses the philosophical, logical, and pragmatic problems of building agent systems. A distinction is made between *practical cognition* — the ability to make and execute plans, from *epistemic cognition* — the knowledge required to make and execute plans. The paper traces the history of planning in AI and describes the OSCAR architecture that combines the two aspects of cognition using a defeasible reasoner. The solution to the frame problem — one of the central issues in planning — is addressed within the OSCAR architecture.

3.2 Symbolic Micro-Theories of Rational Agency

The seminal work in the symbolic theories of rational agency is the formalization of intention by Cohen and Levesque [5]. They introduced a temporal multi-modal logic that captured some of the essential characteristics of intention and action as described by Bratman. This was followed by a number of papers by Rao and Georgeff detailing the interactions between the three attitudes of beliefs, desires, and intentions, and exploring various aspects of Bratman's theory [17]. This led to an active group of researchers working on the foundational aspects of what came to be known as BDI agents and various extensions to the three primary attitudes.

The paper by van der Hoek *et. al.* provides a comprehensive coverage of symbolic modal logic based approaches to formalizing rational agency. The theory integrates informational and motivational attitudes, and also considers the static and dynamic aspects of these attitudes. It uses dynamic logic, suitably extended, to capture an expressive theory of action and capability. On top of this action theory is a rich variety of informational attitudes comprising *default*, *communicated*, *observational beliefs*, and *knowledge*. The authors then look at the dynamics of how these attitudes are created and can change over time by integrating these attitudes with the well known theories of belief revision [8]. This leads to some interesting characterisations of belief changes, such as, *seeing is believing*, *hearing is believing*, and *jumping to conclusions is believing*. With respect to motivational attitudes the authors once again present a comprehensive set of attitudes, such as *wishes*, *goals*, *commitments*, and *intentions*.

In his paper *Know-How*, Singh gives an introduction and overview to the ways that know-how has been formalised in theories of agency. He focusses in particular on his own formalism, cast in a branching-time temporal framework. He contrasts this to other related formalisms, such as Moore's well-known theory of ability, and the various philosophical theories of ability.

The papers in this section demonstrate the power of formal analysis in understanding folk-psychological notions and making them precise. This precision results in a shared understanding of some of the abstract concepts of rational agency and helps in developing more practical systems.

3.3 Theories of Interaction

The primary concern of the papers in the previous section was to formalize the notion of rational agency as seen from within a *single* agent. Although the agent was capable of performing actions and communicating with other agents, the communication or more generally the interaction with other agents was not a major concern. The next set of papers address these issues by looking at the interaction between agents. The first three papers describe symbolic approaches to interaction, while the last one takes a quantitative approach to rational interaction amongst agents.

Grosz and Kraus provide a detailed account of collaborative behaviour in their paper titled *The Evolution of Shared Plans*. The approach extends some of the single-agent notions introduced in the earlier section to modeling the collaborative activity of multiple agents. Central to their work is the notion of *shared plans*: a representation of the joint intention of a group of agents. They formalize this notion and provide an architecture for collaborative agent design. This paper sets the foundation for analyzing the interactions between agents — the primary focus of the next two papers.

In his paper entitled *Speech Acts for Dialogue Agents*, David Traum provides an excellent introduction to the field of speech acts. Starting from the early philosophical work of Austin and Searle he moves on to the impetus provided by early AI research into speech acts by Allen, Cohen, and Perrault. What clearly emerges from this early work is the close relationship between theories of (individual) rational agency as discussed in the earlier section and the role of speech acts in changing the mental attitudes of agents involved in an interaction. Traum then discusses some of the more recent dynamic logical and non-monotonic approaches to speech acts, and the extensions of speech acts to cover *dialogues* (combination of speech acts used to achieve a certain purpose). Using a multi-agent logic, in the tradition of the work discussed in the previous section (c.f. Kraus and Grosz), Traum extends the notion of speech acts and dialogues. In particular, he uses personal commitment and the notion of *obligations* to overcome the stringent mutual-belief requirements for joint activity.

In her paper on *Communication among Rational Agents*, Afsaneh Haddadi addresses a similar problem by defining three different levels of commitments between agents — *potential for cooperation*, *pre-commitment*, and *commitment* of agents to one another. However, unlike Traum these notions are defined purely in terms of the beliefs, desires, intentions, and actions of single agents, i.e., built on top of single-agent mental attitudes. Also, a number of primitive communication acts can be defined in this manner, and primitive communication acts can be combined to form communication plans or dialogues.

Both these papers demonstrate the close link between theories of individual and joint rational agency and interaction between multiple agents. Also, this line of research has led to an active interest in artificial communication languages, such as KQML, ACL, etc [15].

In contrast to the symbolic approaches of Traum and Haddadi, Piotr Gmytrasiewicz takes a quantitative decision-theoretic approach to rational interaction between multiple agents. In particular, he provides a framework for capturing the agent's knowledge about the environment and of other agents in the domain. The framework lends itself to recursive modeling, allowing agents to express and make rational choices of actions based on its model of the other agents' actions. Piotr gives a decision-theoretic characterisation of the speech acts "inform" and "promise" using his modeling and intentional messages. This paper is a vivid illustration of how quantitative and qualitative theories of interaction can support each other.

3.4 Architectures and Languages

The final section of the book moves from philosophical and formal concerns to more pragmatic concerns. The issue of how one goes about implementing such agents is considered. However, rather than merely describing implemented systems, this section consists of a collection of papers that integrate the philosophical and formal foundations of the earlier sections with architectures and programming languages that reflect these concerns.

The paper by Rebecca Thomas, *A Survey of Agent-Oriented Programming*, surveys a class of programming languages that exploit the micro-theories of rational agency. Starting from AGENT0, a simple agent-oriented language developed by Yoav Shoham, the paper goes on to describe the language PLACA (Planning Communicating Agents), which allows agent programs to be written that can plan and achieve their goals.

The paper by Lespérance, Levesque, and Reiter on *A Situation Calculus Approach to Modeling and Programming Agents* introduces a language called Golog and its concurrent version, ConGolog. Golog is a logic programming language that reasons about actions. The close relationship between building agents and the enterprise of planning (cf. Pollock) is once again explored by Lesperance *et al.* One of the more interesting aspects of Golog is its strong link to Situation Calculus and the way it addresses the frame problem [16].

The papers in this section give the reader a glimpse of what is to come in the future from the community of agent researchers, namely tools and languages that help program and model complex behaviours at a richer and more abstract level than is permitted by existing software tools and paradigms.

4 RATIONAL AGENTS: FUTURE DIRECTIONS

The field of rational agents is very much in its infancy. This collection brings together some of the foundational work in this area. It does not cover some of the agent-based systems that have been built to capture the essence of rationality [19]. Not included in this collection are also a growing number of applications that these systems are being tested against — ranging from sophisticated military, telecommunications, aviation, electricity distribution, and robotic applications [13].

However, except for a couple of exceptions (the paper by Lesperance and colleagues being one of them) the connection between the foundational work on rational agency and their implementations are tenuous. There seems to be a great deal of confusion regarding the role of theory and how to build multi-agent systems that exhibit the properties of rationality discussed in the various theories. There are some who even question whether the theories have anything useful to say about practical systems. Leaving aside the controversial issue of whether a formal description of rationality is useful or not, there seems to be two ways in which theories of rationality can be used to understand the building of rational systems:

- To use the formal language as not only a *description* of the agents, their environment and how they interact with each other, but also as a formal specification for the *execution* of such systems.
- To use the formal language for *describing* and *verifying* a set of abstract behaviours that are generated by an implemented system and not use it as a language for execution.

Lack of a clear distinction between the two has led to criticisms of both approaches. Pursuing the former aim leaves one open to the criticism that the language is not expressive enough to capture a rich set of behaviours, while pursuing the latter leaves one open to the criticism that the all powerful and expressive language is not implementable. While it is too early for the community to focus on one in preference to the other, the issues to be addressed with respect to the two aims are quite different.

Research aimed at the first goal should aim to provide a declarative and formal specification language, an operational semantics as to how such a specification can be executed, and explore the limits of the classes of applications or more precisely the types of behaviours that can be modeled using their specification language. Obviously, the more expressive the language and more complex the behaviours that can be captured in this way the better is the formalism.

On the other hand, research aimed at the second goal should start from an implemented system or language, provide an abstraction of the system in terms of an operational semantics, provide a representation language that captures important properties of the system, and provide some mechanism (e.g., model-checking or theorem proving) for verifying the behaviour of such an implemented system. The variants of temporal logic and their use in verifying safety and liveness properties of concurrent programs fall under this category [7, 17].

As the rational agents field matures, one would hope that researchers pursue both these goals and develop tools and techniques that are both theoretically sound and useful in practice. We hope that this collection will facilitate a move in this direction.

5 ACKNOWLEDGEMENTS

We would like to extend our thanks to all the authors in this volume for their care in producing first-rate manuscripts, and their patience while the volume was in production. In addition, we would like to thank Stuart Russell, Danny Bobrow, and *Artificial Intelligence* journal for their permission to reprint Stuart's article, which originally appeared in *Artificial Intelligence* volume 94, July 1997.

Anand S. Rao (Mitchell Madison Group, Australia)
Michael Wooldridge (University of London, UK)

REFERENCES

[1] K. Binmore. *Fun and Games: A Text on Game Theory*. D. C. Heath and Company: Lexington, MA, 1992.

[2] Mark S. Boddy and Thomas Dean. Deliberation scheduling for problem solving in time-constrained environments. *Artificial Intelligence*, 67(2):245–285, 1994.

[3] M. E. Bratman. *Intentions, Plans, and Practical Reason*. Harvard University Press: Cambridge, MA, 1987.

[4] M. E. Bratman, D. J. Israel, and M. E. Pollack. Plans and resource-bounded practical reasoning. *Computational Intelligence*, 4:349–355, 1988.

[5] P. R. Cohen and H. J. Levesque. Intention is choice with commitment. *Artificial Intelligence*, 42:213–261, 1990.

[6] P. R. Cohen and H. J. Levesque. Rational interaction as the basis for communication. In P. R. Cohen, J. Morgan, and M. E. Pollack, editors, *Intentions in Communication*, pages 221–256. The MIT Press: Cambridge, MA, 1990.

[7] E. A. Emerson. Temporal and modal logic. In J. van Leeuwen, editor, *Handbook of Theoretical Computer Science*, pages 996–1072. Elsevier Science Publishers B.V.: Amsterdam, The Netherlands, 1990.

[8] P. Gärdenfors. *Knowledge in Flux*. The MIT Press: Cambridge, MA, 1988.
[9] M. P. Georgeff. Planning. *Annual Review of Computer Science*, 2:359–400, 1987.
[10] M. P. Georgeff and A. L. Lansky. Reactive reasoning and planning. In *Proceedings of the Sixth National Conference on Artificial Intelligence (AAAI-87)*, pages 677–682, Seattle, WA, 1987.
[11] M. P. Georgeff and A. S. Rao. A profile of the Australian AI Institute. *IEEE Expert*, 11(6):89–92, December 1996.
[12] D. Harel. Dynamic logic. In D. Gabbay and F. Guenther, editors, *Handbook of Philosophical Logic Volume II — Extensions of Classical Logic*, pages 497–604. D. Reidel Publishing Company: Dordrecht, The Netherlands, 1984. (Synthese library Volume 164).
[13] N. R. Jennings, K. Sycara, and M. Wooldridge. A roadmap of agent research and development. *Autonomous Agents and Multi-Agent Systems*, 1(1):7–38, 1998.
[14] H. J. Levesque, P. R. Cohen, and J. H. T. Nunes. On acting together. In *Proceedings of the Eighth National Conference on Artificial Intelligence (AAAI-90)*, pages 94–99, Boston, MA, 1990.
[15] J. Mayfield, Y. Labrou, and T. Finin. Evaluating KQML as an agent communication language. In M. Wooldridge, J. P. Müller, and M. Tambe, editors, *Intelligent Agents II (LNAI Volume 1037)*, pages 347–360. Springer-Verlag: Berlin, Germany, 1996.
[16] J. McCarthy and P. J. Hayes. Some philosophical problems from the standpoint of artificial intelligence. In B. Meltzer and D. Michie, editors, *Machine Intelligence 4*. Edinburgh University Press, 1969.
[17] A. S. Rao and M. Georgeff. Decision procedures of BDI logics. *Journal of Logic and Computation*, 8(3):293–344, 1998.
[18] S. Russell and P. Norvig. *Artificial Intelligence: A Modern Approach*. Prentice-Hall, 1995.
[19] M. Wooldridge and N. R. Jennings. Intelligent agents: Theory and practice. *The Knowledge Engineering Review*, 10(2):115–152, 1995.

S. RUSSELL

RATIONALITY AND INTELLIGENCE

ABSTRACT: The long-term goal of our field is the creation and understanding of intelligence. Productive research in AI, both practical and theoretical, benefits from a notion of intelligence that is precise enough to allow the cumulative development of robust systems and general results. The concept of *rational agency* has long been considered a leading candidate to fulfill this role. This paper outlines a gradual evolution in the formal conception of rationality that brings it closer to our informal conception of intelligence and simultaneously reduces the gap between theory and practice. Some directions for future research are indicated.

1 ARTIFICIAL INTELLIGENCE

AI is a field whose ultimate goal has often been somewhat ill-defined and subject to dispute. Some researchers aim to emulate human cognition, others aim at the creation of intelligence without concern for human characteristics, and still others aim to create useful artifacts without concern for abstract notions of intelligence.

This variety is not *necessarily* a bad thing, since each approach uncovers new ideas and provides fertilization to the others. But one can argue that, since philosophers abhor a definitional vacuum, many of the damaging and ill-informed debates about the feasibility of AI have been about definitions of AI to which we as AI researchers do not subscribe.

My own motivation for studying AI is to create and understand intelligence as a general property of systems, rather than as a specific attribute of humans. I believe this to be an appropriate goal for the field as a whole, and it certainly includes the creation of useful artifacts—both as a spin-off and as a focus and driving force for technological development. The difficulty with this "creation of intelligence" view, however, is that it presupposes that we have some productive notion of what intelligence is. Cognitive scientists can say "Look, my model correctly predicted this experimental observation of human cognition," and artifact developers can say "Look, my system is saving lives/megabucks," but few of us are happy with papers saying "Look, my system is intelligent." This difficulty is compounded further by the need for theoretical scaffolding to allow us to design complex systems with confidence and to build on the results of others. "Intelligent" must be given a definition that can be related directly to the system's input, structure, and output. Such a definition must also be *general*. Otherwise, AI subsides into a smorgasbord of fields—intelligence as chess playing, intelligence as vehicle control, intelligence as medical diagnosis.

In this paper, I shall outline the development of such definitions over the history of AI and related disciplines. I shall examine each definition as a predicate P that

M. Wooldridge and A. Rao (eds.), Foundations of Rational Agency, 11–33.

can be applied, supposedly, to characterize systems that are intelligent. For each P, I shall discuss whether the statement "Look, my system is P" is interesting and at least sometimes true, and the sort of research and technological development to which the study of P-systems leads.

I shall begin with the idea that intelligence is strongly related to the capacity for successful behaviour—the so-called "agent-based" view of AI. The candidates for formal definitions of intelligence are as follows:

- P_1: *Perfect rationality*, or the capacity to generate maximally successful behaviour given the available information.
- P_2: *Calculative rationality*, or the in-principle capacity to compute the perfectly rational decision given the initially available information.
- P_3: *Metalevel rationality*, or the capacity to select the optimal combination of computation-sequence-plus-action, under the constraint that the action must be selected by the computation.
- P_4: *Bounded optimality*, or the capacity to generate maximally successful behaviour given the available information and computational resources.

All four definitions will be fleshed out in detail, and I will describe some results that have been obtained so far along these lines. Then I will describe ongoing and future work under the headings of calculative rationality and bounded optimality.

I shall be arguing that, of these candidates, bounded optimality comes closest to meeting the needs of AI research. There is always a danger, in this sort of claim, that its acceptance can lead to "premature mathematization," a condition characterized by increasingly technical results that have increasingly little to do with the original problem—in the case of AI, the problem of creating intelligence. Is research on bounded optimality a suitable stand-in for research on intelligence? I hope to show that P_4, bounded optimality, is more suitable than P_1 through P_3 because it is a real problem with real and desirable solutions, and also because it satisfies some essential intuitions about the nature of intelligence. Some important questions about intelligence can only be formulated and answered within the framework of bounded optimality or some relative thereof. Only time will tell, however, whether bounded optimality research, perhaps with additional refinements, can generate enough theoretical scaffolding to support significant practical progress in AI.

2 AGENTS

Until fairly recently, it was common to define AI as the computational study of "mental faculties" or "intelligent systems," catalogue various kinds, and leave it at that. This doesn't provide much guidance. Instead, one can define AI as the problem of designing systems that *do the right thing*. Now we just need a definition for "right."

This approach involves considering the intelligent entity as an *agent*, that is to say a system that senses its environment and acts upon it. Formally speaking, an agent is defined by the mapping from percept sequences to actions that the agent instantiates. Let **O** be the set of percepts that the agent can observe at any instant, and **A** be the set of possible actions the agent can carry out in the external world (including the action of doing nothing). Thus the *agent function* $f : \mathbf{O}^* \to \mathbf{A}$ defines how an agent behaves under all circumstances. What counts in the first instance is what the agent does, not necessarily what it thinks, or even whether it thinks at all. This initial refusal to consider further constraints on the internal workings of the agent (such as that it should reason logically, for example) helps in three ways: first, it allows us to view such "cognitive faculties" as planning and reasoning as occurring *in the service of* finding the right thing to do; second, it encompasses rather than excludes the position that systems can do the right thing without such cognitive faculties [1, 4]; third, it allows more freedom to consider various specifications, boundaries, and interconnections of subsystems.

The agent-based view of AI has moved quickly from workshops on "situatedness" and "embeddedness" to mainstream textbooks [38, 12] and buzzwords in Newsweek. *Rational* agents, loosely speaking, are agents whose actions make sense from the point of view of the information possessed by the agent and its goals (or the task for which it was designed). Rationality is a property of actions and does not specify—although it does constrain—the process by which the actions are selected. This was a point emphasized by Simon [46], who coined the terms *substantive rationality* and *procedural rationality* to describe the difference between the question of *what* decision to make and the question of *how* to make it. That Rod Brooks's 1991 Computers and Thought lecture was titled "Intelligence without Reason" (see also [5]) emphasizes the fact that reasoning is (perhaps) a *derived* property of agents that might, or might not, be a good implementation scheme to achieve rational behaviour. Justifying the cognitive structures that many AI researchers take for granted is not an easy problem.

One other consequence of the agent-based view of intelligence is that it opens AI up to competition from other fields that have traditionally looked on the embedded agent as a natural topic of study. Control theory is foremost among these, but evolutionary programming and indeed evolutionary biology itself also have ideas to contribute.[1] The prevalence of the agent view has also helped the field move towards solving real problems, avoiding what Brooks calls the "hallucination" problem that arises when the fragility of a subsystem is masked by having an intelligent human providing input to it and interpreting its outputs.

[1] I view this as a very positive development. AI is a field defined by its problems, not its methods. Its principal insights—among them the learning, use, and compilation of explicit knowledge in the service of decision making—can certainly withstand the influx of new methods from other fields. This is especially true when other fields are simultaneously embracing the insights derived within AI.

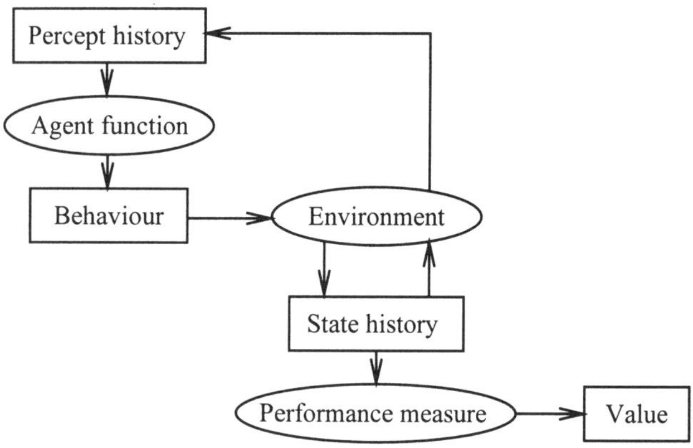

Figure 1. The agent receives percepts from the environment and generates a behaviour which in turn causes the environment to generate a state history. The performance measure evaluates the state history to arrive at the value of the agent.

3 PERFECT RATIONALITY

Perfect rationality constrains an agent's actions to provide the maximum expectation of success given the information available. We can expand this notion as follows (see Figure 1). The fundamental inputs to the definition are the environment class $\mathbf{E}$ in which the agent is to operate and the performance measure U which evaluates the sequence of states through which the agent drives the actual environment. Let $V(f, \mathbf{E}, U)$ denote the expected value according to U obtained by an agent function f in environment class $\mathbf{E}$, where (for now) we will assume a probability distribution over elements of $\mathbf{E}$. Then a perfectly rational agent is defined by an agent function f_{opt} such that

$$f_{\text{opt}} = \text{argmax}_f V(f, \mathbf{E}, U)$$

This is just a fancy way of saying that the best agent does the best it can. The point is that perfectly rational behaviour is a well-defined function of $\mathbf{E}$ and U, which I will call the *task environment*. The problem of computing this function is addressed below.

The theoretical role of perfect rationality in AI is well-described by Newell's paper on the Knowledge Level [35]. Knowledge-level analysis of AI systems relies on an assumption of perfect rationality. It can be used to establish an upper bound on the performance of any possible system, by establishing what a perfectly rational agent would do given the same knowledge.

Although the knowledge that a perfectly rational agent has determines the actions that it will take given its goals, the question of where the knowledge comes from is not well understood. That is, we need to understand rational learning as well as rational action. In the logical view of rationality, learning has received almost no attention—indeed, Newell's analysis precludes learning at the knowledge level. In the decision-theoretic view, Bayesian updating provides a model for rational learning, but this pushes the question back to the prior [6]. The question of rational priors, particularly for expressive representation languages, remains unsettled.

Another aspect of perfect rationality that is lacking is the development of a suitable body of techniques for the specification of utility functions. In economics, many results have been derived on the decomposition of overall utility into attributes that can be combined in various ways [26], yet such methods have made few inroads into AI (but see [50, 2]). We also have little idea how to specify utility over time, and although the question has been raised often, we do not have a satisfactory understanding of the relationship between goals and utility.

The good thing about perfectly rational agents is that if you have one handy, you prefer it to any other agent. Furthermore, if you are an economist you can prove nice results about economies populated by them; and if you want to design distributed intelligent systems, assuming perfect rationality on the part of each agent makes the design of the interaction mechanisms much easier. The bad thing is that the theory of perfect rationality does not provide for the analysis of the internal design of the agent: one perfectly rational agent is as good as another. The *really* bad thing, as pointed out by Simon, is that perfectly rational agents do not exist. Physical mechanisms take time to process information and select actions, hence the behaviour of real agents cannot immediately reflect changes in the environment and will generally be suboptimal.

4 CALCULATIVE RATIONALITY

Before discussing calculative rationality, it is necessary to introduce a distinction between the agent function and the *agent program*. In AI, an agent is implemented as a program, which I shall call l, running on a machine, which I shall call M. An agent program receives as input the current percept, but also has internal state that reflects, in some form, the previous percepts. It outputs actions when they have been selected. From the outside, the behaviour of the agent consists of the selected actions *interspersed with inaction* (or whatever default actions the machine generates).

Calculative rationality is displayed by programs that, *if executed infinitely fast*, would result in perfectly rational behaviour. Unlike perfect rationality, calculative rationality is a requirement that can be fulfilled by many real programs. Also unlike perfect rationality, calculative rationality is not necessarily a desirable property. For example, a calculatively rational chess program will choose the "right" move, but may take 10^{50} times too long to do so.

The pursuit of calculative rationality has nonetheless been the main activity of theoretically well-founded research in AI. In the early stages of the field, it was important to concentrate on "epistemological adequacy" before "heuristic adequacy"—that is, capability in principle rather than in practice.[2] Calculative rationality has been the mainstay of both the logical and the decision-theoretic traditions. In the logical tradition, the performance measure accepts behaviours that achieve the specified goal in all cases and rejects any others. Thus Newell [35] defines rational actions as those that are guaranteed to achieve one of the agent's goals. Logical planning systems, such as theorem-provers using situation calculus, satisfy the conditions of calculative rationality under this definition. In the decision-theoretic tradition, the design of calculatively rational agents has largely gone on outside AI—for example, in stochastic optimal control theory [27]. Representations have usually been very impoverished (state-based rather than sentential) and solvable problems have been either very small or very specialized. Within AI, the development of probabilistic networks or belief networks has opened up many new possibilities for agent design, providing in many cases an exponential reduction in representational complexity. Systems based on influence diagrams (probabilistic networks with action and value nodes added) satisfy the decision-theoretic version of calculative rationality.

In practice, neither the logical nor the decision-theoretic traditions can avoid the intractability of the decision problems posed by the requirement of calculative rationality. One response is to rule out sources of exponential complexity in the representations and reasoning tasks addressed, so that calculative and perfect rationality coincide—at least, if we ignore the little matter of polynomial-time computation. This position was expounded in two fascinating Computers and Thought lectures given by Hector Levesque in 1985 [31, 30] and by Henry Kautz in 1989. The accompanying research results on tractable sublanguages are perhaps best seen as indications of where complexity may be an issue rather than as a solution to the problem of complexity. The idea of restricting expressiveness was strongly opposed by Doyle and Patil [14], who pointed out that it also restricts the applicability of the representation and inference services designed under such constraints.[3]

In the area of distributed AI, the system designer has control over that part of each agent's environment that involves negotiations with other agents. Thus, one possible way to control complexity is to constrain the negotiation problem so that optimal decisions can be made easily. For example, the Clarke Tax mechanism can be used to ensure that the best policy for each agent is simply to state its preferences truthfully [15]. Of course, this approach does not necessarily result in optimal behaviour by the *ensemble* of agents; nor does it solve the problem of complexity in interacting with the rest of the environment.

[2] Perhaps not coincidentally, this decision was taken before the question of computational intractability was properly understood in computer science.

[3] Doyle and Patil propose instead the idea of "rational management of inference." Representation systems "should be designed to offer a broad mix of services varying in cost and quality" and should take into account "the costs and benefits [of computations] as perceived by the system's user." That is, they suggest a solution based on rational metareasoning, as discussed in Section 5.

The most common response to complexity has been to use various speedup techniques and approximations in the hope of getting reasonable behaviour. AI has developed a very powerful armoury of methods for reducing complexity, including the decomposition of state representations into sentential form; sparse representations of environment models (as in STRIPS operators); solution decomposition methods such as partial-order planning and abstraction; approximate, parameterized representations of value functions for reinforcement learning; compilation (chunking, macro-operators, EBL etc.); and the application of metalevel control. Although some of these methods can retain guarantees of optimality and are effective for moderately large problems that are well structured, it is inevitable that intelligent agents will be unable to act rationally in all circumstances. This observation has been a commonplace since the very beginning of AI. Yet systems that select suboptimal actions fall outside calculative rationality *per se*, and we need a better theory to understand them.

5 METALEVEL RATIONALITY

Metalevel rationality, also called Type II rationality by I. J. Good [18], is based on the idea of finding an optimal tradeoff between computational costs and decision quality. Although Good never made his concept of Type II rationality very precise—he defines it as "the maximization of expected utility *taking into account deliberation costs*—it is clear that the aim was to take advantage of some sort of *metalevel architecture* to implement this tradeoff. Metalevel architecture is a design philosophy for intelligent agents that divides the agent into two (or more) notional parts. The *object level* carries out computations concerned with the application domain—for example, projecting the results of physical actions, computing the utility of certain states, and so on. The *metalevel* is a second decision-making process whose application domain consists of the object-level computations themselves and the computational objects and states that they affect. Metareasoning has a long history in AI, going back at least to the early 1970s (see [41] for historical details). One can also view selective search methods and pruning strategies as embodying metalevel expertise concerning the desirability of pursuing particular object-level search operations.

The theory of *rational metareasoning* formalizes the intuition that the metalevel can "do the right thinking." The basic idea is that object-level computations are actions with costs (the passage of time) and benefits (improvements in decision quality). A rational metalevel selects computations according to their expected utility. Rational metareasoning has as a precursor the theory of *information value* [23]—the notion that one can calculate the decision-theoretic value of acquiring an additional piece of information by simulating the decision process that would be followed given each possible outcome of the information request, thereby estimating the expected improvement in decision quality averaged over those outcomes. The application to computational processes, by analogy to information-gathering, seems to have originated with Matheson [32]. In AI, Horvitz [21, 22], Breese and Fehling [3], and Russell and Wefald [40, 41, 42] all showed how the idea of value of computation could solve the basic problems of real-time decision making.

The work done with Eric Wefald looked in particular at search algorithms, in which the object-level computations extend projections of the results of various courses of actions further into the future. For example, in chess programs, each object-level computation expands a leaf node of the game tree. The metalevel problem is then to select nodes for expansion and to terminate search at the appropriate point. The principal problem with metareasoning in such systems is that the local effects of the computations do not *directly* translate into improved decisions, because there is also a complex process of propagating the local effects at the leaf back to the root and the move choice. It turns out that a general formula for the value of computation can be found in terms of the "local effects" and the "propagation function," such that the formula can be instantiated for any particular object-level system (such as minimax propagation), compiled, and executed efficiently at runtime. This method was implemented for two-player games, two-player games with chance nodes, and single-agent search. In each case, the same general metareasoning scheme resulted in efficiency improvements of roughly an order of magnitude over traditional, highly-engineered algorithms.

Another general class of metareasoning problems arises with *anytime* [10] or *flexible* [21] algorithms, which are algorithms designed to return results whose quality varies with the amount of time allocated to computation. The simplest type of metareasoning trades off the expected increase in decision quality for a single algorithm, as measured by a *performance profile*, against the cost of time [45]. A greedy termination condition is optimal if the second derivative of the performance profile is negative. More complex problems arise if one wishes to build complex real-time systems from anytime components. First, one has to ensure the *interruptibility* of the composed system—that is, to ensure that the system as a whole can respond robustly to immediate demands for output. The solution is to interleave the execution of all the components, allocating time to each component so that the total time for each complete iterative improvement cycle of the system doubles at each iteration. In this way, we can construct a complex system that can handle arbitrary and unexpected real-time demands exactly as if it knew the exact time available in advance, with just a small (≤ 4) constant factor penalty in speed [43]. Second, one has to allocate the available computation optimally among the components to maximize the total output quality. Although this is NP-hard for the general case, it can be solved in time linear in program size when the call graph of the components is tree-structured [52]. Although these results are derived in the relatively clean context of anytime algorithms with well-defined performance profiles, there is reason to expect that the general problem of robust real-time decision-making in complex systems can be handled in practice.

Over the last few years, an interesting debate has emerged concerning the nature of metaknowledge and metareasoning. TEIRESIAS [9] established the idea that explicit, domain-specific metaknowledge was an important aspect of expert system creation. Thus, metaknowledge is a sort of "extra" domain knowledge, over and above the object-level domain knowledge, that one has to add to an AI system to get it to work well. On the other hand, in the work on rational metareasoning described above, it is clear that *the metatheory describing the effects of computations is domain-independent* [41, 17]. In principle, no additional domain knowledge is needed to assess the benefits of a computation. In practice, metareasoning from first principles can be very expensive. To avoid this, the results of metalevel analysis for particular domains can be compiled into domain-specific metaknowledge, or such knowledge can be learned directly from experience [41, Chapter 6], [34]. This view of emerging "computational expertise" leads to a fundamental insight into intelligence—namely, that there is an interesting sense in which *algorithms are not a necessary part of AI systems*. Instead, one can imagine a general process of rationally guided computation interacting with properties of the environment to produce more and more efficient decision making. To my mind, this way of thinking finesses one major puzzle of AI: if what is required for AI is incredibly devious and superbly efficient algorithms far surpassing the current best efforts of computer scientists, how did evolution (and how will machine learning) ever get there?

Significant open problems remain in the area of rational metareasoning. One obvious difficulty is that most systems to date have adopted a *myopic* strategy—a greedy, depth-one search at the metalevel. Obviously, the problem of optimal selection of computation *sequences* is at least as intractable as the underlying object-level problem. Nonetheless, sequences must be considered because in some cases the value of a computation may not be apparent as an improvement in decision quality until further computations have been done. This suggests that techniques from reinforcement learning could be effective, especially as the "reward function" for computation—that is, the improvement in decision quality—is easily available to the metalevel *post hoc*. Other possible areas for research include the creation of effective metalevel controllers for more complex systems such as abstraction hierarchy planners, hybrid architectures, and so on.

Although rational metareasoning seems to be a useful tool in coping with complexity, the concept of metalevel rationality as a formal framework for resource-bounded agents does not seem to hold water. The reason is that, since metareasoning is expensive, it cannot be carried out optimally. The history of object-level rationality has repeated itself at the metalevel: perfect rationality at the metalevel is unattainable and calculative rationality at the metalevel is useless. Therefore, a time/optimality tradeoff has to be made for metalevel computations, as for example with the myopic approximation mentioned above. Within the framework of metalevel rationality, however, there is no way to identify the appropriate tradeoff of time for metalevel decision quality. Any attempt to do so via a metametalevel simply results in a conceptual regress. Furthermore, it is entirely possible that in some environments, the most effective agent design will do no metareasoning at all, but will simply respond to circumstances. These considerations suggest that the right approach is to step outside the agent, as it were; to refrain from micro-managing the individual decisions made by the agent. This is the approach taken in bounded optimality.

6 BOUNDED OPTIMALITY

The difficulties with perfect rationality and metalevel rationality arise from the imposition of constraints on things (actions, computations) that the agent designer does not directly control. Specifying that *actions* or *computations* be rational is of no use if no real agents can fulfill the specification. The designer controls the *program*. In [39], the notion of *feasibility* for a given machine is introduced to describe the set of all agent functions that can be implemented by some agent program running on that machine. This is somewhat analogous to the idea of computability, but is much stricter because it relates the operation of a program on a formal machine model with finite speed to the actual temporal behaviour generated by the agent.

Given this view, one is led immediately to the idea that optimal feasible behaviour is an interesting notion, and to the idea of finding the program that generates it. Suppose we define $Agent(l, M)$ to be the agent function implemented by the program l running on machine M. Then the bounded optimal program l_{opt} is defined by

$$l_{\text{opt}} = \text{argmax}_{l \in \mathcal{L}_M} V(Agent(l, M), \mathbf{E}, U)$$

where $\mathcal{L}_M$ is the finite set of all programs that can be run on M. This is P_4, bounded optimality.

In AI, the idea of bounded optimality floated around among several discussion groups interested in the general topic of resource-bounded rationality in the late 1980s, particularly those at Rockwell (organized by Michael Fehling) and Stanford (organized by Michael Bratman). The term "bounded optimality" seems to have been originated by Eric Horvitz [22], who defined it informally as "the optimization of computational utility given a set of assumptions about expected problems and constraints on resources."

Similar ideas have also surfaced recently in game theory, where there has been a shift from consideration of optimal decisions in games to a consideration of optimal decision-making programs. This leads to different results because it limits the ability of each agent to do unlimited simulation of the other, who is also doing unlimited simulation of the first, and so on. Even the requirement of computability makes a significant difference [33]. Bounds on the complexity of players have also become a topic of intense interest. Papadimitriou and Yannakakis [36] have shown that a collaborative equilibrium exists for the iterated Prisoner's Dilemma game if each agent is a finite automaton with a number of states that is less than exponential in the number of rounds. This is essentially a bounded optimality result, where the bound is on space rather than speed of computation.

Philosophy has also seen a gradual evolution in the definition of rationality. There has been a shift from consideration of *act utilitarianism*—the rationality of individual acts—to *rule utilitarianism*, or the rationality of general policies for acting. The requirement that policies be feasible for limited agents was discussed extensively by Cherniak [8] and Harman [19]. A philosophical proposal generally consistent with the notion of bounded optimality can be found in the "Moral First Aid Manual" [13]. Dennett explicitly discusses the idea of reaching an optimum within the space of feasible decision procedures, using as an example the Ph.D. admissions procedure of a philosophy department. He points out that the bounded optimal admissions procedure may be somewhat messy and may have no obvious hallmark of "optimality"—in fact, the admissions committee may continue to tinker with it since bounded optimal systems may have no way to recognize their own bounded optimality.

In work with Devika Subramanian, the general idea of bounded optimality has been placed in a formal setting so that one can begin to derive rigorous results on bounded optimal programs. This involves setting up completely specified relationships among agents, programs, machines, environments, and time. We found this to be a very valuable exercise in itself. For example, the "folk AI" notions of "real-time environments" and "deadlines" ended up with definitions rather different than those we had initially imagined. From this foundation, a very simple machine architecture was investigated in which the program consists of decision procedures of fixed execution time and decision quality. In a "stochastic deadline" environment, it turns out that the utility attained by running several procedures in sequence until interrupted is often higher than that attainable by any single decision procedure. That is, it is often better first to prepare a "quick and dirty" answer before embarking on more involved calculations in case the latter do not finish in time.

The interesting aspect of these results, beyond their value as a demonstration of nontrivial proofs of bounded optimality, is that they exhibit in a simple way what I believe to be a major feature of bounded optimal agents: the fact that the pressure towards optimality within a finite machine results in more complex program structures. Intuitively, efficient decision-making in a complex environment requires a software architecture that offers a wide variety of possible computational options, so that in most situations the agent has at least some computations available that provide a significant increase in decision quality.

One possible objection to the basic model of bounded optimality outlined above is that solutions are not *robust* with respect to small variations in the environment or the machine. This in turn would lead to difficulties in analysing complex system designs. Theoretical computer science faced the same problem in describing the running time of algorithms, because counting steps and describing instruction sets exactly gives the same kind of fragile results on optimal algorithms. The $O()$ notation was developed to deal with this and provides a much more robust way to describe complexity that is independent of machine speeds and implementation details. This robustness is also essential in allowing complexity results to develop cumulatively. In [39], the corresponding notion is asymptotic bounded optimality (ABO). As with classical complexity, we can define both average-case and worst-case ABO, where "case" here means the environment. For example, worst-case ABO is defined as follows:

Worst-case asymptotic bounded optimality
an agent program l is timewise (or spacewise) worst-case ABO in $\mathbf{E}$ *on* M *iff*

$$\exists k, n_0\ \forall l', n\ \ n > n_0 \ \Rightarrow\ V^*(Agent(l, kM), \mathbf{E}, U, n) \geq V^*(Agent(l', M), \mathbf{E}, U, n)$$

where kM *denotes a version of* M *speeded up by a factor* k *(or with* k *times more memory) and* $V^*(f, \mathbf{E}, U, n)$ *is the minimum value of* $V(f, E, U)$ *for all* E *in* $\mathbf{E}$ *of complexity* n.

In English, this means that the program is basically along the right lines if it just needs a faster (larger) machine to have worst-case behaviour as good as that of any other program in all environments.

Another possible objection to the idea of bounded optimality is that it simply shifts the intractable computational burden of metalevel rationality from the agent's metalevel to the designer's object level. Surely, one might argue, the designer now has to solve offline all the metalevel optimization problems that were intractable when online. This argument is not without merit—indeed, it would be surprising if the agent design problem turns out to be easy. There is however, a significant difference between the two problems, in that the agent designer is presumably creating an agent for an entire class of environments, whereas the putative metalevel agent is working in a specific environment. That this can make the problem *easier* for the designer can be seen by considering the example of sorting algorithms. It may be very difficult indeed to sort a list of a trillion elements, but it is relatively easy to design an asymptotically optimal algorithm for sorting. In fact, the difficulties of the two tasks are unrelated. The unrelatedness would still hold for BO as well as ABO design, but the ABO definitions make it a good deal clearer.

It can be shown easily that worst-case ABO is a generalization of asymptotically optimal algorithms, simply by constructing a "classical environment" in which classical algorithms operate and in which the utility of the algorithm's behaviour is a decreasing positive function of runtime if the output is correct and zero otherwise. Agents in more general environments may need to trade off output quality for time, generate multiple outputs over time, and so on. As an illustration of how ABO is a useful abstraction, one can show that under certain restrictions one can construct *universal* ABO programs that are ABO for any time variation in the utility function, using the doubling construction from [43]. Further directions for bounded optimality research are discussed below.

7 WHAT IS TO BE DONE?

This section describes some of the research activities that will, I hope, help to turn bounded optimality into a creative tool for AI system design. First, however, I shall describe work on calculatively rational systems that needs to be done in order to enrich the space of agent programs.

7.1 Components for Calculative Rationality

As mentioned above, the correct design for a rational agent depends on the task environment—the "physical" environment and the performance measure on environment histories. It is possible to define some basic properties of task environments that, together with the complexity of the problem, lead to identifiable requirements on the corresponding rational agent designs [38, Ch. 2]. The principal properties are whether the environment is *fully observable* or *partially observable*, whether it is *deterministic* or *stochastic*, whether it is *static* (i.e., does not change except when the agent acts) or *dynamic*, and whether it is *discrete* or *continuous*. Although crude, these distinctions serve to lay out an agenda for basic research in AI. By analysing and solving each subcase and producing calculatively rational mechanisms with the required properties, theoreticians can produce the AI equivalent of bricks, beams, and mortar with which AI architects can build the equivalent of cathedrals. Unfortunately, many of the basic components are currently missing. Others are so fragile and non-scalable as to be barely able to support their own weight. This presents many opportunities for research of far-reaching impact.

The logicist tradition of goal-based agent design, based on the creation and execution of guaranteed plans, is firmly anchored in fully observable, deterministic, static, and discrete task environments. (Furthermore, tasks are usually specified as logically defined goals rather than general utility functions.) This means that agents need keep no internal state and can even execute plans without the use of perception.

The theory of optimal action in stochastic, partially observable environments goes under the heading of *POMDPs* (Partially Observable Markov Decision Problems), a class of problems first addressed in the work of Sondik [47] but almost completely unknown in AI until recently [7]. Similarly, very little work of a fundamental nature has been done in AI on dynamic environments, which require real-time decision making, or on continuous environments, which have been largely the province of geometry-based robotics. Since most real-world applications are partially observable, nondeterministic, dynamic, and continuous, the lack of emphasis is somewhat surprising.

There are, however, several new bricks under construction. For example, dynamic probabilistic networks (DPNs) [11] provide a mechanism to maintain beliefs about the current state of a dynamic, partially observable, nondeterministic environment, and to project forward the effects of actions. Also, the rapid improvement in the speed and accuracy of computer vision systems has made interfacing with continuous physical environments more practical. In particular, the application of Kalman filtering [24], a widely used technique in control theory, allows robust and efficient tracking of moving objects; DPNs extend Kalman filtering to allow more general representations of world state. Reinforcement learning, together with inductive learning methods for continuous function representations such as neural networks, allow learning from delayed rewards in continuous, nondeterministic environments. Recently, Parr and Russell [37], among others, have had some success in applying reinforcement learning to partially observable environments. Finally, learning methods for static and dynamic probabilistic networks with hidden variables (i.e., for partially observable environments) may make it possible to acquire the necessary environment models [29, 44].

The Bayesian Automated Taxi (a.k.a. BATmobile) project [16] is an attempt to combine all these new bricks to solve an interesting application problem, namely driving a car on a freeway. Technically, this can be viewed as a POMDP because the environment contains relevant variables (such as whether or not the Volvo on your left is intending to change lanes to the right) that are not observable, and because the behaviour of other vehicles and the effects of one's own actions are not exactly predictable. In a POMDP, the optimal decision depends on the joint probability distribution over the entire set of state variables. It turns out that a combination of real-time vision algorithms, Kalman filtering, and dynamic probabilistic networks can maintain the required distribution when observing a stream of traffic on a freeway. The BATmobile currently uses a hand-coded decision tree to make decisions on this basis, and is a fairly safe driver (although probably far from optimal) on our simulator. We are currently experimenting with lookahead methods to make approximately rational decisions, as well as supervised learning and reinforcement learning methods.

As well as extending the scope of AI applications, new bricks for planning under uncertainty significantly increase the opportunity for metareasoning to make a difference. With logical planners, a plan either does or does not work; it has proved very difficult to find heuristics to measure the "goodness" of a logical plan that does not guarantee success, or to estimate the likelihood that an abstract logical plan will have a successful concrete instance. This means that it is very hard to identify plan elaboration steps that are likely to have high value. In contrast, planners designed to handle uncertainty and utility have built-in information about the likelihood of success and there is a continuum from hopeless to perfect plans. Getting metareasoning to work for such systems is a high priority. It is also important to apply those methods such as partial-order planning and abstraction that have been so effective in extending the reach of classical planners.

7.2 Directions for Bounded Optimality

Ongoing research on bounded optimality aims to extend the initial results of [39] to more interesting agent designs. In this section, I will sketch some design dimensions and the issues involved in establishing bounded optimality results.

The general scheme to be followed involves defining a virtual machine M that runs programs from a class $\mathcal{L}_M$. Typically, programs will have a "fixed part" that is shared across some subclass and a "variable part" that is specific to the individual program. Then comparisons are made between the best programs in different subclasses for the same machine. For example, suppose M is a machine capable of running any feedforward neural network. $\mathcal{L}_M$ consists of all such networks, and we might be interested in comparing the subclasses defined by different network topologies, while within each subclass individual programs differ in the weights on the links of the network. Thus, the boundary between machine and program depends to some extent on the range of comparisons that the designer wishes to consider.

At the most general level of analysis, the methodology is now quite straightforward: choose a machine, choose a program that runs on the machine, then dump the resulting agent into a class of environments **E**. The program with the best performance is bounded optimal for M in **E**. For example, M is an IBM PC with a C compiler; $\mathcal{L}_M$ consists of C programs up to a certain size; the environment consists of a population of human chess opponents; the performance measure is the chess rating achieved; the bounded optimal program is the one with the highest rating.

This rather blunt and unenlightening approach has no doubt occurred to many engaged in the construction of chess programs. As stated, the problem is ridiculously hard to solve and the solution, once found, would be very domain-specific. The problem is to define a research agenda for bounded optimality that provides a little more guidance and generality. This can be done by exploiting structure in the definition of the problem, in particular the orthogonality of time and content, and by using more sophisticated agent designs, particularly those that incorporate mechanisms for adaptation and optimization. In this way, we can prove bounded optimality results for more general classes of task environments.

Mechanisms for optimization

Modular design using a hierarchy of components is commonly seen as the only way to build reliable complex systems. The components fulfill certain behavioural specifications and interact in well-defined ways. To produce a composite bounded-optimal design, the optimization problem involves allocating execution time to components [52] or arranging the order of execution of the components [39] to maximize overall performance. As illustrated earlier in the discussion of universal ABO algorithms, the techniques for optimizing temporal behaviour are largely orthogonal to the *content* of the system components, which can therefore be optimized separately. Consider, for example, a composite system that uses an anytime inference algorithm over a belief network as one of its components. If a learning algorithm improves the accuracy of the belief network, the performance profile of the inference component will improve, which will result in a reallocation of execution time that is guaranteed to improve overall system performance. Thus, techniques such as the doubling construction and the time allocation algorithm in [52] can be seen as domain-independent tools for agent design. They enable bounded optimality results that do not depend on the specific temporal aspects of the environment class. As a simple example, we might prove that a certain chess program design is ABO for all time controls ranging from blitz to full tournament play.

The results obtained so far for optimal time allocation have assumed a static, offline optimization process with predictable component performance profiles and fixed connections among components. One can imagine far more subtle designs in which individual components must deal with unexpectedly slow or fast progress in processing and changing needs for information from other components. This might involve exchanging computational resources among components, establishing new interfaces, and so on. This is more reminiscent of a computational market, as envisaged by Wellman [51], than of the classical subroutine hierarchies, and would offer a useful additional level of abstraction in system design.

Mechanisms for adaptation

In addition to combinatorial optimization of the structure and temporal behaviour of an agent, we can also use learning methods to improve the design:

- The *content* of an agent's knowledge base can of course be improved by inductive learning. In [39], it is shown that approximately bounded optimal designs can be guaranteed with high probability if each component is learned in such a way that its output quality is close to optimal among all components of a given execution time. Results from computational learning theory, particularly in the *agnostic learning* model [25], can provide learning methods with the required properties. The key additional step is to analyze the way in which slight imperfection in each component carries through to slight imperfection in the whole agent.

- *Reinforcement learning* can be used to learn value information such as utility functions. Recent results [49] provide convergence guarantees for reinforcement learning with a fairly broad class of function approximators. One can use such learning methods for metalevel information, e.g., the value of computation. In [41, Chapter 6], this is shown to be an effective technique. Formal results on convergence to optimal control of search would be of great interest. Further work is needed, however, since current theorems assume a stationary distribution that generates the agent's experiences whereas an agent that is improving its search control will presumably be exploring different populations of experiences over time.

- *Compilation* methods such as explanation-based learning can be used to transform an agent's representations to allow faster decision making. Several agent architectures including SOAR [28] use compilation to speed up all forms of problem solving. Some nontrivial results on convergence have been obtained by Tadepalli [48], based on the observation that after a given amount of experience, novel problems for which no solution has been stored should be encountered only infrequently.

Presumably, an agent architecture can incorporate all these learning mechanisms. One of the issues to be faced by bounded optimality research is how to prove convergence results when several adaptation and optimization mechanisms are operating simultaneously. A "quasistatic" approach, in which one mechanism reaches convergence before the other method is allowed to take its next step, seems theoretically adequate but not very practical.

Offline and online mechanisms

It is possible to distinguish between *offline* and *online* mechanisms for constructing bounded-optimal agents. An offline construction mechanism is not itself part of the agent and is not the subject of bounded optimality constraints. Let C be an offline mechanism designed for a class of environments $\mathbf{E}$. Then a typical theorem will say that C operates in a specific environment $E \in \mathbf{E}$ and returns an agent design that is ABO (say) for E—that is, an environment-specific agent.

In the online case, the mechanism C is considered part of the agent. Then a typical theorem will say that the agent is ABO for all $E \in \mathbf{E}$. If the performance measure used is indifferent to the transient cost of the adaptation or optimization mechanism, the two types of theorems are essentially the same. On the other hand, if the cost cannot be ignored—for example, if an agent that learns quickly is to be preferred to an agent that reaches the same level of performance but learns more slowly—then the analysis becomes more difficult. It may become necessary to define asymptotic equivalence for "experience efficiency" in order to obtain robust results, as is done in computational learning theory.

It is worth noting that one can easily prove the value of "lifelong learning" in the ABO framework. An agent that devotes a constant fraction of its computational resources to learning-while-doing cannot do worse, in the ABO sense, than an agent that ceases learning after some point. If some improvement is still possible, the lifelong learning agent will always be preferred.

Fixed and variable computation costs

Another dimension of design space emerges when one considers the computational cost of the "variable part" of the agent design. The design problem is simplified considerably when the cost is fixed. Consider again the task of metalevel reinforcement learning, and to make things concrete let the metalevel decision be made by a Q function mapping from computational state and action to value. Suppose further that the Q function is to be represented by a neural net. If the topology of the neural net is fixed, then all Q functions in the space have the same execution time. Consequently, the optimality criterion used by the standard Q-learning process coincides with bounded optimality, and the equilibrium reached will be a bounded-optimal configuration.[4] On the other hand, if the topology of the network is subject to alteration as the design space is explored, then the execution time of the different Q-functions varies. In this case, the standard Q-learning process will not necessarily converge to a bounded-optimal configuration. A different adaptation mechanism must be found that takes into account the passage of time and its effect on utility.

Whatever the solution to this problem turns out to be, the important point is that the notion of bounded optimality helps to distinguish adaptation mechanisms that will result in good performance from those that will not. Adaptation mechanisms derived from calculative rationality will fail in the more realistic setting where an agent cannot afford to aim for perfection.

Fully variable architectures

The discussion so far has been limited to fairly sedate forms of agent architecture in which the scope for adaptation is circumscribed to particular functional aspects such as metalevel Q functions. However, an agent must in general deal with an environment that is far more complex than itself and that exhibits variation over time at all levels of granularity. Limits on the size of the agent's memory may imply that almost complete revision of the agent's mental structure is needed to achieve high performance. For example, one can imagine that a simple rule-based agent living through cycles of winter and summer may have to discard all of its summer rules as winter approaches, and then relearn them from scratch the following year. Such situations may engender a rethinking of some of our notions of agent architecture and optimality, and suggest a view of agent programs as dynamical systems with various amounts of compiled and uncompiled knowledge and internal processes of inductive learning, forgetting, and compilation.

Towards a grammar of AI systems

The approach that seems to be emerging for bounded optimality research is to divide up the space of agent designs into "architectural classes" such that in each class the structural variation is sufficiently limited. Then ABO results can be obtained either by analytical optimization within the class or by showing that an empirical adaptation process results in an approximately ABO design. Once this is done, it should be possible to compare architecture classes directly, perhaps to establish asymptotic dominance of one class over another. For example, it might be the case that the inclusion of an appropriate "macro-operator formation" or "greedy metareasoning" capability in a given architecture will result in an improvement in behaviour in the limit of very complex environments—that is, one cannot compensate for the exclusion of the capability by increasing the machine

[4] A similar observation was made by Horvitz and Breese [20] for cases where the object level is so restricted that the metalevel decision problem can be solved in constant time.

speed by a constant factor. A central tool in such work will be the use of "no-cost" results where, for example, the allocation of a constant fraction of computational resources to learning or metareasoning can do no harm to an agent's ABO prospects.

Getting all these architectural devices to work together smoothly is an important unsolved problem in AI and must be addressed before we can make progress on understanding bounded optimality within these more complex architectural classes. If the notion of "architectural device" can be made sufficiently concrete, then AI may eventually develop a *grammar* for agent designs, describing the devices and their interrelations. As the grammar develops, so should the accompanying ABO dominance results.

8 SUMMARY

I have outlined directions for formally grounded AI research based on bounded optimality as the desired property of AI systems. This perspective on AI seems to be a logical consequence of the inevitable philosophical "move" from optimization over actions or computations to optimization over programs. I have suggested that such an approach should allow synergy between theoretical and practical AI research of a kind not afforded by other formal frameworks. In the same vein, I believe it is a satisfactory formal counterpart of the informal goal of creating intelligence. In particular, it is entirely consistent with our intuitions about the need for complex structure in real intelligent agents, the importance of the resource limitations faced by relatively tiny minds in large worlds, and the operation of evolution as a design optimization process. One can also argue that bounded optimality research is likely to satisfy better the needs of those who wish to emulate human intelligence, because it takes into account the limitations on computational resources that are presumably responsible for most of the regrettable deviation from perfect rationality exhibited by humans.

Bounded optimality and its asymptotic cousin are, of course, nothing but formally defined properties that one may want systems to satisfy. It is too early to tell whether ABO will do the same kind of work for AI that asymptotic complexity has done for theoretical computer science. Creativity in design is still the prerogative of AI researchers. It may, however be possible to systematize the design process somewhat and to automate the process of adapting a system to its computational resources and the demands of the environment. The concept of bounded optimality provides a way to make sure the adaptation process is "correct."

My hope is that with these kinds of investigations, it will eventually be possible to develop the conceptual and mathematical tools to answer some basic questions about intelligence. For example, *why* do complex intelligent systems (appear to) have declarative knowledge structures over which they reason explicitly? This has been a fundamental assumption that distinguishes AI from other disciplines for agent design, yet the answer is still unknown. Indeed, Rod Brooks, Hubert

Dreyfus, and others flatly deny the assumption. What is clear is that it will need *something like* a theory of bounded optimal agent design to answer this question.

Most of the agent design features that I have discussed here, including the use of declarative knowledge, have been conceived within the standard methodology of "first build calculatively rational agents and then speed them up." Yet one can legitimately doubt that this methodology will enable the AI community to discover all the design features needed for general intelligence. The reason is that no conceivable computer will ever be remotely close to approximating perfect rationality for even moderately complex environments. Perfect rationality is, if you like, a "Newtonian" definition for intelligent agents whereas the real world is a particle accelerator. It may well be the case that agents based on improvements to calculatively rational designs are *not even close* to achieving the level of performance that is potentially achievable given the underlying computational resources. For this reason, I believe it is imperative not to dismiss ideas for agent designs that do not seem at first glance to fit into the "classical" calculatively rational framework. Instead, one must attempt to understand the potential of the bounded optimal configurations within the corresponding architectural class, and to see if one can design the appropriate adaptation mechanisms that might help in realizing these configurations.

As mentioned in the previous section, there is also plenty of work to do in the area of making more general and more robust "bricks" from which to construct AI systems for more realistic environments, and such work will provide added scope for the achievement of bounded optimality. In a sense, under this conception AI research is the same now as it always should have been.

ACKNOWLEDGEMENTS

An earlier version of this paper appeared in the Proceedings of the Fourteenth International Joint Conference on Artificial Intelligence, published by IJCAII. That paper drew on previous work with Eric Wefald [41] and Devika Subramanian [39]. The latter work contains a more rigorous analysis of many of the concepts presented here. Thanks also to Michael Wellman, Michael Fehling, Michael Genesereth, Russ Greiner, Eric Horvitz, Henry Kautz, Daphne Koller, and Bart Selman for many stimulating discussions and useful suggestions on the topic of bounded rationality. The research was supported by NSF grants IRI-8903146, IRI-9211512 and IRI-9058427, by a UK SERC Visiting Fellowship.

Stuart Russell (University of California Berkeley, USA)

REFERENCES

[1] Philip E. Agre and David Chapman. Pengi: an implementation of a theory of activity. In *Proceedings of the Tenth International Joint Conference on Artificial Intelligence (IJCAI-87)*, pages

268–272, Milan, Italy, August 1987. Morgan Kaufmann.

[2] Fahiem Bacchus and Adam Grove. Graphical models for preference and utility. In *Proceedings of the Eleventh Conference on Uncertainty in Artificial Intelligence (UAI-95)*, pages 3–10, Montreal, 1995. Morgan Kaufmann.

[3] J. S. Breese and M. R. Fehling. Control of problem-solving: Principles and architecture. In R. D. Shachter, T. Levitt, L. Kanal, and J. Lemmer, editors, *Uncertainty in Artificial Intelligence 4*. Elsevier/North-Holland, Amsterdam, London, New York, 1990.

[4] R. A. Brooks. Engineering approach to building complete, intelligent beings. *Proceedings of the SPIE—The International Society for Optical Engineering*, 1002:618–625, 1989.

[5] R. A. Brooks. Intelligence without representation. *Artificial Intelligence*, 47(1–3):139–159, 1991.

[6] Rudolf Carnap. *Logical Foundations of Probability*. University of Chicago Press, Chicago, Illinois, 1950.

[7] A. R. Cassandra, L. P. Kaelbling, and M. L. Littman. Acting optimally in partially observable stochastic domains. In *Proceedings of the Twelfth National Conference on Artificial Intelligence (AAAI-94)*, pages 1023–1028, Seattle, Washington, August 1994. AAAI Press.

[8] C. Cherniak. *Minimal Rationality*. MIT Press, Cambridge, Massachusetts, 1986.

[9] Randall Davis. Meta-rules: reasoning about control. *Artificial Intelligence*, 15(3):179–222, December 1980.

[10] Thomas Dean and Mark Boddy. An analysis of time-dependent planning. In *Proceedings of the Seventh National Conference on Artificial Intelligence (AAAI-88)*, pages 49–54, St. Paul, Minnesota, 21–26 August 1988. Morgan Kaufmann.

[11] Thomas Dean and Keiji Kanazawa. A model for reasoning about persistence and causation. *Computational Intelligence*, 5(3):142–150, 1989.

[12] Thomas L. Dean, John Aloimonos, and James F. Allen. *Artificial Intelligence: Theory and Practice*. Benjamin/Cummings, Redwood City, California, 1995.

[13] Daniel C. Dennett. The moral first aid manual. Tanner lectures on human values, University of Michigan, 1986.

[14] J. Doyle and R. S. Patil. Two theses of knowledge representation: language restrictions, taxonomic classification, and the utility of representation services. *Artificial Intelligence*, 48(3):261–297, April 1991.

[15] E. Ephrati and J. S. Rosenschein. The clarke tax as a consensus mechanism among automated agents. In *Proceedings of the Ninth National Conference on Artificial Intelligence (AAAI-91)*, volume 1, pages 173–178, Anaheim, California, July 1991. AAAI Press.

[16] Jeff Forbes, Tim Huang, Keiji Kanazawa, and Stuart Russell. The BATmobile: Towards a Bayesian automated taxi. In *Proceedings of the Fourteenth International Joint Conference on Artificial Intelligence (IJCAI-95)*, Montreal, Canada, August 1995. Morgan Kaufmann.

[17] M. L. Ginsberg and D. F. Geddis. Is there any need for domain-dependent control information? In *Proceedings of the Ninth National Conference on Artificial Intelligence (AAAI-91)*, volume 1, pages 452–457, Anaheim, California, July 1991. AAAI Press.

[18] I. J. Good. Twenty-seven principles of rationality. In V. P. Godambe and D. A. Sprott, editors, *Foundations of Statistical Inference*, pages 108–141. Holt, Rinehart, Winston, Toronto, 1971.

[19] Gilbert H. Harman. *Change in View: Principles of Reasoning*. MIT Press, Cambridge, Massachusetts, 1983.

[20] E.J. Horvitz and J.S. Breese. Ideal partition of resources for metareasoning. Technical Report KSL-90-26, Knowledge Systems Laboratory, Stanford University, Stanford, California, February 1990.

[21] E. J. Horvitz. Problem-solving design: Reasoning about computational value, trade-offs, and resources. In *Proceedings of the Second Annual NASA Research Forum*, pages 26–43, Moffett Field, California, 1987. NASA Ames Research Center.

[22] E. J. Horvitz. Reasoning about beliefs and actions under computational resource constraints. In Laveen N. Kanal, Tod S. Levitt, and John F. Lemmer, editors, *Uncertainty in Artificial Intelligence 3*, pages 301–324. Elsevier/North-Holland, Amsterdam, London, New York, 1989.

[23] Ronald A. Howard. Information value theory. *IEEE Transactions on Systems Science and Cybernetics*, SSC-2:22–26, 1966.

[24] R. E. Kalman. A new approach to linear filtering and prediction problems. *Journal of Basic Engineering*, pages 35–46, March 1960.

[25] M. Kearns, R. Schapire, , and L. Sellie. Toward efficient agnostic learning. In *Proceedings of the Fifth Annual ACM Workshop on Computational Learning Theory (COLT-92)*, Pittsburgh, Pennsylvania, July 1992. ACM Press.

[26] Ralph L. Keeney and Howard Raiffa. *Decisions with Multiple Objectives: Preferences and Value Tradeoffs*. Wiley, New York, 1976.

[27] P.R. Kumar and Pravin Varaiya. *Stochastic systems: Estimation, identification, and adaptive control*. Prentice-Hall, Englewood Cliffs, New Jersey, 1986.

[28] J. E. Laird, P. S. Rosenbloom, and A. Newell. Chunking in Soar: the anatomy of a general learning mechanism. *Machine Learning*, 1:11–46, 1986.

[29] S. L. Lauritzen. The EM algorithm for graphical association models with missing data. *Computational Statistics and Data Analysis*, 19:191–201, 1995.

[30] H. J. Levesque and R. J. Brachman. Expressiveness and tractability in knowledge representation and reasoning. *Computational Intelligence*, 3(2):78–93, May 1987.

[31] H. J. Levesque. Making believers out of computers. *Artificial Intelligence*, 30(1):81–108, October 1986.

[32] J. E. Matheson. The economic value of analysis and computation. *IEEE Transactions on Systems Science and Cybernetics*, SSC-4(3):325–332, 1968.

[33] N. Megiddo and A. Wigderson. On play by means of computing machines. In Joseph Y. Halpern, editor, *Theoretical Aspects of Reasoning about Knowledge: Proceedings of the 1986 Conference (TARK-86)*, pages 259–274, Monterey, California, March 19–22 1986. IBM and AAAI, Morgan Kaufmann.

[34] Steven Minton. Is there any need for domain-dependent control information? a reply. In *Proceedings of the Thirteenth National Conference on Artificial Intelligence (AAAI-96)*, volume 1, pages 855–862, Portland, Oregon, August 1996. AAAI Press.

[35] Allen Newell. The knowledge level. *Artificial Intelligence*, 18(1):82–127, 1982.

[36] C. H. Papadimitriou and M. Yannakakis. On complexity as bounded rationality. In *Symposium on Theory of Computation (STOC-94)*, 1994.

[37] Ronald Parr and Stuart Russell. Approximating optimal policies for partially observable stochastic domains. In *Proceedings of the Fourteenth International Joint Conference on Artificial Intelligence (IJCAI-95)*, Montreal, Canada, August 1995. Morgan Kaufmann.

[38] Stuart J. Russell and Peter Norvig. *Artificial Intelligence: A Modern Approach*. Prentice-Hall, Englewood Cliffs, New Jersey, 1995.

[39] Stuart J. Russell and Devika Subramanian. Provably bounded-optimal agents. *Journal of Artificial Intelligence Research*, 3, May 1995.

[40] Stuart J. Russell and Eric H. Wefald. On optimal game-tree search using rational meta-reasoning. In *Proceedings of the Eleventh International Joint Conference on Artificial Intelligence (IJCAI-89)*, pages 334–340, Detroit, Michigan, August 1989. Morgan Kaufmann.

[41] Stuart J. Russell and Eric H. Wefald. *Do the Right Thing: Studies in Limited Rationality*. MIT Press, Cambridge, Massachusetts, 1991.

[42] Stuart J. Russell and Eric H. Wefald. Principles of metareasoning. *Artificial Intelligence*, 49(1–3):361–395, May 1991.

[43] S. Russell and S. Zilberstein. Composing real-time systems. In *Proceedings of the Twelfth International Joint Conference on Artificial Intelligence (IJCAI-91)*, Sydney, August 1991. Morgan Kaufmann.

[44] Stuart Russell, John Binder, Daphne Koller, and Keiji Kanazawa. Local learning in probabilistic networks with hidden variables. In *Proceedings of the Fourteenth International Joint Conference on Artificial Intelligence (IJCAI-95)*, pages 1146–52, Montreal, Canada, August 1995. Morgan Kaufmann.

[45] Herbert A. Simon. A behavioral model of rational choice. *Quarterly Journal of Economics*, 69:99–118, 1955.

[46] Herbert A. Simon. Rational choice and the structure of the environment. In *Models of Bounded Rationality*, volume 2. MIT Press, Cambridge, Massachusetts, 1958.

[47] E. J. Sondik. *The Optimal Control of Partially Observable Markov Decision Processes*. PhD thesis, Stanford University, Stanford, California, 1971.

[48] Prasad Tadepalli. A formalization of explanation-based macro-operator learning. In *Proceedings of the Twelfth International Joint Conference on Artificial Intelligence (IJCAI-91)*, pages 616–622, Sydney, August 1991. Morgan Kaufmann.

[49] John N. Tsitsiklis and Benjamin Van Roy. An analysis of temporal-difference learning with function approximation. Technical Report LIDS-P-2322, Laboratory for Information and Decision Systems, Massachusetts Institute of Technology, 1996.

[50] Michael P. Wellman. Reasoning about preference models. Technical Report MIT/LCS/TR-340, Laboratory for Computer Science, MIT, Cambridge, Massachusetts, 1985.

[51] Michael P. Wellman. A market-oriented programming environment and its application to distributed multicommodity flow problems. *Journal of Artificial Intelligence Research*, 1(1):1–23, 1994.

[52] Shlomo Zilberstein and Stuart Russell. Optimal composition of real-time systems. *Artificial Intelligence*, 83, 1996.

A. SLOMAN

WHAT SORT OF ARCHITECTURE IS REQUIRED FOR A HUMAN-LIKE AGENT?

ABSTRACT: This paper is about how to give human-like powers to complete agents. For this the most important design choice concerns the overall architecture. Questions regarding detailed mechanisms, forms of representations, inference capabilities, knowledge etc. are best addressed in the context of a global architecture in which different design decisions need to be linked. Such a design would assemble various kinds of functionality into a complete coherent working system, in which there are many concurrent, partly independent, partly mutually supportive, partly potentially incompatible processes, addressing a multitude of issues on different time scales, including asynchronous, concurrent, motive generators. Designing human like agents is part of the more general problem of understanding design space, niche space and their interrelations, for, in the abstract, there is no one optimal design, as biological diversity on earth shows.

1 INTRODUCTION

A complete functioning agent, whether biological, or simulated in software, or implemented in the form of a robot, needs an integrated collection of diverse but interrelated capabilities, i.e. an architecture. At present, most work in AI and Cognitive Science addresses only components of such an architecture (e.g. vision, speech understanding, concept formation, rule learning, planning, motor control, etc.) or mechanisms and forms of representation and inference (logic engines, condition-action rules, neural nets, genetic algorithms) which might be used by many components. While such studies can make useful contributions it is important to ask, from time to time, how everything can be put together, and that requires the study of architectures.

Analysing possible architectures is closely related to the task of defining an ontology for mental objects, states and processes (percepts, beliefs, desires, attitudes, intentions, moods, emotions, character, inferences, learning, etc.). Ideas about the ontology can help to guide design choices. However, exploring an architecture can reveal unexpected features of the ontology it is capable of supporting, and that can feed back into new ideas about ontologies and design requirements. So the processes of theorising, designing, implementing and experimenting are related in a cyclic fashion.

At present I do not think we know much about the space of possible architectures, and our ideas regarding the ontology to be supported by such an architecture are still very primitive (having advanced little beyond folk psychology, though that's as good a starting place as any). So we are not yet in a position to choose

M. Wooldridge and A. Rao (eds.), Foundations of Rational Agency, 35–52.

one architecture, or even a sub-class. So all such work must remain exploratory and speculative for the time being, including the work reported here.

2 WHAT IS AN ARCHITECTURE?

What do I mean by "architecture"? A fully functioning system has architectures at different levels of abstraction, corresponding to different implementation layers, e.g. there is the architecture of an underlying physical mechanism (Turing machine, von Neumann machine, dataflow machine, neural net, chemical control mechanism, etc.), the architecture of a complex algorithm (e.g. a parsing algorithm which has components that handle different types of sub-structure in the input), the architecture of an integrated collection of concurrent software modules (e.g. the architecture of an operating system, or the architecture of a factory control system). When computer scientists talk about architecture they often mean to refer to the structure of the lowest level physical mechanism. There is a more important notion of architecture for our purposes, which is closer to what we mean by the architecture of a building, or a large organisation. This refers to the large scale functional decomposition: it is the concept of architecture that might be used by a software engineer, or systems analyst.

Besides differences in levels of abstraction or implementation, there are differences in types of functionality. A human-like agent needs to be able to perform a large and diverse collection of tasks, both externally (finding and consuming food, avoiding predators, building shelters, making tools, finding mates, etc.) and internally (interpreting sensory data, generating motives, evaluating motives, selecting motives, creating plans, storing information for future use, making inferences from new or old information, detecting inconsistencies, monitoring plan execution, monitoring various kinds of internal processing, noticing resemblances, creating new concepts and theories, discovering new rules, noticing new possibilities, etc.).

At present we do not know much about the range of internal tasks performed by the human architecture since neither observation of behaviour, nor introspection nor neurophysiological studies can give direct insight into most of what is going on in abstract virtual machines (for reasons indicated below). Nevertheless we can start our exploration from our best current hunches gleaned from all these sources.

3 THERE IS NO UNIQUE DESIGN FOR INTELLIGENCE

Even if the list of internal capabilities given above is a good start, we must not assume that all intelligent agents will have the same collection. Different kinds of agents may have different subsets. Even among humans there is enormous diversity, especially if we consider extreme cases, such as Newton, Mozart, and idiot savants. Within an individual the collection of capabilities is not fixed either, as is clear both from observation of young children and studies of aging.

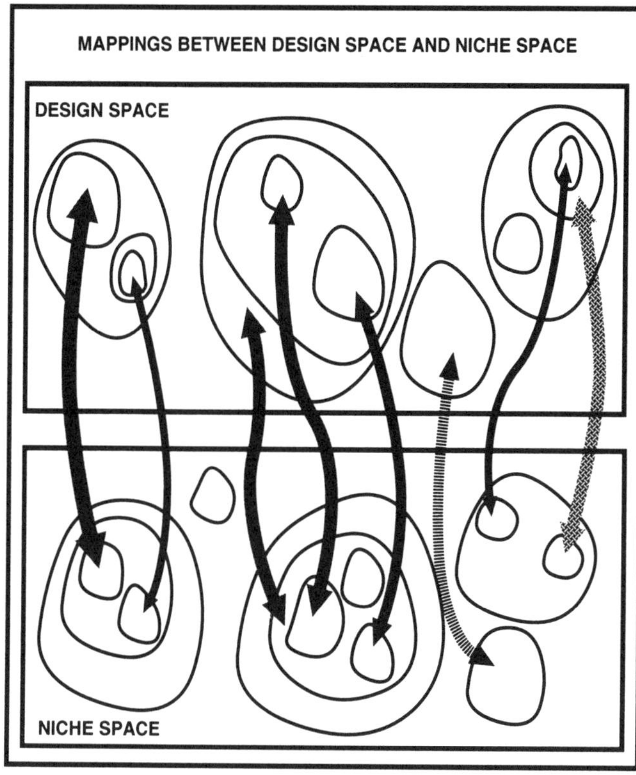

Figure 1. Mappings between design space and niche space

Thus we should not assume that an intelligent agent has a fixed architecture: part of the processes of learning and development may include changes to the architecture, for instance development of major new collections of capabilities and development of new links between old capabilities. Some individuals seem to go on developing and extending their architectures longer than others. It may turn out that one of the *most* important features of a human architecture, a source of much of its power, is the potential for self modification and the consequential diversification within a cooperating community.

4 DESIGN SPACE AND NICHE SPACE

For any collection of capabilities (i.e. for each set of requirements for a design) we can consider the designs that might implement such capabilities. In general there will not be unique design solutions. I have summarised this in [10, 12, 14] by suggesting that we need to explore a space of possible designs for behaving systems (design space) and a space of possible sets of requirements (niche space) and the mappings between the two. It is not to be expected that there is any one "right" architecture. As biological diversity demonstrates, many different architectures can be successful, and in different ways. There are different "niches" (sets of requirements and constraints) for which architectures can be evaluated and compared, and such evaluations will not generally yield a Yes/No decision, but rather an analysis of trade-offs, often involving several dimensions of comparison. This comment does not imply that the spaces are smooth continua without any sharp boundaries: on the contrary, both are likely to have many significant discontinuities (as should be obvious from the structure of the space of designs for software systems) and part of our task is to understand the nature of those discontinuities. the two spaces and their relationships. The variety of types of arrows is intended to show that there are different kinds and degrees of match between a region in design space and a region in niche space.

5 TRAJECTORIES IN DESIGN SPACE AND NICHE SPACE

One task for AI and related disciplines is to investigate possible trajectories in design space and in niche space, i.e. possible transformations from one design to another or from one niche to another. This involves exploring and analysing possible forms of development, adaptation and learning within individuals and also possible types of evolutionary change.

Some changes occur within continuous regions of design space and niche space (e.g. smooth increases in speed of processing), while other trajectories cross discontinuities, e.g. introducing a notation or mechanism that (in principle) allows construction of nested symbolic structures of unbounded depth, going from a system of propositional logic to full predicate logic with quantifiers, or going from a purely reactive architecture to one that includes deliberative capabilities (described below).

There are some types of changes that can happen within a single individual, such as the changes from frog spawn to tadpole to adult frog, or the change from helpless human infant to naughty child, to sophisticated quantum physicist. Other types of trajectories in design space are not possible within an individual, but require evolution across gradually changing generations, or, in the case of artifacts, major re-engineering. For example, I suspect that there is no environmental manipulation that can transform a frog's egg into a giraffe. I do not know whether some sequence of evolutionary pressures could lead from a frog to a giraffe, possibly via regression to a simpler form (a common ancestor).

Whether any self-modifying artificial information processing system could start with the ability to write computer programs in assembly language and somehow extend itself by inventing languages like Algol, Simula67, Lisp, C++, Prolog, etc. or by inventing a new type of operating system for itself, remains an open research question, linked to other questions about mechanisms underlying human creativity.

Since all organisms form part of the environment for other organisms (including others of the same species) evolution in the design of one can constitute evolution in the niche for another, and *vice versa.* A study of which forms of co-evolution are and are not possible would be an essential part of the study of trajectories.

Another kind of trajectory is the evolution of a culture, i.e. the collection of concepts, knowledge, skills, norms, ideals, etc. shared (to varying degrees) among members of a community. There seem to be forms of learning that are possible in a culture but not in an individual (e.g. because they take too long to be achieved in one lifetime, or because they essentially involve interactions between individuals, such as social and political developments). Another way of thinking about this is to regard an enduring society as a particular form of self-modifying agent with a complex distributed architecture.

A different sort of question is whether a particular design permits instances to be assembled ready made in a laboratory or whether they would have to grow themselves. It may be physically impossible to assemble directly mechanisms that are capable of supporting certain kinds of functional architectures (e.g. assembling a fully functional adult human brain), because of the 3-D structural intricacies. This does not rule out the possibility of *growing* one in a laboratory, using sophisticated developmental and learning processes. But those are long term research issues, on which we can reserve judgement.

Whether a *software* equivalent to an adult human brain could be assembled in a fully functional form is another question. The answer may turn out to be "yes" in theory but "no" in practice, if the system is to be implemented in physical mechanisms and operate within human-like constraints of weight, physical size, speed of operation, and energy consumption. These are all questions on which opinions will differ until more research has been done.

6 MUST DESIGNS BE INTELLIGIBLE?

Another question on which there is disagreement is whether the provision of a large set of capabilities, such as those listed above, necessarily involves the creation of an *intelligible* design, with identifiable components performing separate tasks, or whether the functionality could sometimes (or always?) emerge only in a very complex and incomprehensible fashion from myriad interacting components.

For example, experimenters using genetic algorithms to evolve neural nets to control a robot sometimes create networks that work, but which seem to be impossible to understand (not unlike some legacy software which has grown over many years of undisciplined development).

This is related to the question whether a niche (i.e. a set of requirements) will always decompose into a collection of distinct capabilities which can be served by distinct components of a design, or whether there is always so much intricate "cross-talk" between requirements and between elements of designs that clean, intelligible, modular solutions will turn out to be impossible, except in relatively trivial cases.[1]

Even if designs are unintelligible at one level of description, there may be higher level descriptions of important features which can be discovered if only we develop the right sets of concepts. Cohen and Stewart [3] suggest that this emergence of higher level order is a feature of all complex systems, including biological systems.

7 HOW CAN AN ARCHITECTURE BE EVALUATED?

Evaluation of an architecture (or a generic design for a family of related architectures) can take different forms, depending on one's interests.

For instance, someone with a practical objective would be primarily interested in observable performance. This could include multiple dimensions of evaluation, involving input-output mappings, speed, running costs, generality, precision, accuracy, adaptability.

A much discussed (and maligned) criterion is the Turing test. The main point to note about this is that it corresponds to a tiny subset of niche space (even if interesting regions of design space are potentially relevant, as Turing claimed, at least implicitly). For someone interested in designs that fit other regions of niche space, the Turing test would be of limited value: a machine that passed the Turing test with flying colours might not be able to learn to fly an airliner safely, or to interpret the sensory information and control the movements of a robot.

Arguing about which performance criterion is correct is just silly: different criteria will be relevant to different scientific and engineering goals.

The task of designing a system satisfying observable performance criteria may lead to a concern with *internal* processes. For instance, whether a system can modify its performance by changing its strategies when things go wrong will depend on what sorts of internal monitoring, analysis and evaluation are possible, and what sorts of short term and long term internal self-modification are possible. This in turn will depend on the forms of representation and inference available, and the generative power of the internal building blocks.

Someone with a biological or psychological orientation, rather than practical engineering objectives, will have different criteria for evaluating models, for instance requiring a fairly close correspondence with information-processing states, and possibly even neural mechanisms, within the organism being modelled. Detecting such a correspondence, or lack of it, may be very difficult, especially when the objective is to achieve a correspondence at a high level of abstraction compatible with significant differences in physical construction and differences in observable behaviour (just as different human beings sharing many design features will differ in their behaviour and capabilities).

A more general and ambitious scientific concern would be not just the evaluation of any particular model, or the study of any particular type of organism, but rather the comparative study of different architectures and their relationships

[1] I have argued against certain sorts of modularity in vision, in [8].

to different niches. This could also include an interest in possibilities for change: i.e. a study of possible trajectories in design-space and niche-space, as described above. In particular questions about the *power* of an architecture may need to distinguish the power of the system at any particular time and the potential for increased power through learning and self-modification: consider the difference between a newborn human infant and other newborn mammals which walk, find the mother's nipple, and even run with the herd shortly after birth.

8 DESIGNS FOR A NEW PHILOSOPHY

This comparative analysis of types of designs and niches and their relationships is very close to old philosophical problems about the nature of mind, intentionality, consciousness, etc.

One difference is that whereas older philosophers used to ask questions like: "What is a mind?" or "What are the necessary and/or sufficient conditions for something to be conscious?" we can now ask "How many different kinds of minds are there and how do they differ in their architectures and their capabilities?" These questions unify philosophy, psychology, biology and AI. (Though we must resist any temptation to assume that the concept of a mind is initially clear, or that there are sharp boundaries between things with and things without minds!)

In philosophy, there is a long tradition of linking the possession of mental states (beliefs, desires, intentions, etc.) with rationality. This tradition has recently manifested itself in Dennett's notion of the "intentional stance" and Newell's "Knowledge level", both of which require that actions be explainable in terms of beliefs and desires as if the agent were rational. However from our broader standpoint we can explore a variety of more or less "rational" architectures and assess them from different standpoints. E.g. for genes to perpetuate themselves it may be *essential* that agents sometimes behave in a manner that is not rational from the agent's viewpoint. There are many ways in which exploring design space can shed light on philosophical problems.

9 IS THE TASK TOO HARD?

Given the enormous diversity in both design space and niche space and our limited understanding of both, one reaction is extreme pessimism regarding our ability to gain significant insights. My own attitude is cautious optimism: let us approach the study from many different directions and with many different methodologies and see what we can learn. Even the discovery that a particular approach does not get very far is an advance in knowledge.

In particular, the Cognition and Affect group at Birmingham has been trying to use a combination of philosophical analysis, critical reflection on shared common sense knowledge about human capabilities, analysis of strengths and especially weaknesses in current AI systems, and where appropriate hints from biology, psychology, psychiatry and brain science, to guide a combination of speculation and exploratory implementation (e.g. using the general-purpose Sim_agent toolkit [16]). The implementations inevitably lag far behind the speculation! The rest of this paper illustrates some of the speculation regarding functional decomposition [2]. I have speculated elsewhere about the diversity of forms of representation required in systems with human-like intelligence [3].

10 "BROAD" AGENT DESIGNS

For now, let us ignore most of the types and levels of architecture and focus mainly on the highest level functional architecture: the global organisation of a collection of coexisting, interacting, capabilities, each of which may be described at a high level of abstraction, for instance, receiving or collecting information from the environment, analysing such information, interpreting the information; making plans to modify the environment, modifying the environment, monitoring modifications; generating new motivators, assessing motivators, working out costs and benefits of motivators, assessing likelihood of success, deciding whether to accept or reject them; monitoring internal processes, evaluating internal processes, modifying internal processes; and many more, concerned with different time-scales, different spheres of influence, different purposes. (Not all purposes need ultimately be those of the agent: e.g. much of animal behaviour serves the needs of a community, or a gene-pool, rather than the individual.)

This focus on the problem of combining a large number of diverse kinds of functionality, each of which may not (at first) be specified or modelled in much depth, has been dubbed the "broad and shallow" approach by the OZ group at Carnegie Mellon University [1].

11 THREE LEVELS OF CONTROL

Within this framework I'd like to offer some speculations about the gross features of the human information processing architecture. These speculations are prompted by reflection on (a) many facts about human capabilities, (b) considerations regarding evolution of intelligence and (c) engineering design considerations inspired by reflection on limitations of current AI systems.

A brain is, above all, an information processing control system. I'd like to suggest that there are three rather different sorts of control, which might have evolved at different times.

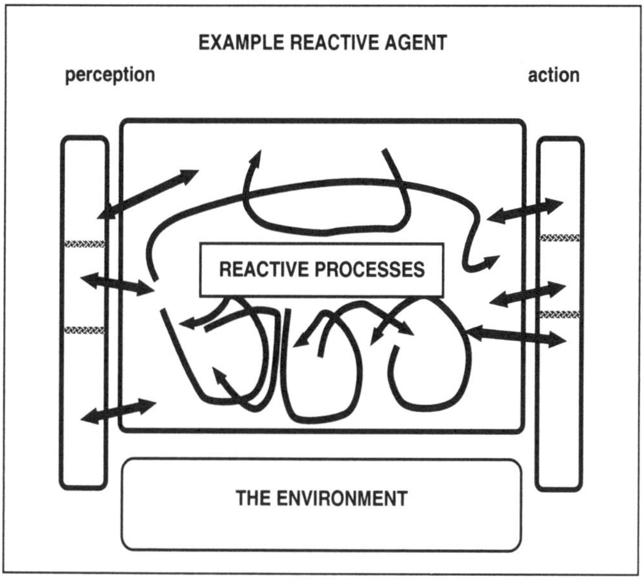

Figure 2. An architecture for a reactive agent

11.1 A reactive subsystem

The first sort has been the focus of a lot of interest in recent years, in connection with "reactive" agents. In a purely reactive agent (or one sort of reactive agent) information is acquired through external sensors and internal monitors and propagates through and around the system, and out to effectors of various kinds, as indicated roughly in Figure 2.

This leaves open the possibility of some effects being counterbalanced by opposing tendencies, or some of the outputs of sub-components being gated or inhibited by others. Many different relatively unintelligent mechanisms of conflict resolution can fit into a reactive system. What a purely reactive system cannot do is explicitly construct representations of alternative possible actions, evaluate them and choose between them, all in advance of performing them.

Processes occur in parallel in a reactive system because there are dedicated coexisting circuits. I presume there are many organisms like that (e.g. insects), and older, more primitive parts of the human brain are also like that.

In human beings, and possibly other animals, there are forms of learning, or rather training, that extend the capabilities of the reactive sub-mechanism. Thus we can distinguish designs for reactive systems that are largely static (apart from dynamic tuning of feedback loops perhaps), and designs that are extendable, pos-

[2] Reported in several previous papers [15, 7, 8, 9, 10, 2, 14, 17]. Compare [5].

[3] E.g. see [6, 11, 13]

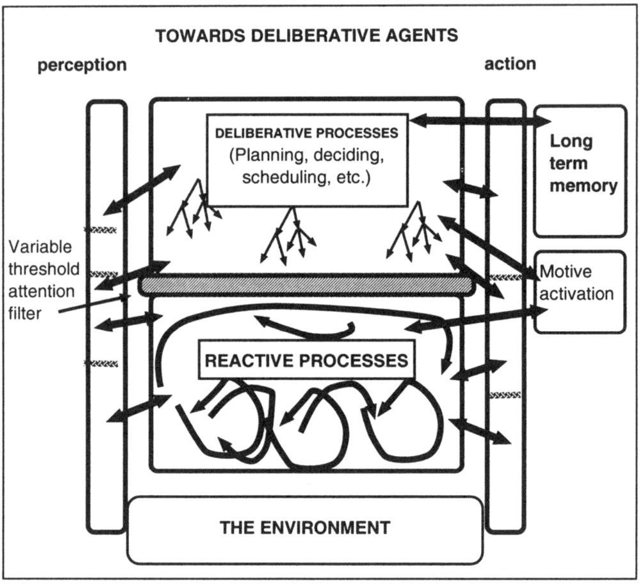

Figure 3. A hybrid reactive and deliberative agent

sibly under the control of other mechanisms within the global architecture.

11.2 A deliberative subsystem

One of the major characteristics of a reactive system as conceived here is that all responses, whether internal or external, happen as soon as their triggering conditions are satisfied (provided that the response is not inhibited as a result of another reactive mechanism.) This principle of automatic triggering is independent of how the system is implemented, e.g. whether it uses a collection of neural networks, or condition-action rules in a symbolic rule interpreter, or something like procedure calls in a programming language, or just a hard-wired circuit.

If such a system is well matched to its niche, the fact that it is relatively inflexible and unintelligent is of no concern. It could be that insects are like this. Perhaps those mammals (e.g. deer) which are born with sophisticated capabilities that enable them to run with the herd also have an essentially reactive control system.

Such a system can break down when the pre-designed collections of conditions for triggering responses are confronted with new situations for which no appropriate responses are available. This is typical of the sort of niche that requires our second main type of control architecture, a “deliberative” architecture which is able to assemble new combinations of actions to cope with novel contexts, as indicated roughly in Figure 3.

In general the space of such combinations is explosive in its complexity[4], and that means that if the new combinations have to be tried out by acting on them a very large number of experiments will be required, which may be both time consuming and very dangerous. So it is beneficial if the search can be done hypothetically, using some kind of model which is evaluated internally.

That sort of niche requires designs that include a type of memory in which temporary structures can be created, evaluated and then tried out. It may require storage of a number of different temporary structures, e.g. alternative plans that have to be compared in some way prior to selection. (This is the core difference between a deliberative and a purely reactive system.)

The processes which create, modify, compare, evaluate, and select such new structures may themselves be implemented using more primitive reactive systems, which unlike the previous ones are primarily concerned with operations on an internal world rather than operations on the environment, though the result of their manipulations can be improved ability to operate on the environment.

This kind of deliberative mechanism, by definition, does not have pre-allocated resources for various functional capabilities: rather it is using a general subsystem to create and evaluate new capabilities including some which are then rejected.

There are many implications of this. In particular, because the same facility is being re-used for different sub-tasks, questions about resource limitations arise, which are not relevant to reactive systems where dedicated circuits exist for the different sub-capabilities. Other obvious questions arise, such as whether and how these newly created structures can be stored and retrieved in similar contexts in future.

Yet another problem is whether the re-activation of a previously constructed plan necessarily makes use of the same mechanisms as create new solutions to problems, so that it is not possible then to use the deliberative mechanism to solve a new problem while one of its previous products is being used.

A possible solution is to transfer newly constructed solutions to the reactive subsystem, where they can in future be run in parallel with new deliberative processes. This seems to be a feature of many kinds of human learning, including familiar examples such as learning to drive a car, learning to read text or sight read music, becoming a fluent programmer, learning many sporting skills.

The diagrams above and below are intended to indicate that perceptual subsystems and action subsystems can develop different levels of abstraction corresponding to the different requirements of the more central architectures to which they are connected. For instance a deliberative architecture may require the perception of abstract "affordances" in the environment in order to perceive the possibility of a particular step in a plan that is being considered. Moreover social agents and agents that interact with other animals may need to be able to infer complex internal mental states from observed facial expressions posture, gestures, etc. Our

[4]If K choices have to be made from N types of components there will be of the order of N^K possible combinations.

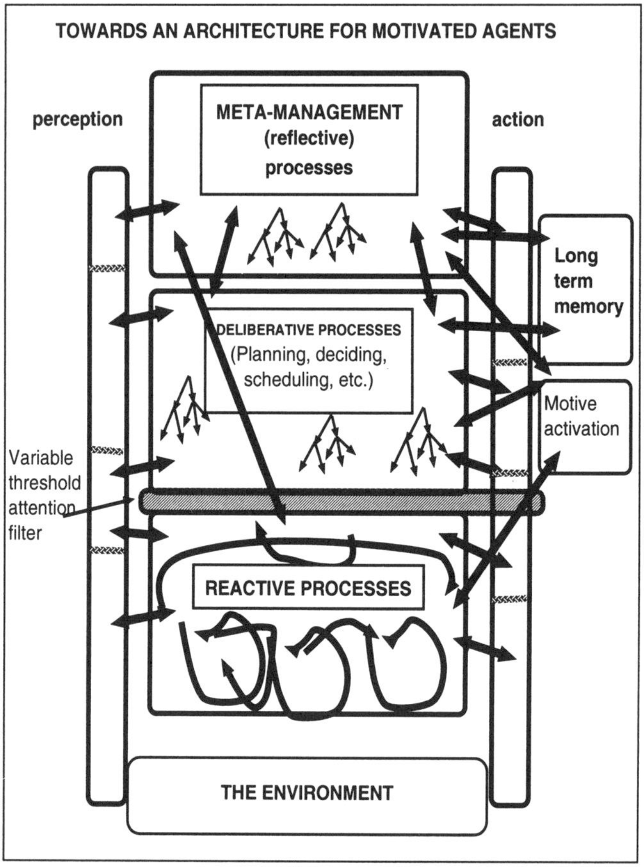

Figure 4. Adding a meta-management layer

ability to see a face as happy or sad or threatening would be examples.

In previous papers my colleagues and I (largely inspired by [5]) have been exploring some of the consequences of the division of labour between a reactive system and a deliberative system, including the implications of concurrent triggering of new motives by the reactive system, sometimes when the deliberative system is overloaded, necessitating some sort of "attention filter" to protect processes that are urgent, important and difficult. Some emotional states can be interpreted as arising out of "perturbances" in such an architecture [17].

11.3 A meta-management subsystem

The third sort of control system, which we have previously described as a meta-management system (e.g. [2, 14, 17]) is concerned with monitoring and control of the deliberative mechanism, as indicated in Figure 4.

The idea is that just as a reactive system may suffer from excessive rigidity in a changing environment, so may a deliberative mechanism. In particular since the environment of the deliberative system is in part the internal architecture of the agent, and since that environment changes as the products of the deliberative system are stored and made available for future use, it is very likely that what works in the early stages of an agent's development may not be very good at much later stages. For this and other reasons it would be useful for internal monitoring mechanisms to be able to keep records of processes, problems, decisions taken by the deliberative mechanism, and perform some kind of evaluation, relative to high level long term generic objectives of the agent (some of which might be determined genetically, and some of which might be learnt in some way, including possibly being absorbed from a culture).[5]

Generic objectives could include such things as not failing in too many tasks, not allowing the achievement of one goal to interfere with other goals, not wasting a lot of time on problems that turn out not to be solvable, not using a slow and resource-consuming strategy if it turns out that a faster or more elegant method is available, and detecting possibilities for structure sharing among actions.

Although such a meta-management system may have a lot in common with a deliberative sub-system, the point of making the distinction is that the deliberative mechanisms *could* exist without the kinds of self-monitoring and self-assessing capabilities just described. In fact, I conjecture that comparative studies will show that that is the case in many animals. Moreover just as deliberative mechanisms can vary in their scope and sophistication so also can meta-management mechanisms.

It might be argued that if meta-management is needed then so also is meta-meta-management, and so on. However, the three kinds of subsystems may suffice if the kinds of self-monitoring and self-modifying capabilities which I have ascribed to the third layer can be applied to itself. We then need no new kind of subsystem.

There are many unanswered questions. For example, experience with computing systems suggests that it is difficult or impossible for everything to be monitored: in fact in the limiting case that would produce an infinite regress of monitoring mechanisms. It may also be the case that there are incompatibilities between the requirement for certain processes to be internally monitored and the requirement for them to run fast on dedicated circuits. This could imply, for example, that the self-monitoring mechanisms used for meta-management cannot have direct access to all the details of the workings of the reactive system.

To overcome this, special additional circuits within the reactive system might be used to transfer information about low level processes to deliberative and meta-management processes which can use it for high level evaluations of current activities. Such "internal perception" mechanisms could simplify and abstract, if that suffices for the job, in which case higher levels will have access only to in-

[5]For more on reasons for self-monitoring see [4].

complete and possibly misleading information about what is going on, not unlike senior management in a large organisation!

These design problems are relevant to a lot of contemporary discussions about consciousness, qualia, and the role of introspection. My own view is that the vast majority of what is written on such topics (even by distinguished scientists) is of dubious value because it has not been based on an implementable theory of the architecture which could support the concepts used by the discussants. (I am not restricting consideration only to computational implementations.)

12 FURTHER QUESTIONS

The sort of discussion presented here needs to be combined with the more familiar AI research on formalisms and algorithms. It could well turn out that quite different formalisms are suited to the different tasks. Different formalisms and ways of manipulating them may require the existence of different kinds of representational media.

In particular a reactive subsystem may be able to use forms of representation and control which are not suited to a deliberative system, including, in the extreme case, hard-wired circuits and reflexes. If so that raises interesting problems about what happens when as a result of training new structures created by the deliberative system get implanted (or transplanted?) to the reactive subsystem.

Is the very old idea that some forms of learning are a bit like compiling from a high level to a low level language supported by this?

Alternatively might it be that the very information structure that is created by a deliberative mechanism can also be used by a reactive system, but in a far less flexible (though speedy) fashion?

Too often it seems that debates about mechanisms and formalisms (e.g. logical notations vs neural nets) are conducted in a spirit in which issues of partisanship, or fashion, have more influence than scientific considerations. I suspect that by asking how all the various components can be put together into complete working systems we may be able to make more progress with such problems and even learn that instead of having to choose between apparently incompatible options we have to use both, but in different parts of the system. In short, debates about which sorts of formalisms are best should be replaced by investigations mapping formalisms to tasks, within the more general study of relations between designs and niches.

13 OTHER ASPECTS OF THE ARCHITECTURE

Claiming that an architecture has reactive, deliberative and meta-management subsystems does not imply that each of these is a monolithic mechanism, or that everything in the architecture must fit neatly into one of these categories.

Perception is an interesting example. In an agent whose complete architecture is reactive, perceptual mechanisms will use fixed algorithms for analysing their input and determining what should be sent on to other parts of the system. Where the architecture includes a deliberative component, however, a perceptual system could have a dual role, namely both feeding information directly into the reactive subsystem and also collaborating with the deliberative system when it constructs and evaluates alternative possible action plans. A chess-player working out what move to make will often find it useful to stare at the board and use it as an extension of short term memory (though a more advanced player can do this all internally). Similarly an animal considering how to pick something up, or which route to take across a cluttered environment, may find that the problem is easier to solve while the environment is visible, again because the perceptual structures form part of the re-usable short term memory structure required for creating and evaluating options.

The often rediscovered fact that humans use spatial representations for solving many kinds of problems, including some very abstract problems, may be a manifestation of the overlap between a spatial perception mechanism and the deliberative mechanism. On the other hand, the visual feedback that allows smooth and rapid movement of a hand to pick up a cup could be an example of a deep connection between spatial perception and some reactive mechanisms.

If all this is correct, perceptual mechanisms are neither entirely in the reactive subsystem nor entirely in the deliberative subsystem. Similar comments could apply to the motor output system, if the reactive subsystem sometimes controls it and at other times the deliberative subsystem takes over, or if both can be simultaneously involved in different aspects of the control of behaviour, e.g. thinking about phrasing and dynamics by performing a well-rehearsed piece of music.

A different sort of point concerns the question whether within the perceptual system there is a need for a distinction between reactive and deliberative subsystems. It may be that the perception of complex structures (e.g. hearing grammatical sentence structures, or seeing a complex piece of machinery) requires some ambiguities of parsing or local interpretation to be resolved by temporary construction of alternatives which are compared. If so, a perceptual mechanism may need to include something analogous to deliberative mechanisms, though possibly tailored specifically to the tasks and forms of representation in that mode of perception. (This was taken for granted in much AI vision research in the 1960s and 1970s, but later went out of fashion.)

14 MOTIVATION

I have hinted that new motives can be generated asynchronously in different parts of the system. How all these motives are managed is a complex topic that has not been investigated much in AI[6].

In psychology and neuroscience, I have the impression that much of the study of motivation, emotions and related states and processes, has assumed that humans

[6]Though see [2] and references therein.

are essentially the same as other animals, such as rats. This assumption may be misleading. Motivational processes in an agent whose deliberative mechanisms can explicitly represent the long term future may have significant additional complexity compared with the processes that occur in a rat, for example. Can the latter feel humiliated, guilty, awe-struck or driven by a long term ambition?

Agents that can learn through positive and negative reinforcement will have their motivational mechanisms linked to their learning mechanisms so that rewards and punishment bring about changes. Agents that also include meta-management, i.e. agents that are capable of monitoring, evaluating, and modifying high level aspects of their own internal processes, will be capable of having very abstract types of motivation that simply could not occur in agents with simpler architectures, for instance the desire to be an honest and generous person.

The three layers support very different sorts of mental processes, including motivational and emotional processes, a fact that has not been noticed among emotion theorists, leading to a plethora of different definitions of "emotion" and unrelated theories of emotion, which appear to contradict one another, but are actually talking about different things. Our diagnosis is that those who stress emotions based on the limbic system and observable in rats and most other animals are studying effects of the reactive layer. Those who stress emotions such as apprehension, disappointment and relief, related to phases in the execution of plans, are studying effects of the deliberative layer. By contrast poets, novelists and those who study emotions involving loss of control of thought processes (e.g. our work on grief and perturbances [17]), are studying processes involving the reflective, or meta-management layer.

There is much more to be said about motivation, moods, character, personality, and the like. In particular, requirements for concurrency and independence of various subsystems can lead to a variety of kinds of states in which subsystems disturb one another, possibly producing less than optimal global performance. Some human emotional states, including states that are too sophisticated to occur in rats, may be like that.

Some AI researchers believe that it should be the goal of AI to design agents that overcome human limitations while displaying all their strengths. This may not be possible if some of the limitations are inevitable consequences of the mechanisms and architectures required to produce those strengths.

15 CONCLUSION

I have tried to outline a methodology which takes account of the existence of niche space and design space and their relationships.

I have also tried to illustrate the application of this methodology to the analysis of a particular class of designs and niches, showing how this might be achieved using an architecture which (among other things) has reactive, deliberative and meta-management components (a trio that may correspond loosely to old and familiar concepts from philosophy, psychology and common sense).

What I have not done is to spell out examples of complete working architectures to show what kinds of ontologies for mental states and processes they support and how well they can explain sophisticated aspects of human mentality. This is ongoing work; for more on our work see the project ftp directory:

`ftp://ftp.cs.bham.ac.uk/pub/groups/cog_affect`

16 ACKNOWLEDGEMENTS

This research was funded in part by: the Renaissance Trust, the UK Joint Council initiative on Cognitive Science and HCI, and DRA Malvern. I have received much help from colleagues and students at Birmingham, including Luc Beaudoin, Chris Complin, Darryl Davis, Glyn Humphreys, Brian Logan, Riccardo Poli, Christian Paterson, Ed Shing, Tim Read and Ian Wright.

This is a slightly expanded version of an invited paper presented in August 1996 at the Cognitive Modeling Workshop at AAAI96, Portland Oregon.

Aaron Sloman (University of Birmingham, UK)

REFERENCES

[1] J. Bates, A. B. Loyall, and W. S. Reilly. Broad agents. In *Paper presented at AAAI spring symposium on integrated intelligent architectures*, 1991. (Available in SIGART BULLETIN, 2(4), Aug. 1991, pp. 38–40).

[2] L.P. Beaudoin. *Goal processing in autonomous agents*. PhD thesis, School of Computer Science, The University of Birmingham, 1994.

[3] J. Cohen and I. Stewart. *The collapse of chaos*. Penguin Books, New York, 1994.

[4] J. McCarthy. Making robots conscious of their mental states. In *AAAI Spring Symposium on Representing Mental States and Mechanisms*, 1995.

[5] H. A. Simon. Motivational and emotional controls of cognition, 1967. Reprinted in *Models of Thought*, Yale University Press, 29–38, 1979.

[6] A. Sloman. Interactions between philosophy and ai: The role of intuition and non-logical reasoning in intelligence. In *Proc 2nd IJCAI*, London, 1971. Reprinted in *Artificial Intelligence*, pp 209-225, 1971, and in J.M. Nicholas, ed. *Images, Perception, and Knowledge*. Dordrecht-Holland: Reidel. 1977.

[7] A. Sloman. Motives mechanisms and emotions'. *Emotion and Cognition*, 1(3):217–234, 1987. Reprinted in M.A. Boden (ed), *The Philosophy of Artificial Intelligence*, 'Oxford Readings in Philosophy' Series, Oxford University Press, 231–247, 1990.

[8] A. Sloman. On designing a visual system (towards a gibsonian computational model of vision). *Journal of Experimental and Theoretical AI*, 1(4):289–337, 1989.

[9] A. Sloman. Prolegomena to a theory of communication and affect. In A. Ortony, J. Slack, and O. Stock, editors, *Communication from an Artificial Intelligence Perspective: Theoretical and Applied Issues*, pages 229–260. Springer, Heidelberg, Germany, 1992.

[10] A. Sloman. Prospects for ai as the general science of intelligence. In A. Sloman, D. Hogg, G. Humphreys, D. Partridge, and A. Ramsay, editors, *Prospects for Artificial Intelligence*, pages 1–10. IOS Press, Amsterdam, 1993.

[11] A. Sloman. Semantics in an intelligent control system. *Philosophical Transactions of the Royal Society: Physical Sciences and Engineering*, 349(1689):43–58, 1994.

[12] A. Sloman. Exploring design space and niche space. In *Proceedings 5th Scandinavian Conference on AI, Trondheim*, Amsterdam, 1995. IOS Press.
[13] A. Sloman. Musings on the roles of logical and non-logical representations in intelligence. In Janice Glasgow, Hari Narayanan, and Chandrasekaran, editors, *Diagrammatic Reasoning: Computational and Cognitive Perspectives*, pages 7–33. MIT Press, 1995.
[14] A. Sloman. What sort of control system is able to have a personality, 1995. Available at URL ftp://ftp.cs.bham.ac.uk/pub/groups/cog_affect/ Aaron.Sloman.vienna.ps.Z, (Presented at Workshop on Designing personalities for synthetic actors, Vienna, June 1995).
[15] A. Sloman and M. Croucher. Why robots will have emotions. In *Proc 7th Int. Joint Conf. on AI*, Vancouver, 1981.
[16] A. Sloman and R. Poli. Sim_agent: A toolkit for exploring agent designs. In Mike Wooldridge, Joerg Mueller, and Milind Tambe, editors, *Intelligent Agents Vol II (ATAL-95)*, pages 392–407. Springer-Verlag, 1996.
[17] I.P. Wright, A. Sloman, and L.P. Beaudoin. Towards a design-based analysis of emotional episodes. *Philosophy Psychiatry and Psychology*, 3(2):101–126, 1996.

J. L. POLLOCK

PLANNING AGENTS

1 INTRODUCTION

Agents are entities that act upon the world. Rational agents are those that do so in an intelligent fashion. What is essential to such an agent is the ability to select and perform actions. Actions are selected by planning, and performing such actions is a matter of plan execution. So the essence of a rational agent is the ability to make and execute plans. This constitutes *practical cognition*. In order to perform its principal function of practical cognition, a rational agent must also be able to acquire the knowledge of the world that is required for making and executing plans. This is done by *epistemic cognition*. Rational agents embedded in a realistically complicated world (e.g., human beings) may devote more time to epistemic cognition than to practical cognition, but even for such agents, epistemic cognition is in an important sense subservient to practical cognition.

This general view of rational agents has become enshrined in what are called *BDI architectures*. These endow agents with beliefs, desires (goals), and intentions (plans). The first implemented BDI architecture was PRS [5] . Other interesting architectures of this same genus are represented by HOMER [26] and IRMA [2] . A useful survey can be found in [28] . The OSCAR architecture [16,18] is also of this same general structure. Indeed, the structure is the standard one assumed in philosophical theories of rationality, although philosophers have usually focussed on intentions to perform individual actions rather than multi-step plans. [1]

The AI literature on rational agents is full of references to *autonomous agents*, which are agents that learn. It is a commonly held view that only autonomous agents are "real" rational agents, because others are merely running a program—doing what their makers tell them to do—rather than figuring out for themselves what to do. This view seems to me to be incoherent. First, agents that learn are not somehow magically making up their own program—they are running a learning program. Second, I see no reason to believe that learning is in any way essential to an agent's performing in a sophisticated or flexible manner. The purpose of machine learning is to enable the agent to acquire general knowledge about the world, but that same knowledge can be acquired by *reasoning* from perceptual input if the agent is equipped with the ability to perform inductive and abductive reasoning. I

[0]This work was supported by NSF grant no. IRI-9634106

[1]See my [17,18] for a discussion of this difference.

M. Wooldridge and A. Rao (eds.), Foundations of Rational Agency, 53–79.

have elsewhere talked about *Q&I (quick and inflexible) modules*,[2] which are special purpose modules enabling an agent to perform certain cognitive tasks rapidly (they are "quick") provided the environment cooperates by satisfying appropriate constraints (they are "inflexible"). The kind of learning provided by machine learning programs is best viewed as a Q&I module for inductive and abductive reasoning. Such learning *might* be faster than explicit inductive or abductive reasoning, but that remains to be shown. There is in principle nothing that an agent running a machine learning program can do that an agent equipped with the ability to perform sophisticated epistemic reasoning cannot do. So I think the emphasis on autonomy is a red herring. The emphasis should instead be on planning and plan-execution, and on the kinds of epistemic cognition that the latter require.

Planning is one of the oldest subfields of AI, and it has been a very active subfield. Work in planning has traditionally proceeded in what is essentially a bottom up fashion. "Theories of planning" have been constructed on the basis of logical and mathematical theories of search. The construction of planners has then been viewed as an engineering task that proceeds by applying the basic science provided by these theories. This approach has led to valuable insights, but the resulting planners are in many ways crude. This is for a variety of reasons. First, planning theorists have been quick to identify planning with a certain kind of search, but that leads directly to the view that a correct plan is *guaranteed* to achieve its goals (provided, of course, that the background beliefs of the planning agent are not faulty). Human planning is not like that at all. Plans do sometimes fail because they are based upon false beliefs, but it is much more common for them to fail because they are based upon *incomplete beliefs*. When planning in the real world, we never know everything that is or might be relevant to a planning problem, so there is no way we could conceivably construct plans that are guaranteed to work. That does not stop us from planning. To accommodate such planning within a theory of planning, we need a more sophisticated view of what plans are and what the criteria of adequacy should be for good plans. In the last few years, a number of researchers have begun to address these problems, but so far nothing has emerged to replace the elegant but simplistic view of planning as search that has dominated work on planning.

To make planning more tractable, planning theorists have traditionally built their planners on the basis of some simplifying assumptions. These include most notably the assumptions (1) that the planner has all the relevant knowledge that it needs at the time it begins to plan, and (2) that nothing of importance happens in the world except as a result of the agent's actions. The second assumption is actually a corollary of the first, because exogenous events would be events about which the planner lacks knowledge. Planners based upon these assumptions have substantially advanced our knowledge of the processes by which plans can be discovered or generated, but it is universally acknowledged that these assumptions must eventually be jettisoned and planners constructed that do not depend upon

[2] [19], chapter one.

them. Much recent work has been devoted to relaxing these assumptions, but this work is still in its relative infancy.

We have learned much from the bottom up approach to planning, but the problem of building a sophisticated planning agent is probably too difficult to solve by treating it as a pure engineering problem. We should also approach planning from the top down. The inadequacies of current planners are most easily revealed by looking at the kind of sophisticated planning that human beings perform, trying to describe various aspects of it precisely, and then addressing the question of how an artificial agent might be constructed that is capable of performing such sophisticated planning. An essential aspect of this approach is that planning is viewed as something performed by a planning agent, and planning is integrated into the overall cognitive architecture of the agent rather than being a stand-alone module.

2 INTERLEAVING PLANNING AND REASONING

The first AI planner [6] did its work by reasoning deductively about plans. But almost all subsequent AI planners have worked algorithmically rather than by reasoning. You give them the necessary information and the goal and they run a program that searches the space of possible plans (relative to the given information) until it finds a plan whose execution is guaranteed to achieve the goal. The algorithmic approach to planning has been dictated largely by the failure of AI to solve the frame problem.

Although this is not its official formulation,[3] we can think of the frame problem as the problem of constructing a reasoner that can reason correctly about the consequences of actions. Ideally, a rational agent should come to a planning problem with some general knowledge about the world and some specific knowledge about the current situation, and then it should construct a plan by reasoning about what actions will achieve various results. But, of course, this is impossible if the agent is unable to reason correctly about the consequences of actions. To circumvent this problem, virtually all AI planning systems are based upon some variant of the STRIPS representation of actions.[4] On the STRIPS representation, the human operator does the reasoning about the consequences of actions, and then builds the results of that reasoning into the description of the actions, thus absolving the artificial agent from performing the reasoning it is unable to do. This has generally been accomplished by associating a list of *preconditions* and an *add-list* and *delete-list* with each action. The add-list is the list of literals that will be made true by performing the action when the preconditions are satisfied, and the delete-list is the list of literals that will cease to be true as a result of performing the action when the preconditions are satisfied.

This way of avoiding the frame problem is obviously a kludge. We really want to build agents that do all the relevant reasoning. However, it is a helpful kludge

[3] See my [19] for a discussion of how exactly the frame problem should be formulated.
[4] See [3].

for the purpose of constructing useful planning systems. Unfortunately, its power is limited. First, note that it is important that the add- and delete-lists consist of literals. This is because the planning agent does no reasoning about the consequences of actions. If we were to put, say, a conjunction into the add-list, the agent would be unable to conclude that the conjuncts will be made true as well.[5] This restriction to literals means that the designing of a planning system cannot encode relevant knowledge in a straightforward way, e.g., as a set of first-order formulas. Existing knowledge must be carefully instantiated and refined for use by the planner.

Partially because of these knowledge representation problems, the STRIPS-representation encounters the *qualification problem* and the *ramification problem*. The qualification problem can be regarded as the problem of making the preconditions sufficiently inclusive to ensure that the add- and delete-lists really consist of *guaranteed consequences* of the action given the preconditions.[6] In a realistically complex environment, this desideratum is almost impossible to achieve. We can list consequences that will *normally* ensue when an action is performed, but it is virtually impossible to list consequences that will result no matter what. In the real world, unexpected eventualities can foil even the most sophisticated planning.

This aspect of the qualification problem is only a problem insofar as the task of planning is viewed as that of generating a plan that is *guaranteed* to succeed. Human beings do not view planning in that way. We search for plans that are *likely* to succeed. If we revise our aspirations for AI planners similarly, then this aspect of the qualification problem becomes much less serious.

There remains a residue of the qualification problem that is an obstacle to planning even if we do not insist that our plans are guaranteed to succeed. The world-knowledge that we want to bring to bear on a planning problem tends to be fairly general. But on the STRIPS representation, the list of preconditions and the add- and delete-lists must consist of literals. These are generated by instantiating our general knowledge in various ways. However, there will usually be infinitely many different ways of instantiating our general knowledge. Suppose, for example, I am making a simple plan about how to feed my cat. My general knowledge allows me to infer that I can usually do this by opening a can of catfood, placing the contents in a bowl, and placing the bowl before my cat. However, my general knowledge also allows me to infer that this will not work if my cat is chasing a mouse, or comatose, or frightened by the noise of the dishwasher, etc. These qualifications can be made more complex without limit. Each produces a different STRIPS action (because the preconditions, add- and delete-lists are part of the action). Which

[5]Recently, some planners have been built that do a limited amount of reasoning about consequences and hence are able to use a more expressive formalism. For example, both UCPOP [13,27] and PRODIGY [25] are able to reason about disjunctive goals and universally general goals. However, they are based upon algorithms requiring that the effects of an action be computable, so the logical consequence relation must be decidable. That rules out a truly expressive formalism like full first-order logic.

[6]The qualification problem originated with [12], where it was formulated more generally as a problem for axiomatizing real-world domain in the situation calculus.

actions should be provided to the planner?

The ramification problem is closely related to this residue of the qualification problem. The ramification problem is the problem of deciding which consequences of actions to build into the STRIPS actions. For example, consider the consequences of striking a match. If we think about it for a while, we can enumerate among the effects such things as displacing air around the match, marginally depleting the ozone layer, raising the temperature of the earth's atmosphere, illuminating Alpha Centauri a little, making that match unavailable for future use, etc. These are effects that we typically do not care about, and so we do not reason about them. But this does not mean that we can omit them from the add- and delete-lists with impunity, because occasionally we might care about one of them. What this illustrates is that which consequences we build into a STRIPS-action must be determined by which planning problem we are trying to solve. For example, I would not ordinarily think about the fact that feeding my cat results in his gaining a small amount of weight, but if he is precariously balanced on a board that will collapse if he gains that weight, then my planning must take that into account and include a step to reinforce the board or move the cat.

Human planners solve both these problems by reasoning. They take account of only those qualifying circumstances that they think are likely to occur, and only those consequences that are relevant to their present interests. On the STRIPS-representation, this puts a very heavy burden on the reasoning of the human operator, and means in effect that the planner must be re-tailored for each planning problem. Unfortunately, it is generally impossible to decide what STRIPS-actions to provide for the planner without first solving the planning problem. For example, when reasoning by goal-regression, a human planner considers what he or she might do to achieve a certain goal, and instantiates his or her general knowledge in ways that will answer this question. What is important here is that the planning is interleaved with epistemic reasoning aimed at finding appropriate information for use in the planning, and the course of the epistemic reasoning is directed by where the agent is in its planning. In effect, requiring the human operator of an AI planner to perform this epistemic reasoning in advance requires him or her to already know how the planning problem will be solved, but this makes the planner superfluous.

I take it that what the qualification and ramification problems illustrate is the importance of interleaving planning and epistemic reasoning, and performing the epistemic reasoning in a way that is directed by or focused by the practical considerations involved in the planning. The epistemic reasoner must be *interest-driven* in the sense that it tries to answer particular queries that are passed to it by practical cognition. These will be factual queries of relevance to the planning problem, and will arise *in the course of* planning. We cannot assume that the planner comes to the planning problem equipped with precisely the knowledge it needs to solve the planning problem (and without the ability to do further reasoning) without ignoring essential aspects of planning and cognition.

3 SCHEMATIC PLANS

What I have been arguing is that epistemic cognition and planning cannot be separated and performed by different unrelated cognitive modules. The ramification and qualification problems illustrate that one reason for this is that realistic planners cannot be supposed to come to their planning problems equipped with all the knowledge they need to solve them. Instead, *in the course of planning*, they must adopt epistemic interest in various questions whose answers impinge on the planning process, and undertake epistemic investigations aimed at answering those questions. This means that *the course of planning effects the course of epistemic cognition*. The converse is true as well. Plans do not spring full-blown, with all their details specified, from a single planning session. This is because, no matter how much epistemic cognition we have performed, we never have all the relevant knowledge. We cannot know all the details relevant to even so simple a plan as walking across the room (e.g., will the cat run underfoot?). Human beings accommodate this by adopting plans that are to varying degrees schematic, and then try to make them more precise as further knowledge is acquired. This has not gone overlooked in AI planning. One aspect of it is captured by hierarchical planning.[7] In hierarchical planning, plans are constructed using "coarse-grained operations" like FLY_TO_LA or GET_A_PHD_IN_COMPUTER_SCIENCE or MAKE_A_CUP_OF_COFFEE or WALK_ACROSS_THE_ROOM. On the basis of general knowledge, we think it likely that we will be able to perform these operations. However, in order to perform these operations we must engage in further planning. We may not be able to do that until we acquire further knowledge that is not currently available. For example, in order to fly to LA I must buy a plane ticket, and in order to do that I must acquire knowledge about flight schedules. By constructing a plan using the coarse-grained operator FLY_TO_LA, I am able to go ahead and plan without first acquiring knowledge about flight schedules. When such knowledge later becomes available, I can refine my plan to take account of it.

Hierarchical planning captures only one aspect of the schematicity of human planning. Another respect in which human plans are schematic is scheduling. Most practical AI planners separate planning and scheduling. First they adopt a plan, and then they construct a schedule for when to perform the plan-steps. This is fairly obviously inadequate. The acceptability of a plan is certainly dependent upon our being able to schedule it in a reasonable way. Thus far I have not remarked on the fact that human planning is in some sense decision-theoretic. I have observed that we do not require plans to be guaranteed to achieve their goals—we only require it to be likely that they will do so. But even that is an oversimplification. We weigh the value of the goals discounted by the probability of success against the expected cost of executing the plan. This yields an expected value for the plan, and we decide whether to adopt a plan by comparing its expected value with that of competing plans.[8] It is apparent that the schedule that is attached to

[7]Hierarchical planning was introduced by Sacerdotti [21,22].

[8]See chapter five of my [18] for a discussion of the logic of the reasoning involved in the compari-

a plan can profoundly affect its expected value. Thus the schedule must be an integral part of the plan when it is being evaluated for adoption.

Although the schedule must be part of the plan, human beings rarely schedule the steps of a plan with much precision. For example, I may plan to do something *this afternoon*. Rarely do I plan to do something at 3:07:13 PM. This is not because I am being intellectually sloppy. It is good rational policy. First, most small differences in scheduling do not affect the expected value of a plan to any appreciable degree. Second, by leaving unimportant scheduling decisions indeterminate, one maintains flexibility. An agent residing in a realistically complicated world will continually receive new information through perception, and some of this new information will present new opportunities that provide new goals and call for new planning. Ideally, planning for new goals should disrupt previous planning as little as possible. This is, to my mind, the principal virtue of *least commitment planning*.[9] By constructing plans that do not make unnecessary commitments, either to the ordering of plan-steps, or to the times of execution of plan-steps, or to precisely how to execute coarse-grained plan-steps, we maximize the likelihood that the plans will not conflict with new planning.

I have argued that it is often good rational policy to construct schematic plans. But of course, at some point the schematicity must be eliminated. In order to execute a plan step, the agent must decide precisely how and when to do it. It was once common to suppose that planning and plan execution were separate processes, the latter being essentially simple—something like running a computer program. As planning theorists have come to appreciate that planning is generally based on incomplete and sometimes inaccurate information, they have also come to realize that plan execution is not so mechanical. At the very least it involves monitoring the course of the plan execution and replanning if things do not go as expected. But in fact, it involves more than that. The schematicity of plans generally extends into the period of plan execution. For example, suppose I plan to walk into the kitchen and fix a cup of coffee. My plan may initially consist of three steps: (1) walk to the kitchen door, (2) walk from there to the vicinity of the espresso machine, (3) prepare a cup of espresso. Each step is highly schematic. In executing the first step, I start walking in the direction of the kitchen door, but I do not plan precisely where I am going to place my feet before I begin the walk. Instead, the details of each step are determined on the fly as obstacles appear in my course or in response to my own sense of balance and knowledge of how I must place my feet in order to avoid falling down. Once I find myself in the vicinity of the espresso machine, I may then and only then decide on the order in which I perform the various operations required to make espresso. For example, whether I grind the beans first or put water in the boiler first may depend upon just where I find myself, in what order the various operations occur to me, or mere whim. As far as scheduling, I may never construct a precise schedule. At some point I decide to do something

son.

[9]See [27] for a general discussion of least commitment planning.

now, and then after I have done that I decide to do something else *now*, and so on.[10]

4 PLANNING BY REASONING

The upshot of the previous sections is that epistemic cognition, planning, and plan execution must all be interleaved. The course of each affects the course of the others. Perception and epistemic cognition based upon it gives rise to knowledge of opportunities, and this gives rise to goals. Adopting goals initiates planning for their achievement. Planning poses queries that are passed to epistemic cognition in an effort to acquire the knowledge needed for planning. Plan execution requires monitoring the world (epistemic cognition) and refinement, revision, or abortion of the plan as execution proceeds. Thus a realistic rational agent must perform all of these tasks in unison, not sequentially.

As I remarked above, virtually all contemporary planners are algorithmic planners. Plans are found by running a search algorithm rather than by reasoning. It is extremely difficult to see how to interleave algorithmic planning and epistemic cognition. An algorithmic planner depends upon already having all the information it needs, and if some of that information changes during the course of the plan search, it appears that an algorithmic planner will simply have to start all over again. The only obvious way to interleave planning and epistemic cognition is to do the planning by reasoning, just as human beings do.

In order to build a planner that plans by reasoning we must solve three problems. First, we must have an adequate reasoner. AI has many automated deductive reasoners, but deductive reasoning does not provide an adequate basis for epistemic cognition. This has been acknowledged in philosophical epistemology for the last thirty years, and I hope I am right in thinking that this is now generally appreciated in AI as well. If not, all I can do is point my AI colleagues to the philosophical literature on the topic.[11]

There is a large AI literature on non-deductive reasoning. The earlier literature goes under the name "nonmonotonic logic", and more recently the term "defeasible reasoning" (taken from philosophy) has become popular. The importance of defeasible reasoning has been stressed by many important AI theorists, but there has not been a consensus on how defeasible reasoning should work and how it can be implemented. Without an implemented defeasible reasoner, we cannot implement a system of epistemic cognition, and without that we cannot implement a planner that plans by reasoning.

Given an automated defeasible reasoner, the second problem that must be solved

[10]HOMER [26] is an agent architecture that is designed to accommodate this observation about schematicity and plan execution. HOMER constructs hierarchical plans and produces subsidiary plans for executing high level operators as additional knowledge is acquired during the course of plan execution.

[11]As a start, I would suggest my own [15,18].

is the construction of a general agent architecture that incorporates the system of defeasible reasoning and integrates it with planning and epistemic cognition. The third problem is that planning by reasoning runs afoul of the frame problem. Without a solution to the frame problem, we will be unable to build an artificial agent that is able to construct plans by reasoning.

Fortunately, I do not think that any of these problems are as serious as they once were. The OSCAR architecture for rational agents is based upon a general purpose defeasible reasoner, and is designed to integrate epistemic and practical cognition. I cannot claim that there is general agreement that OSCAR does it right, but I am convinced that OSCAR at least does it for the most part right, and perhaps OSCAR's success at providing the inference engine required for many AI purposes will inspire increasing confidence in OSCAR's correctness. That still leaves the frame problem as an obstacle to building planners that plan by reasoning. Fortunately, the frame problem is no longer so mysterious as it once was. In the last couple of years several authors have proposed solutions.[12] In particular, I have proposed a solution that is implemented in OSCAR and enables reasonably efficient reasoning about the consequences of actions. So OSCAR provides a basis upon which to at least begin the process of building agents that plan by reasoning. In the next section I will give a brief description of the OSCAR architecture, and in the following section I will sketch my proposed solution to the frame problem. Then I will return to the question of how to build a planning agent.

5 THE OSCAR ARCHITECTURE

OSCAR is an architecture for rational agents, based upon the world's only implemented general-purpose defeasible reasoner.[13] The architecture divides cognition into epistemic cognition (cognition about what to believe) and practical cognition (cognition about what to do). Epistemic cognition is interest-driven in the sense that it proceeds bidirectionally. The reasoner reasons forwards from perceptual input, and backwards from queries it is trying to answer. Initially, queries are posed by practical cognition, and then as reasoning proceeds, new epistemic interests are generated by reasoning backwards from the initial ones.

The architecture for epistemic cognition is diagramed in figure 1. Reasoning begins with perceptual input and ultimate-epistemic-interests. The latter are queries that are passed to epistemic cognition from practical cognition, without being derived from other interests by backwards reasoning. Conclusions are stored as nodes ("inference-nodes") in the *inference-graph*, which records both inference-relations between conclusions and defeat-relations between conclusions. Interests are stored in the *interest-graph*, which records their inference-relations. Conclusions have

[12] See Shoham [24], Gelfond and Lifschitz [4], Lin and Reiter [9,10], and Shanahan [23].

[13] The theory underlying OSCAR is described in detail in my [18]. The implementation is described in The OSCAR Manual. Both the manual and the implemented system can be downloaded from http://www.u.arizona.edu/~pollock/.

degrees of justification attached to them (numbers between 0 and 1) and interests have degrees of interest. These numbers are used in three ways. First, degrees of justification are used in adjudicating conflicts and computing defeat statuses. Second, they are used in interest discharge. The higher the degree of interest, the more serious the matter at issue, and so the more justified a conclusion must be to settle the matter. So for a conclusion to discharge an interest, its degree of justification must be at least as great as the degree of interest. Backwards reasoning from an interest creates *interest-links*, which link a set of interests to the original interest. The significance of an interest-link is that if conclusions are drawn discharging the link-interests, then a conclusion discharging the original interest can be inferred from those conclusions. The third way in which degrees of interest and degrees of justification are used is in prioritizing the reasoning. Newly drawn conclusions and newly constructed interests are placed on the *inference-queue*, and then the next step of reasoning is performed by retrieving the first member of the inference-queue and processing it. The ordering of the inference-queue is based, among other things, on the degrees of justification and the degrees of interest of the items on it. Other things being equal, the more certain a conclusion or the more serious an interest, the higher its priority.

Reasoning is mediated by reasons: forwards-directed reasons mediate forwards-reasoning (reasoning from conclusions to conclusions); backwards-directed reasons mediate backwards-reasoning (reasoning from interests to interests). OSCAR appears to be unique in that its forwards-directed reasons are entirely distinct from its backwards-directed reasons. For example, ADJUNCTION is a backwards-directed reason, instructing the reasoner that if it is interested in $(P \wedge Q)$ then it should become interest in P and in Q. Simplification is a forwards-directed reason instructing the reasoner that if it has concluded $(P \wedge Q)$ then it should infer P and Q from that conclusion. This structure of reasoning makes OSCAR very efficient.[14]

Reasoning proceeds in terms of reasons. *Backwards-reasons* are used in reasoning backwards, and *forwards-reasons* are used in reasoning forwards. Forwards-reasons are data-structures with the following fields:

- reason-name.
- forwards-premises — a list of forwards-premises.
- backwards-premises — a list of backwards-premises.
- reason-conclusions — a list of formulas.

[14]In a recent comparison of OSCAR with more standard resolution-based theorem provers, OSCAR proved to be forty times faster than Otter, a highly regarded state-of-the-art resolution-based theorem prover. The comparison was arranged by Geoff Sutcliffe, librarian of the TPTP library of theorem proving problems. It is noteworthy that OSCAR did so well despite the fact that OSCAR only does theorem proving as a side effect of its general reasoning, and is burdened with a great deal of extra machinery that is required for defeasible reasoning. In addition, Otter is written in C while OSCAR is written in LISP. This should give Otter an advantage of as much as an order of magnitude.

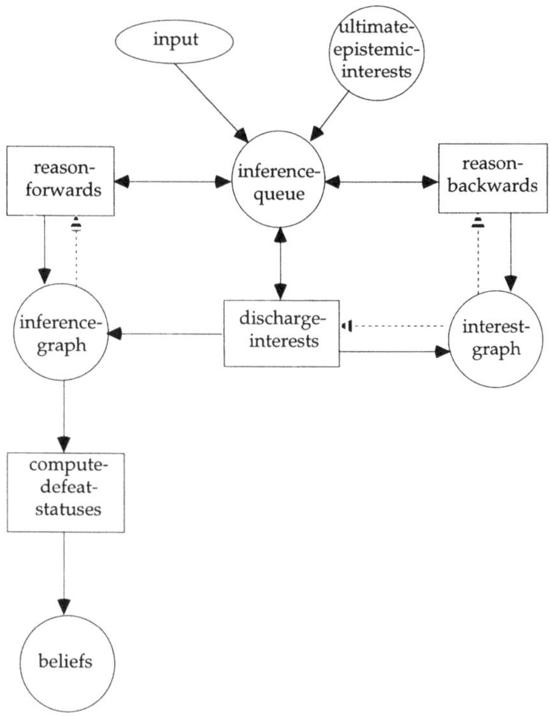

Figure 1. The defeasible reasoner

- conclusions-function — an optional function used for computing the conclusions of an inference made in accordance with the reason. If the function is not supplied, the conclusions are computed by instantiating the reason-conclusions.
- defeasible-rule — T if the reason is a defeasible reason, NIL otherwise.
- reason-variables — variables used in pattern-matching to find instances of the reason-premises.
- reason-strength — a real number between 0 and 1, or an expression containing some of the reason-variables and evaluating to a number.
- reason-description — an optional string describing the reason.

Forwards-premises are data-structures encoding the following information:

- fp-formula — a formula.
- fp-kind — :inference, :percept, or :desire (the default is :inference)

- fp-condition — an optional constraint that must be satisfied by an inference-node for it to instantiate this premise.
- clue? — explained below.

Similarly, *backwards-premises* are data-structures encoding the following information:

- bp-formula
- bp-kind

Backwards-reasons will be data-structures encoding the following information:

- reason-name.
- forwards-premises.
- backwards-premises.
- reason-conclusions — a list of formulas.
- conclusions-function.
- reason-variables — variables used in pattern-matching to find instances of the reason-premises.
- strength — a real number between 0 and 1, or an expression containing some of the reason-variables and evaluating to a number.
- defeasible-rule — T if the reason is a defeasible reason, NIL otherwise.
- reason-condition — a condition that must be satisfied by the sequent of interest before the reason is to be deployed.

Simple forwards-reasons have no backwards-premises, and *simple backwards-reasons* have no forwards-premises. Given inference-nodes that instantiate the premises of a simple forwards-reason, the reasoner infers the corresponding instance of the conclusions. In the same way, given an interest that instantiates the first conclusion of a simple backwards-reason, the reasoner adopts interest in the corresponding instances of the backwards-premises. Given inference-nodes that discharge those interests, an inference is made to the conclusions from those inference-nodes.

In deductive reasoning, with the exception of a rule of reductio ad absurdum, we are unlikely to encounter any but simple forwards- and backwards-reasons.[15] Nevertheless, the use of backwards-premises in forwards-reasons and the use of forwards-premises in backwards-reasons provide an invaluable form of control

[15]This is discussed at greater length in Chapter Two of my [18].

over reasoning. This will be illustrated below. *Mixed* forwards- and backwards-reasons are those with both forwards- and backwards-premises. Given inference-nodes that instantiate the forwards-premises of a mixed forwards-reason, the reasoner does not immediately infer the conclusion. Instead the reasoner adopts interest in the corresponding instances of the backwards-premises, and an inference is made only when those interests are discharged. Similarly, given an interest instantiating the first conclusion of a mixed backwards-reasons, interests are not immediately adopted in the backwards-premises. Interests in the backwards-premises are adopted only when inference-nodes are constructed that instantiate the forwards-premises.

Reasons are most easily defined in OSCAR using the macros DEF-FORWARDS-REASON and DEF-BACKWARDS-REASON:

(def-forwards-reason *symbol*
    :forwards-premises *list of formulas optionally interspersed with expressions of the form (:kind ...) or (:condition ...)*
    :backwards-premises *list of formulas optionally interspersed with expressions of the form (:kind ...) or (:condition ...)*
    :conclusions *list of formulas*
    :conclusions-function
    :strength *number or a an expression containing some of the reason-variables and evaluating to a number.*
    :variables *list of symbols*
    :defeasible? *T or NIL (NIL is the default)*
    :description *an optional string (quoted) describing the reason*)

(def-backwards-reason *symbol*
    :conclusions *list of formulas*
    :conclusions-function
    :forwards-premises *list of formulas optionally interspersed with expressions of the form (:kind ...) or (:condition ...)*
    :backwards-premises *list of formulas optionally interspersed with expressions of the form (:kind ...) or (:condition...)*
    :condition *this is a predicate applied to the binding produced by the target sequent*
    :strength *number or an expression containing some of the reason-variables and evaluating to a number.*
    :variables *list of symbols*
    :defeasible? *T or NIL (NIL is the default)*
    :description *an optional string (quoted) describing the reason*)

The use of these macros will be illustrated in the next two sections.

6 A SOLUTION TO THE FRAME PROBLEM

The OSCAR architecture provides the first cog in the machinery with which I propose to build a planning agent. The second cog must be a solution to the frame problem. I have recently proposed such a solution, and implemented it in OSCAR.[16] The proposal consists of a triumvirate of principles.

The first is a principle of temporal projection. This is a precise version of what has also been called "the common sense principle of inertia". What this principle tells us is that there is a presumption that things do not change. Many authors have suggested that some such principle is a necessary constituent of a solution to the frame problem. Because of its popularity, I will not defend it further here.[17] The version I employ is as follows:

> *TEMPORAL-PROJECTION*
> If P is temporally-projectible and $\Delta t < log(.5)/log(\rho)$ then believing P-at-t is a defeasible reason of strength $2 \cdot (\rho^{\Delta t} - .5)$ for the agent to believe P-at-$(t + \Delta t)$.

ρ is a constant (a probability less than 1.0) that I call *the temporal-decay factor*. It is used in the principle to guarantee that the presumption of nonchange weakens as the time interval increases. The temporal-projectibility constraint is a syntactic constraint that is required for much the same reason projectibility constraints are required in principles of induction. Its main function here is to preclude disjunctions. This is discussed in detail in my [19,20]. This principle can be implemented in OSCAR as follows:

```
(def-backwards-reason TEMPORAL-PROJECTION
    :conclusions "(p at time)"
    :condition (and (temporally-projectible p) (numberp time))
    :forwards-premises
        "(p at time0)"
        (:condition (and (time0 < time*) ((time* - time0) < 693)))
    :variables p time0 time
    :defeasible? T
    :strength (- (* 2 (expt *temporal-decay* (- time time0))) 1)
    :description
    "It is defeasibly reasonable to expect temporally projectible truths to
    remain unchanged.")
```

The frame problem arises in connection with causal reasoning. We want to reason about what changes would be caused by an action. I have argued (in my [19,20]) that the causal relation involved in this reasoning is *causal-sufficiency*, where "A when P is causally sufficient for Q after an interval ε" means "if action A is performed at a time t and P is true at t, then it follows from causal laws that Q is true throughout some open interval beginning at $t + \varepsilon$". The latter can be symbolized as

$$(\forall t)[(\text{A-at-}t \wedge \text{P-at-}t) \equiv> (\exists\delta)Q\text{-throughout-}(t+\varepsilon, t+\varepsilon+\delta)].$$

---

[16] See [19,20].
[17] See my [19] for a full defense of this principle.

where "≡>" is the conditional that results from instantiating causal laws.

Suppose that *A* when *P* is causally sufficient for *Q* after an interval ε. If *A* and *P* are true at time *t*, it follows that *Q* is true at *some* time $t^* > t+\varepsilon$, but it does not follow that *Q* is true at any particular time. This is because $t+\varepsilon$ is not contained in the open-interval $(t+\varepsilon, t+\varepsilon+\delta)$, and as δ is existentially quantified, for any time $t^* > t+\varepsilon$, it could be that$t^* > t+\varepsilon+\delta$. However, for any time $t^* > t+\varepsilon$, it does follow that there is a time t^{**} such that $t^* > t^{**} > t+\varepsilon$ and *Q* is true at t^{**}. Consequently, if *Q* is temporally-projectible, temporal-projection gives us a defeasible reason to expect *Q* to be true at t^*. This reasoning is captured by the following principle:

> <u>*CAUSAL-IMPLICATION*</u>
> If *Q* is temporally projectible, $(t^{**} - (t+\varepsilon)) < log(.5)/log(\rho)$, and $(t+\varepsilon) < t^*$, then "(*A* when *P* is causally sufficient for *Q* after an interval ε) $\wedge$ *A-at-t* $\wedge$ *P-at-t*" is a defeasible reason for "*Q-at-t**."

This is implemented as follows:

```
(def-backwards-reason CAUSAL-IMPLICATION
    :conclusions "(Q at time*))"
    :forwards-premises
    "(A when P is causally sufficient for Q after an interval)"
       (:condition (every #'temporally-projectible (conjuncts Q)))
    "(A at time)"
    (:condition
       (and (eq op 'clopen) ((time + interval) < time*)
       ((time* - (time + interval)) < 693))
    :backwards-premises
    "(P at time)"
    :variables A P Q interval time time* op
    :strength (- (* 2 (expt *temporal-decay* (- time* time))) 1)
    :defeasible? T)
```

At one time, many researchers thought that a principle of temporal projection might be sufficient by itself to solve the frame problem, but that hope was dashed by Steve Hanks and Drew McDermott [7]. They illustrated the difficulty with an example that has become known as the Yale Shooting Problem. The general form of the problem is this. Suppose we have a causal law to the effect that if *P* is true at a time *t* and action *A* is performed at that time, then *Q* will be true shortly thereafter. Suppose we know that *P* is true now, and *Q* false. What should we conclude about the results of performing action *A* in the immediate future? Hanks and McDermott illustrate this by taking *P* to be "The gun is loaded and pointed at Jones", *Q* to be "Jones is dead", and *A* to be the action of pulling the trigger. We suppose (simplistically) that there is a causal law dictating that if the trigger is pulled on a loaded gun that is pointed at someone, that person will shortly be dead.

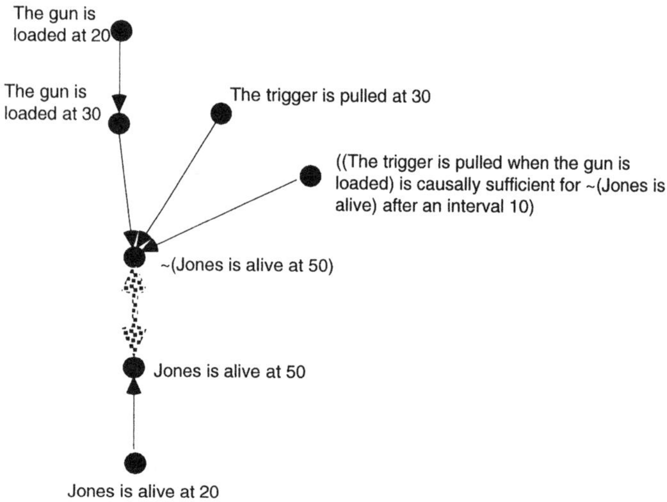

Figure 2. The unsolved Yale shooting problem

Under these circumstances, it seems clear that we should conclude that Jones will be dead shortly after the trigger is pulled.

The difficulty is that all we can infer from what we are given is that when *A* is performed either *P* will no longer be true or *Q* will be true shortly thereafter. Intuitively, we want to conclude (at least defeasibly) that *P* will remain true at the time *A* is performed and *Q* will therefore become true shortly thereafter. But none of our current machinery enables us to distinguish between *P* and *Q*. Because *P* is now true and *Q* is now false, we have a defeasible reason for believing that *P* will still be true when *A* is performed, and we have a defeasible reason for believing that *Q* will still be false shortly thereafter. We know that one of these defeasible conclusions will be false, but we have no basis for choosing between them, so this becomes a case of collective defeat. This is diagramed in figure 2, where the thin arrows indicate inference-relations and the fuzzy arrows indicate defeat-relations. We have equally good defeasible arguments for "Jones is alive at 50" and "Jones is not alive at 50", and no independent source of defeat, so both conclusions are defeated. That, however, is the intuitively wrong answer.

A solution to the Yale Shooting Problem must give us some reason to favor the upper defeasible argument over the lower one in figure 2. My proposal is this. Whenever one event causes another, temporal projection gives us a reason for expecting the change not to occur. Consequently, when reasoning about a causal system, part of the force of describing it as causal must be that the defeasible presumption against the effect occurring is somehow removed. Thus, although we normally expect Jones to remain alive, we do not expect this any longer when he is shot. To remove a defeasible presumption is to defeat it. This suggests that there is some kind of general "causal" defeater for temporal projection.

When we reason about causal mechanisms, we think of the world as "unfolding" temporally, and changes only occur when they are forced to occur by what has already happened. In our example, when A is performed, nothing has yet happened to force a change in P, so we conclude defeasibly that P remains true. But given the truth of P, we can then deduce that at a slightly later time, Q will become true. Thus when causal mechanisms force there to be a change, we conclude defeasibly that the change occurs in the later states rather than the earlier states. This seems to be part of what we mean by describing something as a causal mechanism. Causal mechanisms are systems that force changes, where "force" is to be understood in terms of temporal unfolding.[18] The idea is that in defeating a temporal projection, we can use things that have already happened but not things that are about to happen. I propose to capture this intuition with the following defeater for temporal projection:

> *CAUSAL-UNDERCUTTER*
> Where $t_0 < t_1$ and $(t_1 + \epsilon) < t$, "A-at-$t_1 \wedge Q$-at-$t_1 \wedge (A$ when Q is causally sufficient for $\neg P$ after an interval ϵ)" is a defeasible undercutting defeater for the inference from P-at-t_0 to P-at-t by temporal-projection.

This can be implemented as follows:[19]

```
(def-backwards-undercutter CAUSAL-UNDERCUTTER
    :defeatee *temporal-projection*
    :forwards-premises
        "(A when Q is causally sufficient for ¬p after an interval interval)"
        "(A at time1)"
        (:condition (and (time0 ≤ time1) ((time1 + interval) < time)))
    :backwards-premises
        "(Q at time1)"
    :variables A Q p time0 time time1 interval
    :defeasible? T)
```

With the addition of this principle, the Yale Shooting Problem is solved as in figure 3.

7 REASONING DEFEASIBLY ABOUT PLANS

I have argued that planning should be done by reasoning rather than algorithmically. The two most obvious difficulties for producing a planner that works in this

[18] This intuition is reminiscent of Shoham's [24] "logic of chronological ignorance", although unlike Shoham, I propose to capture the intuition without modifying the structure of the system of defeasible reasoning. This is also related to the proposal of Gelfond and Lifschitz [4]. This same idea underlies my analysis of counterfactual conditionals in Pollock [14].

[19] DEF-BACKWARDS-UNDERCUTTER is a variant of DEF-BACKWARDS-REASON that computes the form of the conclusion as the form of the undercutting defeater for the reason listed.

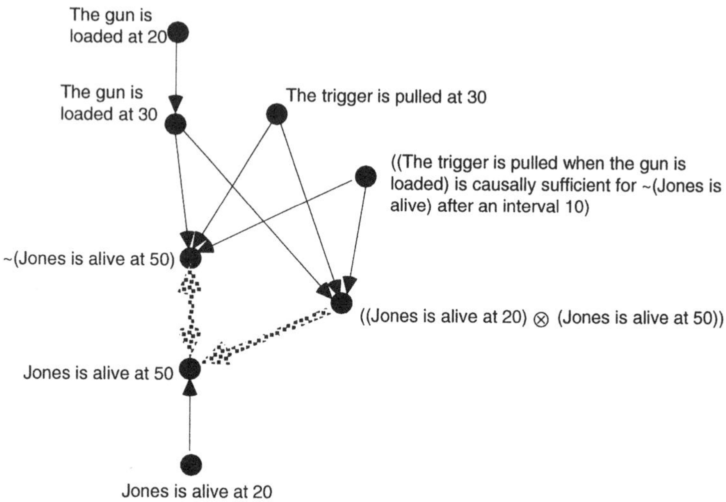

Figure 3. The solved Yale shooting problem

way are that (1) we require an agent architecture that can provide the requisite inference-engine, and (2) if an agent based upon this architecture is to be able to reason about plans, we must build a solution to the frame problem into the reasoning. Sections five and six sketched my implemented proposals for meeting these two difficulties. Does this mean that we are now in a position to build a planning agent?

Unfortunately, the answer is "Not quite." Several large problems remain. The first is that in reasoning about plans, a planning agent must reason about what *would happen if* an action *were* performed. This is *counterfactual* or *subjunctive* reasoning. Section six is about *indicative* causal reasoning—reasoning about what will be the case if an action *is* performed. The difference is a subtle one, but the literature on counterfactual or subjunctive conditionals makes it apparent that it is also an important one.[20] Before we are in a position to build a planning agent, we must incorporate the solution to the frame problem into an implemented account of subjunctive reasoning. This is a task that I have not yet undertaken. When I do, I propose to pursue an account of subjunctive reasoning based upon my theory of subjunctive conditionals in my [14].

Even though we lack an implementation of subjunctive reasoning, we can still make some progress in exploring how planning might be done by reasoning. For this purpose, let us write subjunctive conditionals in the form "P $\gg$ Q". Although we lack a full theory of how to reason about these conditionals, a few principles are obvious. For example, a version of simplification holds for the consequents of these conditionals:

[20] See Lewis [8], and Pollock [14].

```
(def-forwards-reason SIMPLIFY-≫
    :forwards-premises "(P ≫ (Q ∧ R))"
    :conclusions "(P ≫ Q)" "(P ≫ R)"
    :variables P Q R)
```

So for present purposes I will allow myself to make use of these conditionals even without a complete theory of precisely how they work.

Let us turn to planning then. I will sketch the construction of a goal-regression planner within OSCAR. The plans produced by this planner are essentially the same as those produced by SNLP [11]. *Plans* will be data-structures with the following fields:

- plan-steps — *a list of plan-nodes*
- plan-goals — *the list of goals the plan aims to achieve*
- a set of ordering constraints.

Plans will be nonlinear, in the sense that the plan-steps may be only partially ordered. The only ordering imposed is that required by causal-links and that contained in the explicit ordering constraints. The interpretation of a nonlinear plan is that every linearization of the plan (i.e., every total ordering of the plan-steps consistent with the explicit ordering constraints and causal-links) is expected to achieve the goals of the plan.

Plan-nodes will be data-structures with the following fields:

- plan-node-action
- plan-node-goals — *the list of goals the plan-node aims to achieve*
- call-set — *the list of causal-links having the node as their target.*

Finally, *causal-links* will be data-structures with the following fields:

- causal-link-root—*a plan-node*
- causal-link-goal
- causal-link-target—*a plan-node.*

The causal-link "$n - g \rightarrow n^*$" signifies that the root n aims to achieve the goal g which is a precondition for the target n^* to achieve whatever goal it aims to achieve.

OSCAR will find plans by attempting to answer queries of the form (?p)(plan-for p *goal*), meaning "find a p such that p is a plan for *goal*". The fundamental operation of goal-regression is provided by a backwards-reason:

```
(def-backwards-reason GOAL-REGRESSION
     :conclusions "(plan-for plan goal)"
     :condition (interest-variable plan)
     :backwards-premises
          "((precondition ∧ action) ≫ goal)"
          "(plan-for subplan precondition)"
          "(define plan (extend-plan action goal subplan))"
     :variables precondition action goal plan subplan)
```

The "define" premise works like a "let", temporarily binding the variable *subplan*. The function (*extend-non-linear-plan action goal subplan precondition*) constructs a new plan-node having *action* as its action and *precondition* as its goal, and then constructs a new plan by adding this plan-node to *subplan*.

The simplest case of goal-regression planning occurs when the goal is already true. In that case we conclude that it is achieved by a null-plan, which is a plan with no plan-steps. This reasoning is accomplished by the following backwards-reason:

```
(def-backwards-reason NULL-PLAN
     :conclusions "(plan-for plan goal)"
     :condition (interest-variable plan)
     :backwards-premises
          "goal"
          "(define plan (null-plan goal))"
     :variables goal plan)
```

null-plan goal) is a plan with no plan-steps and with a goal-node that is a plan-node having *goal* as its plan-node-action, and having a single causal-link *start* – goal $\rightarrow$ goal.

Goal-regression planning is made complicated by conjunctive goals. When faced with a conjunction of goals, humans plan separately for the individual goals, and then try to merge the plans. However, that does not always work because the plans for the individual goals may interfere with each other. In decision-theoretic planning, separate plans can interfere subtly by affecting each other's expected values, but at this stage I am only dealing with non-decision-theoretic planning. In this case, the way one plan interferes with another is by *clobbering one of its causal-links*. Suppose *plan*2 contains the causal-link $n - g \rightarrow n^*$. This means that g is a subgoal required for node n^* to do its job, and g is achieved in *plan*2 by node n. *Plan*1 clobbers this causal-link iff *plan*1 contains a subplan p consisting of nodes from *plan*1 where the final step of p could (consistent with the ordering constraints) be executed between n and n^*, and p is a plan for $\neg g$ This means that there is a linearization of the plan that results from combining *plan*1 and *plan*2 in which node n makes g true, and then the last node of p is executed and makes g false again before it is used by node n^*.

How should we deal with this problem? It is tempting to suppose that when planning for conjunctive goals, we should construct plans for the individual goals, and then verify that they do not clobber one another *before* we merge them into a single plan and conclude that the merged plan is a plan for the conjunctive goal. However, this is not the way human planning works. Humans assume defeasibly that the separate plans do not clobber one another, and so infer defeasibly that the merged plan is a good plan for the conjunctive goal. Having made that defeasible inference, they then look for possible clobberings that would defeat it, but they do not regard it as essential to establish that there are no such clobberings before they make the inference. The difference is an important one. The plan might be such that if it is to be executed, execution must begin before the search for clobberings terminates. The agent that assumes defeasibly that the plan is a good one will go ahead and begin execution, while the more conservative agent that requires proof that there are no clobberings will be unable to execute the plan.

Which kind of agent is preferable? It might seem that the conservative agent is the more rational one. But in fact, the conservative agent cannot be made to work. This is related to the difference between algorithmic planners and planners that plan by reasoning. An algorithmic planner makes the assumption that it has all the relevant knowledge before planning begins. Given that assumption, it can compute whether there is any clobbering. But as we have seen, that is an unrealistic assumption. A realistic planning agent will guide its search for knowledge in light of the planning it is doing. In particular, when it constructs two plans and wants to merge them, it will then begin an investigation into whether there is any clobbering. However, this can be an indefinitely complex epistemic investigation. There will not in general be any point at which the agent can stop the investigation and conclude definitively that there is no clobbering. All it can conclude at any given point is that so far it sees no reason to think there is any clobbering. This is just one more example of the fact that a realistic agent that reasons defeasibly can never stop reasoning, because there is always the possibility that further reasoning will reveal heretofore unsuspected defeaters. Of course, we do not want the agent to spend all its time looking for defeaters instead of acquiring new beliefs, so the prioritization of the inference-queue must be adjusted to give the search for defeaters a relatively low (but still significant) priority.

The conclusion to be drawn from this is that the only way to make planning work in a realistically complex environment is by assuming defeasibly that separate plans do not clobber each other's causal-links. This point is of more far ranging significance than might at first be realized. It arose in the context of conjunctive *sub*goals generated by goal-regression. But it arises equally from more general features of planning. Traditional planning systems have often adopted the fiction that all of the goals are given from the beginning, and the planner plans for them all simultaneously. This is certainly not true of human beings. We acquire new goals all the time. When we acquire a new goal, we do not throw out all of our previous plans and begin planning all over again for a single new overarching conjunctive goal that includes both our previous goals and our new goal. No planner could work that way. It is computationally too difficult. Rather, humans plan for the new goal and try to merge that plan with the plans that have already been made for other goals. Artificial agents must work that way as well. Of course, sometimes plans cannot be merged and we may have to modify some of our earlier plans and possibly replan for some of our earlier goals, but the point is that we do not do that unless we have to. What makes this possible is the defeasible assumption that plans can be merged without clobbering.

Planning separately for the different conjuncts of a conjunctive goal is accomplished by using the following backwards-reason:

```
(def-backwards-reason SPLIT-CONJUNCTIVE-GOAL
    :conclusions "(plan-for plan∧ (goal1 ∧ goal2))"
    :condition (interest-variable plan∧)
    :backwards-premises
        "(plan-for plan1 goal1)"
        "(plan-for plan2 goal2)"
        (:condition (or (plan-steps plan1) (plan-steps plan2)))
        "(define plan∧ (merge-plans plan1 plan2 goal1 goal2))"
    :defeasible? t
    :variables goal1 goal2 plan1 plan2 plan∧)
```

Inferences made in accordance with this reason are defeated by finding that the merged plan clobbers one of the causal-links in one of the constituent plans. This is accommodated by adopting the following defeater for split-conjunctive-goal:

```
(def-backwards-undercutter CLOBBER-CAUSAL-LINKS
    :defeatee split-conjunctive-goal
    :backwards-premises
        "(define links (union (call-set (goal-node plan∧)) (unionmapcar #'call-set
            (plan-steps plan∧))))"
        "(plan-clobbers-causal-links plan∧ links)"
    :variables plan∧ links)
```

This is supplemented with principles enabling the reasoner to reason about plans clobbering links.

Nonlinear plans attempt to make as few ordering commitments as possible, for the reasons discussed in section three. But sometimes that is the source of the clobbering, and the clobbering can be avoided by adding ordering constraints. When clobbering is discovered, the following reason attempts to add ordering constraints that avoid the clobbering by promoting or demoting the offending plan-nodes:

```
(def-backwards-reason ADD-ORDERING-CONSTRAINTS
    :conclusions "(plan-for plan∧ (goal1 ∧ goal2) goals)"
    :condition (interest-variable plan∧)
    :forwards-premises
        "(plan-for plan1 goal1)"
        "(plan-for plan2 goal2)"
        "(merged-plan plan plan1 plan2)"
        (:clue? t)
```

```
    "(plan-clobbers-causal-link plan R node link)"
    (:condition
      (or (member link (call-set (goal-node plan)))
          (some #'(lambda (n) (member link (call-set n))) (plan-steps plan))))
    (:clue? t)
    "(define plan∧ (add-not-between node link plan plan1 plan2))"
    (:condition (not (null plan∧)))
:defeasible? t
:variables plan1 plan2 plan∧ plan node link goal1 goal2 R)
```

Discovering the clobbering *occasions* the inference to "(plan-for *plan+ goal goals*)" without being a premise of the inference. This is captured in OSCAR by making the last premise of the reason a *clue*. When a premise is a clue, it must be instantiated by an inference-node before a new conclusion is drawn, but the basis from which the new conclusion is inferred does not include the inference-node instantiating the clue. Note also that this is a degenerate backwards-reason.

Although I have not discussed all of the reasons required for reasoning about nonlinear plans, it may still be useful to consider a simple example. Suppose Horatio is at the library, and wants to know Beethoven's birthday and also what time it is. Suppose Horatio is given the following conditionals:

$$(\forall x) \quad [((\text{at-library } x) \wedge (\text{ask-librarian } x)) \gg (\text{know-beethoven-birthday } x)]$$
$$(\forall x) \quad [((\text{at-clock } x) \wedge (\text{read-clock } x)) \gg (\text{know-time } x)]$$
$$(\forall x) \quad [((\text{at-library } x) \wedge (\text{go-to-clock } x)) \gg ((\text{at-clock } x) \wedge \neg(\text{at-library } x))]$$

Using the reason-schemas described above, OSCAR finds the following plan:

```
PLAN-STEPS:
  (2) (ask-librarian horatio)
     causal-links:
        0– (at-library Horatio)→ 2
  (4) (go-to-clock horatio)
     causal-links:
        0 – (at-library Horatio)→ 4
     ordering constraints:
        4 > 2
  (6) (read-clock horatio)
     causal-links:
        4 – (at-clock Horatio)→ 6
GOAL: ((know-beethoven-birthday Horatio) ∧ (know-time Horatio))
     established by:
        2 → (know-beethoven-birthday Horatio)
        6 → (know-time Horatio)
```

This is, of course, a very simple problem, but this approach to planning works equally well for more complex examples. As remarked above, it produces essentially the same plans as SNLP, but produces them by reasoning rather than algorithmically. The addition of a few more reason-schemas enables OSCAR to plan at the richer level of UCPOP [13]. As we have seen, there is a tremendous advantage to this approach in that it automatically interleaves planning and epistemic reasoning and thereby achieves all the benefits described above. At the same time, there is a cost. Doing anything by reasoning is slow. OSCAR is roughly an order of magnitude slower than a comparable algorithmic planner.[21] However, the slowdown is a linear one, and the planning is still much faster than a human being can accomplish, so the cost does not seem too onerous.

8 SOPHISTICATED PLANNING

The planning system described in the preceding section may actually be useful, but it falls far short of being a solution to the problem of constructing a planning agent that can plan at the level of sophistication of a human being. Although it is a start, the planner described above should be viewed as a proof of concept for the idea of planning by reasoning rather than as a solution to the problem of building a truly sophisticated planner.

This planner does make some significant advances. The most important is that it integrates epistemic cognition and planning. Thus rather than requiring the planner to have all relevant knowledge before planning begins, the course of planning can direct the agent to address various epistemic issues that were not previously addressed and then return to the planning problem once answers have been acquired. Because planning is just part of reasoning, and the reasoning is prioritized by factors relating to its importance, this means, for example, that a planning agent can begin planning for one goal, break that off in the middle when a more important goal arises, and then return to the original planning problem later without losing the work that has already been done.[22] This is an automatic consequence of the OSCAR architecture. It also means that the information the planner uses in planning can come in any logical form. It does not have to be artificially restricted to lists of literals, as in the STRIPS-representation.

So what's missing? Unfortunately, quite a lot. First let me mention a relatively simple omission (at least, simple relative to the others). This is that the planner described above does not construct conditional plans.[23] Human plans are full of conditionals. For example, in planning the route I will drive across town, I may plan to go one way unless the traffic is too heavy, but if the traffic is heavy I will go another way. The problem for building a conditional planner lies in deciding which eventualities should be treated as subgoals and thereby initiate further goal regression and which should be left as conditions. Existing conditional planners tend to handle this in an *ad hoc* manner. The human operator is required to list the

[21] I have used UCPOP as my standard of comparison. See Penberthy and Weld [13], or Weld [27].

[22] This is one of the desiderata Beetz and McDermott [1] stress.

[23] This is to be distinguished from planning with actions having conditional effects. OSCAR handles the latter in the same manner as UCPOP.

eventualities that can go into conditions. Obviously, we want more. It is not even correct to suppose that there is any kind of precise division between eventualities that are treated as subgoals and those that are treated as conditions. How they are treated depends, among other things, on whether we can think of a subplan for achieving them. For instance, while visiting in someone else's home I may spy a dish of particularly succulent candy. This may generate in me the goal of eating a piece, and this may inspire me to form the plan to accept a piece if my host offers it to me. Depending on the circumstances (e.g., how well I know my host), my planning may stop there with no attempt to plan further for the subgoal of getting my host to make such an offer, or I might construct the further plan of looking fondly at the candy and saying, "Um, that looks good!". Whether I can propound a plan like the latter depends entirely on the circumstances.

A much more serious lacuna in the preceding planner is that it is not decision-theoretic. Human planning is profoundly decision-theoretic. We attach values to goals, probabilities to possible outcomes of actions, and expected values to plans. When plans conflict, we must decide which plans to adopt, and we do that in part by comparing their expected values.[24] The decision-theoretic character of plans affects all aspects of planning. My proposal for solving the frame problem must be generalized to accommodate reasoning about the probabilistic consequences of actions. I have not yet even addressed the question of how to reason about subjunctive conditionals, but when such an account is produced it must immediately be generalized to extend it to subjunctive probabilities (probabilities that something *would be* the case if something else *were* the case). The reasons employed in searching for plans must also be revised to take account of decision-theoretic dimensions. Furthermore, the conclusion of the reasoning involved in the plan search will no longer be that the plan discovered is a plan for the goal, but rather that it is a plan for the goal and has a certain expected value. In non-decision-theoretic planning, drawing the conclusion that the plan is a plan for the goal can perhaps be identified with adopting the plan, but that is no longer adequate in decision-theoretic planning. We may conclude that a plan has a certain expected value but should not be adopted because it conflicts with other plans having even higher expected values. Thus we need a whole new tier of rational machinery built on top of the planning reasoning that decides which plans are to be adopted.

An equally serious lacuna is the omission of any serious machinery aimed at scheduling. The planner of the preceding section can do only the most rudimentary scheduling, that in terms of ordering constraints and causal links. Serious scheduling must be combined with decision-theoretic planning, because choices between schedules will often profoundly affect the expected value of a plan, and that is the basis upon which scheduling decisions should be made.

I have talked about how to build agents that plan, but one of the most important properties of planning agents is that they plan as little as possible. Planning is

[24] The logic of this is more complex than planning theorists generally realize. We cannot simply pick the plan with the higher expected value. See chapter six of my [18] for a discussion of this.

computationally difficult, and accordingly both slow and exhaustive of cognitive resources that could be employed elsewhere. Human beings solve simple planning problems by reusing previous plans. Plan-schemas are stored in memory and re-instantiated to handle new planning problems without engaging in new planning. Planning theory has taken this observation to heart, and there is a large literature on plan reuse. There is reason to hope, however, that planning by reasoning will simplify the problem of plan reuse. If the reasoning is such as to generate general beliefs about plans rather than very specific beliefs about fully concrete plans, then those general beliefs can be retained and reused as needed in new planning problems. However, this is a dimension of planning by reasoning that I have not yet explored.

Earlier, I stressed the importance of interleaving planning and plan execution, but I have said nothing further about plan execution. That is a topic that must be explored at length. My general approach to the topic will be that sketched in chapter eight of my [18].

9 OVERVIEW OF A PLANNING AGENT

Rational agents are agents that adopt and execute plans in an intelligent fashion. Such an agent, embedded in a realistically complex world, will reason endlessly, and the bulk of the cognitive work involved in planning and plan-execution will be carried out by reasoning. Reasoning from perceptual input will lead the agent to beliefs that will apprise it of opportunities and lead to the adoption of goals. The adoption of a goal initiates planning. Planning proceeds by posing queries for epistemic cognition regarding how the goal can be achieved. In attempting to find a plan, the agent may be led far afield into epistemic endeavors that are only related to the original planning problem very indirectly. Thus theoretical physics may ultimately emerge from trying to catch lunch. Most (perhaps all) of the complexities of epistemic cognition have their ultimate source in trying to answer the questions that arise in the course of planning and plan-execution. The bulk of cognition will be epistemic cognition, but there remains an ineliminable residue of practical cognition. All epistemic cognition can do is provide us with beliefs about the expected values of plans and about how different plans might be related to one another (and so be in competition in various ways). There are still non-epistemic decisions to be made regarding which plans to adopt.

The most important conclusion of this paper is that planning itself must be done by reasoning in a sophisticated agent. It cannot be done algorithmically, because planning, plan-execution, and epistemic cognition must all be done in unison rather than sequentially. They cannot be performed by isolated modules, because the course of each drives the others and is in turn driven by feedback from the others. A rational agent must be an agent that adopts and executes plans by employing its faculties of epistemic cognition to reason about them.

John Pollock (University of Arizona, Tucson, USA)

REFERENCES

[1] Michael Beetz and Drew McDermott. Local planning of ongoing activities. In Brian Drabble, editor, *Proceedings of the Third International Conference on Artificial Intelligence Planning Systems*. AAAI Press, 1996.
[2] M. Bratman, D. Isreal, and M. Pollack. Plans and resource-bounded practical reasoning. *Computational Intelligence*, 4:349–355, 1988.
[3] R. E. Fikes and N. J. Nilsson. Strips: a new approach to the application of theorem proving to problem solving. *Artificial Intelligence*, 2:189–208, 1971.
[4] Michael Gelfond and Vladimir Lifschitz. Representing action and change by logic programs. *Journal of Logic Programming*, 17:301–322, 1993.
[5] Michael Georgeff and Amy Lansky. Reactive reasoning and planning. In *Proceedings AAAI-87*, pages 677–682, 1987.
[6] C. Green. Application of theorem-proving to problem solving. In *Proceedings IJCAI-69*, pages 219–239, 1969.
[7] Steve Hanks and Drew McDermott. Default reasoning, nonmonotonic logics, and the frame problem. *AAAI-86*, 1986.
[8] David Lewis. *Counterfactuals*. Harvard University Press, Cambrdige, Mass, 1973.
[9] Fangzhen Lin and Raymond Reiter. How to progress a database (and why) i. logical foundations. In *Proceedings of the Fourth International Conference on Principles of Knowledge Representation (KR94)*, pages 425–436, 1994.
[10] Fangzhen Lin and Raymond Reiter. How to progress a database ii: The strips connection. *IJCAI-95*, pages 2001–2007, 1995.
[11] David McAllester and David Rosenblitt. Systematic nonlinear planning. In *Proceedings of AAAI-91*, pages 634–639, 1991.
[12] John McCarthy. Epistemological problems in artificial intelligence. In *Proceedings IJCA-77*, 1977.
[13] J. Scott Penberthy and Daniel Weld. Ucpop: a sound, complete, partial order planner for adl. In *Proceedings 3rd International Conference on Principles of Knowledge Representation and Reasoning*, pages 103–114, 1992.
[14] John Pollock. *The Foundations of Philosophical Semantics*. Princeton University Press, 1984.
[15] John Pollock. *Contemporary Theories of Knowledge*. Rowman and Littlefield, 1987.
[16] John Pollock. Oscar: a general theory of rationality. *Journal of Experimental and Theoretical AI*, 1:209–226, 1990.
[17] John Pollock. New foundations for practical reasoning. *Minds and Machines*, 2:113–144, 1992.
[18] John Pollock. *Cognitive Carpentry*. MIT Press, 1995.
[19] John Pollock. Perceiving and reasoning about a changing world. technical report of the oscar project., 1996. This can be downloaded from http://www.u.arizona.edu/∼pollock/.
[20] John Pollock. Reason in a changing world. In Dov M. Gabbay and Hans Jürgen Ohlbach, editors, *Practical Reasoning: Internaltional Conference on Formal and Applied Practical Reasoning*, pages 495–509, 1996.
[21] E. D. Sacerdotti. The non-linear nature of plans. In *Proceedings IJCAI-75*, 1975.
[22] E. D. Sacerdotti. *A Structure of Plans and Behavior*. Elsevier-North Holland, Amsterdam, 1977.
[23] Murray Shanahan. Robotics and the common sense informatic situation. In *Proceedings of the 12th European Conference on Artificial Intelligence*. John Wiley & Sons, 1996.
[24] Yoav Shoham. *Reasoning about Change*. MIT Press, 1987.
[25] Manuela Veloso, Jaime Carbonell, Alicia Perez, Daniel Borrajo, Eugene Fink, and Jim Blythe. Integratin planning and learning: the prodigy architecture. *Journal of Experimental and Theoretical Artificial Intelligence*, 7, 1995.
[26] S. Vere and T. Bickmore. A basic agent. *Computational Intelligence*, 6:41–60, 1990.
[27] Daniel Weld. An introduction to least commitment planning. *AI Magazine*, 15:27–62, 1994.
[28] M. Wooldridge and N. R. Jennings. Intelligent agents: theory and practice. *The Knowledge Engineering Review*, 10:115–152, 1995.

P. J. GMYTRASIEWICZ

TOWARD RATIONAL INTERACTIONS IN MULTIAGENT DOMAINS

ABSTRACT: We describe a framework that applies the decision-theoretic principle of rationality to designing socially competent agents. By social competence we mean the agents' ability to determine what to do in the presence of other agents, and how to communicate with them. Our framework consists of a method of representing and processing a finite amount of an agent's knowledge to arrive at the rational choice of physical and communicative action. The representation captures the agent's knowledge about the environment and about the other agents, including its knowledge about their states of knowledge, which can include what they know about the other agents, and so on. This reciprocity leads to a recursive nesting of models. Our framework puts forth a representation for these nested models, and uses dynamic programming to solve this representation for the agent's rational choice of action. The same framework is used for communication. Following decision theory we postulate that a rational speaker design a speech act so as to maximally increase the benefit obtained as the result of the interaction. We quantify the gain in the quality of interaction as the expected utility, and we present a method that allows an agent to compute the expected utility of various communicative actions.

1 INTRODUCTION

We adopt the decision-theoretic principle of expected utility maximization as a paradigm for designing autonomous rational agents, and present a method of representing and processing a finite amount of an agent's knowledge to arrive at the rational choice of coordinated action and communication. We follow the theoretical foundations of decision theory since it is the theory that formalizes the rational choice under uncertainty. Intuitively, decision-theoretic rationality postulates that agents do things that are expected to pay off , given what they know about their environment. This postulate is particularly useful when an agent is dealing with a partially unknown environment. It is then that possibilities and likelihoods of various states of affairs have to be weighted and combined with the agent's goals to determine the right way to do things. Decision theory provides a theoretical foundation of this process, and proves [4, 9] that rationality results in optimal choice of an agent's action, given its beliefs about the world and its preferences.

We apply decision-theoretic rationality to a particular case of an agent interacting with other agents. Our main motivation is to take advantage of decision-theoretic treatment of uncertainty; the other autonomous agents are, in principle, at least partially unknown. Therefore, it is necessary to consider and weigh the possibilities of their various states of knowledge, goals and intentions while interacting and communicating with them.

M. Wooldridge and A. Rao (eds.), Foundations of Rational Agency, 81–104.

To help the reader put our work in perspective we should stress that the representations we postulate here are used for the purpose of decision-making in multiagent situations, i.e., we do not postulate a general knowledge-representation and reasoning formalism. Thus, the representations we discuss are invoked only when there is a need for making a decision about which course of action to pursue, and our methods are embedded among many of the other components constituting a full-fledged autonomous agent. These usually include a suitably designed knowledge base[1], sensing and learning routines that update the KB, planning routines that propose alternative courses of action, and so on. This paper will not address any of the difficult challenges posed by the above components; we will concentrate solely on the issue of decision-making, understood as choosing among alternative courses of action generated, say, by a symbolic planning system.

The expected utilities of alternative courses of action are generally assessed based on their results. Intuitively, an agent is attempting to quantify how much better off it would be in a state resulting from it having performed a given action. In a multiagent setting, however, an agent usually cannot anticipate future states of the world unless it can hypothesize the actions of other agents. Therefore, an agent has to model other agents influencing its environment to fully assess the outcomes and the utilities of its own actions. We say that an agent is *coordinating* with other agents precisely when it considers the anticipated actions of others as it chooses its own action.

An agent that is modeling other agents to determine what they are likely to do, however, also should consider the possibility that they are similarly modeling other agents in choosing their actions. To anticipate the action of another agent, therefore, an agent could model how that other agent might be modeling other agents. In fact, this nested modeling could continue on to how an agent is modeling how an agent is modeling how an agent is modeling, and so on.

2 REPRESENTATION

For the purpose of decision-making, it is convenient to represent the decision-making situation an agent is facing. One well known representation method are the belief networks, and their extensions, the influence diagrams [22]. Influence diagrams probabilistically represent the uncertainties of the domain, the ways the agent's alternative actions influence the domain, and the expected utilities of the resulting states. For the purpose of multiagent decision-making, we found it convenient to use a representation that also includes the agent's alternative courses of action, but is more compact since it explicitly contains only the relation between the utilities and the agents' alternative actions. This representation is a payoff matrix, widely used in decision and game theories. Compared to influence diagrams, therefore, the payoff matrices are less explicit, since the uncertain information about the state of the world is compiled and summarized as the utility values;

[1] Our preliminary implementation uses a KB configured as an ontology of object/frames.

however, they do contain the information we wish to reason about, namely, how the other agents influence the expected utilities of the original agent's alternative actions.

In order to anticipate the intentions of other intelligent agents, one may further model their rational choice by representing their decision-making situation with a payoff matrices. Together with the models the other agents may have of still other agents, all of the available information about other agents, about what they know about others, and so on, can be represented as a hierarchy of payoff matrices. We called this representation the Recursive Modeling Method (RMM) [11, 12].

2.1 *Example Multiagent Interaction 1*

To put our description of RMM in concrete terms, we will consider a particular decision-making situation[2] encountered by an autonomous outdoor robotic vehicle, called R_1 (see Figure 1), attempting to coordinate its actions with another robotic vehicle, R_2. Outdoor robotic vehicles have multiple uses, predominantly acting in environments that are too hostile or hazardous for human-controlled vehicles. Among these uses are information gathering—or reconnaissance—activities to assess, for example, the extent of a chemical spill (in the case of an industrial accident), the positions of opposing forces (on a battlefield), or the location of a submerged ship (in the case of an underwater vehicle).

For a ground-based vehicle, gathering large amounts of information depends on moving to vantage points that command a wide view, such as hilltops. Thus, we will assume that a robotic vehicle, whose mission is to gather as much information as it can while minimizing its cost (fuel and/or time consumed), will prefer to move to nearby locations with high elevation. From the perspective of robot R_1, whose point of view we will take in analyzing this situation, two possible vantage points P1 and P2 are worth considering. P2 has a higher elevation and would allow twice as much information to be gathered as P1, and so, the robot is willing to incur greater cost to go to P2. Based on domain-specific knowledge, in this example R_1 expects that gathering information at P2 will be worth incurring a cost of 4 (or, put another way, the information gathered from P2 has an expected value of 4), while the observation from P1 will be worth 2.

R_1 thus has three possible courses of action: it can move to P1 and gather information there (a_1^1); it can move to P2 and gather information there (a_2^1); or it do neither and just sit still (a_3^1).[3] The expected cost (time or energy) to R_1 of pursuing each of these courses of action is proportional to the distance traveled, yielding a cost of 1 for a_1^1, 2 for a_2^1, and 0 for a_3^1. We further assume in this

[2]The two examples here follow ones developed in [12].

[3]These courses of action could have been proposed as plausible by a symbolic planner, and each of them may have to be further elaborated by the robot. While all possible detailed plans for these high-level courses of action could be enumerated and represented in a payoff matrix, it is clearly desirable to include just a few abstract actions or plans. This permits evaluation of choices at a level of detail where the quality of the decision is maximized while the costs of making the decision are minimized [18].

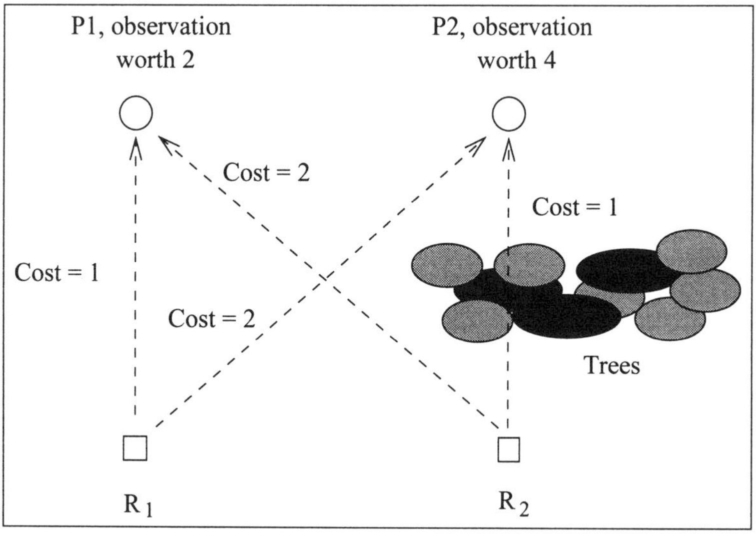

Figure 1. Example Scenario of Interacting Agents

example that each of the robots can make only one observation, and that each of them benefits from *all* information gathered (no matter by which robot), but incurs cost only based on its own actions.[4]

Given the above information, residing in robot R_1's knowledge base, R_1 can build a payoff matrix that summarizes the information relevant to its decision-making situation. The relevant alternative behaviors of R_2 that matter will be labeled a_1^2 through a_3^2, and correspond to R_2's alternative plans of taking the observation from point P1, P2, and staying put or doing something else, respectively. Thus, the entry in the matrix corresponding to R_1's pursuing its option a_1^1 and R_2's pursuing a_2^2 is the payoff for R_1 computed as the total value of the information gathered by both robots from both P1 and P2 minus R_1's own cost: (2+4) - 1 = 5. The payoff to R_1 corresponding to R_1's pursuing a_1^1 and R_2's pursuing a_1^2 is (2+0) - 1 = 1, since the information gathered is worth 2 and redundant observations add no value. All of the payoffs can be assembled in the following payoff matrix:

		R_2		
		a_1^2	a_2^2	a_3^2
	a_1^1	1	5	1
R_1	a_2^1	4	2	2
	a_3^1	2	4	0

In order to arrive at the rational decision as to which of its options to pursue, R_1 has to predict what R_2 will do. If R_2 were to take the observation from the

[4]These assumptions are only for the purpose of keeping our examples simple. In no way do they limit the applicability of our methodology.

point P2, i.e., its a_2^2 option, it would be best for R_1 to observe from P1. But if R_2 decided to stay put, R_1 should observe from the point P2, i.e., pursue its option a_2^1. In general, R_1 might be uncertain as to which action R_2 will take, in which case it should represent its conjecture as to R_2's action as a probability distribution over R_2's possible alternative courses of action. At this point we would like to remark that the above payoff matrix, together with a probability distribution over R_2's actions, comprise a full picture of R_1's decision-making situation, in the sense that all of the relevant features and facts are contained in this representation. Thus, while the fact that, say, R_1 is 10 miles away from P2 is not represented explicitly, it is contained in the payoff matrix through an estimate of the cost of getting to P2. As we mentioned, a more explicit representation of facts like that is left to the agent's general-purpose knowledge base that could use Bayesian networks, frame-based techniques, or the combination of the two.

Since R_1's payoff depends on what R_2 decides to do, R_1 may choose to model R_2's decision-making. If R_1 thinks that R_2 will attempt to maximize its own expected utility, then R_1 can adopt the intentional stance toward R_2 [8], treat R_2 as rational, and model R_2's decision-making situation using payoff matrices. R_2's payoff matrix, if it knows about both observation points, arrived at analogously to R_1's matrix above, has the following form:

		R_1		
		a_1^1	a_2^1	a_3^1
	a_1^2	0	4	0
R_2	a_2^2	5	3	3
	a_3^2	2	4	0

That is not all, though, because R_1 realizes that robot R_2 possibly does not know about the observation point P2 due to the trees located between R_2 and P2.[5] R_1, therefore, has to deal with another source of uncertainty: there are two alternative models of R_2's decision-making situation. If R_2 is unaware of P2, then it will not consider combinations of actions involving a_2^1 or a_2^2 and its payoff matrix is 2×2, as follows:

		R_1	
		a_1^1	a_3^1
R_2	a_1^2	0	0
	a_3^2	2	0

A sensible way for R_1 to deal with its uncertainty as to which of the models of R_2 is correct is to represent its subjective belief as to their likelihood of being correct as a probability distribution. In this example, we assume that R_1, having knowledge about the sensors available to R_2 and assessing the density of the foliage between R_2 and P2, assigns a probability for R_2 seeing through the trees as 0.1.

Let us note that R_2's best choice of action, in each of the intentional models that R_1 has, also depends on what it, in turn, thinks that R_1 will do. Thus, R_1 should,

[5] R_2 not knowing about P2 assumes that R_2 does not have a complete and accurate map of the terrain, or if it does then it cannot locate its own position on that map.

in each of these models, represent how R_2 might model R_1. If it were to model R_1 as rational as well, the nesting of models would continue. If, on the other hand, it were to lack the information needed to build a model of R_1's preferences over joint actions, then the nesting of models would terminate. Perhaps it might have some subjective probabilities over R_1's actions, based on some simplified model of R_1 or on past experiences with R_1. This would mean that the nesting terminates in what we call a *sub-intentional* model. Or, if R_1 is even more of a mystery, then the recursion terminates in what we call a *no-information* model.

To keep this example simple and illustrative, let us make some arbitrary assumptions about how R_1's state of knowledge terminates, as follows: in the case that R_1 supposes that R_2 is of the type[6] that cannot see through the trees, then R_1 knows that R_2 does not know anything about R_1. But in the event that R_2 is of the type that can see through the trees, then R_1 itself has no knowledge in its knowledge base about how it might be modeled by R_2.

While the scenario used here seems relatively simple, we invite the reader to develop his or her own intuitions at this point by considering the problem facing our robot R_1: What is the best course of action, given the information R_1 has about the situation and about R_2? Should R_1 move to P1 and hope that R_2 will cooperate by observing from P2? Or should R_1 go to P2 itself, due to the importance of this observation and in the face of uncertainty as to R_2's behavior? How does the probability of R_2's knowing about P2 influence R_1's choice? We will provide the answers in Section 3.1.

According to our approach in RMM, R_1's knowledge as to the decision-making situation that it faces can be cast into the representation depicted in Figure 2, which we will call the recursive model structure. The top level of this structure is how R_1 sees its own decision-making situation, represented as R_1's payoff matrix. On the second level are the alternative models R_1 can form of R_2, with the alternative branches labeled with the probabilities R_1 assigns to each of the models being correct.

The third level is occupied by no-information models that terminate the recursive nesting in this example. These models represent the limits of the agents' knowledge: The model $No-Info^2$ represents the fact that, in the case when R_2 cannot see P2, R_1 knows that R_2 has no knowledge that would allow it to model R_1. Thus, the uncertainty is associated with R_2, and the model's superscript specifies that the state of no information is associated with its ancestor on the second level of the structure in Figure 2. The $No-Info^1$ model, terminating the other branch of the recursive structure, represents R_1's own lack of knowledge (on the first level of the structure) of how it is being modeled by R_2, if R_2 can see through the trees. In general, the no-information models can represent the knowledge limitations on any level; the limitations of R_1's own knowledge,[7] R_1's knowing the

[6] Our use of this term coincides with the notion of agent's type introduced by Harsanyi in [15].

[7] Note that we assume the agent's introspective ability. This amounts to the agent's being able to detect the lack of statements in its knowledge base that describe beliefs nested deeper than the given level.

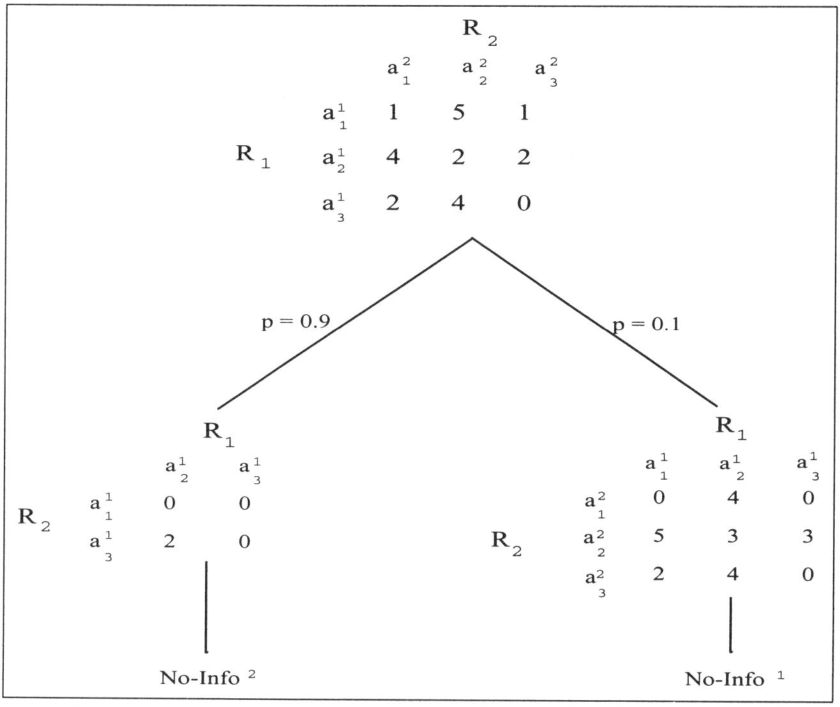

Figure 2. Recursive Model Structure depicting R_1's Knowledge in Example 1.

knowledge limitations of other agents, and so on. For details and a further specification of the no-information models, we refer the reader to [12].

We have to stress that we do not advocate the usage of no-information (or subintentional) models as an *ad hoc* means to terminate the recursive nesting of beliefs. Thus, our formalism does not propose that agents construct the nested models until, say, they think they have enough of them, and terminate them at an arbitrary depth with the no-information models. Rather, we take a knowledge-based view and propose that the representation of the agent's state of knowledge be terminated when, and only when, the limits of the agent's knowledge, residing in its knowledge base, are reached. Thus, if an agent has information and beliefs about other's beliefs nested, say, to the fifteenth level, then the full representation will contain all of these levels, but if an agent's knowledge base does not contain any knowledge nested to more then the second level, then the representation will terminate with the no-information model directly below this level.[8]

[8]Our remarks here take the theoretical point of view of a perfectly rational reasoner that represents and considers *all* of its knowledge during decision-making. This does not preclude that pragmatically motivated simplifications, for example to meet a time constraint, are impossible.

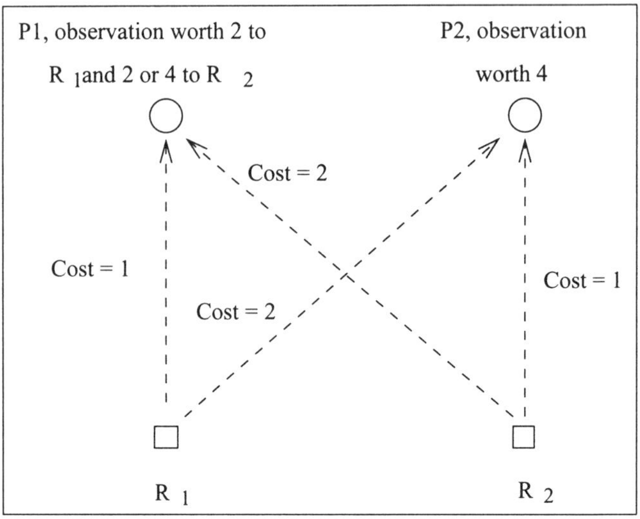

Figure 3. Another Scenario of Interacting Agents.

2.2 Example Multiagent Interaction 2

Let us consider a possible variation of the previous scenario. We will assume that there is no uncertainty about the environment—that R_2 is sure to see P2. However, now there is uncertainty about the agents' properties. R_1 knows that observation point P1 also provides a sunny spot where a vehicle with a solar array can re-charge its batteries. While R_1 has no solar array on board, it believes that R_2 might, and that re-charging is worth 2 to agents. Thus, if R_2 has a solar array, it will value going to P1 as 4 (2 for observing plus 2 for re-charging), but, if it has no array, then R_2 will value P1 the same as R_1 does (at 2). This situation is summarized in Figure 3.

Moreover, R_1 is camouflaged to resemble a bush (R_2 is clearly an agent), and R_1 estimates that it is equally likely that its disguise will fool R_2 as it is that R_2 will correctly recognize R_1 as a robotic vehicle without a solar array. In the case where the disguise fools R_2, the model R_1 has of how R_2 will see R_1 is simply a physical model of a bush. We will not elaborate on the details of physical modeling here beyond assuming, following Dennett (see the discussion in [21]), that R_1 will believe that R_2 will employ the information of the physical state of the perceived bush, and knowledge of the laws of nature, to predict its behavior (in this case the model is likely to predict that the bush will simply remain still). In the event the disguise does not fool R_2, R_1 thinks it would be viewed correctly as a robot, but it has no deeper knowledge about what R_2 would think R_1 thinks about R_2.

Now, for the purpose of decision-making, R_1's information can be represented as depicted in Figure 4. The no-information models used both specify that it is the

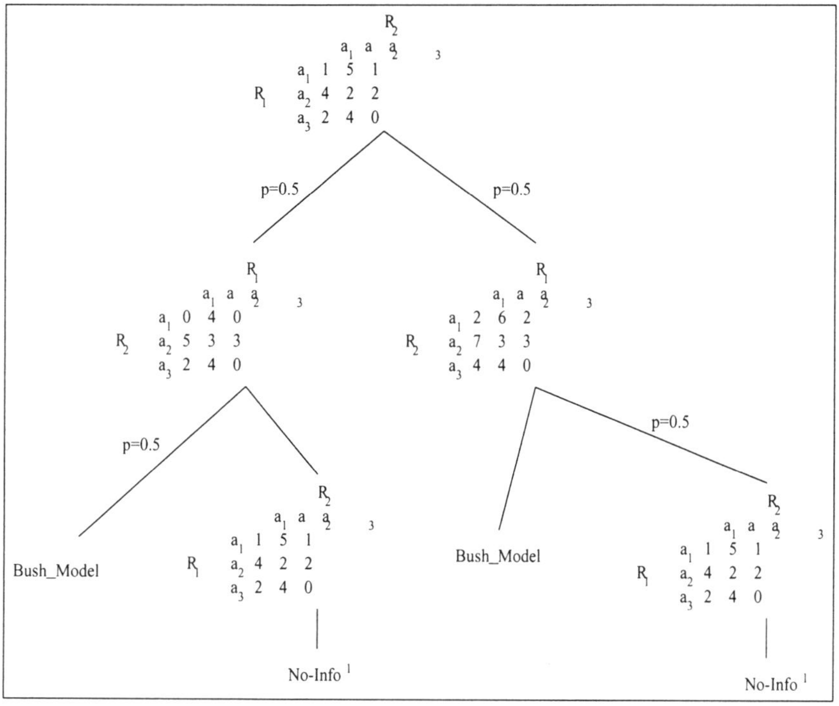

Figure 4. Recursive Model Structure depicting R_1's Knowledge in Example 2.

agent R_1, on level 1, that has run out of information at the third level of recursive modeling.

The above examples do not define our representation formally. The reader is referred to [11, 12] for a rigorous definition.

3 SOLVING RMM USING DYNAMIC PROGRAMMING

The recursive nature of RMM makes it possible to express the rational choice of an agent on a given level of modeling in terms of choices of the agents modeled on deeper levels. Thus, the problem of finding a solution exhibits *optimal substructure*, i.e., the solution to an agent's problem of rational choice involves solving, as subproblems, the rational choices of the other agents involved. Further, the problem exhibits the property of having overlapping subproblems, since some of the models of agents' decision-making, for example in the representation in Figure 4, may be identical and do not have to be recomputed. These properties of the problem make RMM a candidate for a solution using dynamic programming [7].

In case of RMM, the dynamic programming solution traverses the recursive model structure propagating the information bottom-up. The result is an assignment of expected utilities to the agent's alternative actions, based on all of the information the agent has about its own decision-making situation, the other agents' decision-making situations, and so on. The rational agent can then choose an action with the highest expected utility. We illustrate the computations using previously developed examples.

3.1 Solution to Example 1

For illustrative purposes, our solution will assume a simplified case of the no-information models expressed as uniform probability distributions. This approximation[9] is equivalent to assuming that an agent with an unknown model of the other agents will act as if it did not know anything about them. A solution of the case when an agent has information about other agents but it is not known what this information is, is discussed in [12].

As we mentioned, the dynamic programming solution proceeds bottom-up. If R_2 cannot see P2, its expected utilities can be computed from the model in the left branch on Figure 2 as probabilistic sums of the payoffs with uniform probabilities:

$$\begin{aligned} u^{(R_1,1),R_2}_{a^2_1} &= 0.5 \times 0 + 0.5 \times 0 = 0 \\ u^{(R_1,1),R_2}_{a^2_3} &= 0.5 \times 2 + 0.5 \times 0 = 1 \end{aligned}$$

Thus, R_1 knows that if R_2 cannot see point P2 it will remain stationary.

If R_2 can see point P2, then the expected utilities of its actions can be computed from the model in the right branch on Figure 2 as:

$$\begin{aligned} u^{(R_1,1),R_2}_{a^2_1} &= 0.333 \times 0 + 0.333 \times 4 + 0.333 \times 0 = 1.333 \\ u^{(R_1,1),R_2}_{a^2_2} &= 0.333 \times 5 + 0.333 \times 3 + 0.333 \times 3 = 3.667 \\ u^{(R_1,1),R_2}_{a^2_3} &= 0.333 \times 2 + 0.333 \times 4 + 0.333 \times 0 = 3 \end{aligned}$$

Thus, R_1 knows that if R_2 can see P2 it will move toward it and observe from there.

Accounting for the likelihood of R_2 being able to see P2, the probability that it will pursue observation from P1 is zero, the probability that it will pursue observation from P2 is 0.1, and the probability that it will remain stationary is 0.9. These probabilities can now be used to compute the expected utilities of R_1's alternative actions from the top matrix in Figure 2 as:

[9] One of many approximations to perfect rationality possible within the RMM framework. See [12] for other examples and discussion.

$$u^{R_1}_{a^1_1} = .1 \times 5 + .9 \times 1 = 1.4$$
$$u^{R_1}_{a^1_2} = .1 \times 2 + .9 \times 2 = 2$$
$$u^{R_1}_{a^1_3} = .1 \times 4 + .9 \times 0 = 0.4$$

Thus, the best choice for R_1 is to pursue its option a^1_2, that is, to move toward point P2 and make an observation from there. It is the rational coordinated action given R_1's state of knowledge, since the computation included all of the information R_1 has about agent R_2's expected behavior. Intuitively, this means that R_1 believes that R_2 is so unlikely to go to P2 that R_1 should go there itself.

In may be useful at this point to compare the solution just obtained to one that could be arrived using traditional game theory [24, 25]. As we argued in [12], the traditional game-theoretic equilibrium analysis does not apply to this example since there is no common knowledge, i.e., agent R_1 does not know what R_2 knows about what R_1 knows, and so on.[10] However, the solution obtained above happens to coincide with one of two possible solutions that could be arrived at by traditional game-theoretic equilibrium analysis, if a number additional assumptions about what the agents know were made in this case. Thus, if R_1 were to assume that R_2 knows about the point P2, and that R_2 knows that R_1 knows, and so on, then R_1's move toward P2 would a part of the equilibrium in which R_1 goes to P2 and R_2 goes to P1. This shows that the solutions obtained in RMM analysis can coincide with game-theoretic solutions, but that it depends on fortuitous assumptions about agents' knowledge. It is also not difficult to construct a finite state of R_1's knowledge that would result in R_1's rational action to be pursuing observation from P1 and expecting R_2 to observe from P2, which happens to be the other equilibrium point that could be derived if the agents were assumed to have common knowledge [1] about P2. The coincidence would, again, be a matter of making *ad hoc* assumptions about the agents' states of knowledge.

The bottom line is that agents using RMM simply optimize subject to whatever state of knowledge they happen to be in. In game-theoretic literature, this approach has been called a decision-theoretic approach to game theory, and has been advocated in [16] and recently by [1]. It illustrates two main advantages of RMM: First, we avoid the need for unrealistic assumptions about the agents' common and infinite knowledge. Second, we are able to arrive at the rational course of action given any finite state of an agent's knowledge, including epistemic states that need not imply equilibria [1].

[10] A reader that finds the assumption of common knowledge quite natural when, say, they imagine themselves interacting with another person within eye sight, should imagine themselves interacting with R2D2 instead, to realize the assumption of common knowledge may not be as natural as they first thought.

3.2 Solution to Example 2

For R_1's state of knowledge, depicted in Figure 4, the computation is analogous to the one in the previous subsection; it takes us deeper into the nested levels of R_1's knowledge but similarly proceeds bottom-up as in Example 1.[11] R_1 has two alternative models of R_2, but in each case the models below are identical. In each of them we assume that the prediction resulting from the Bush-Model would be that it remains stationary.

The expected utilities of R_1 alternative actions, given the uniform probabilities in the no-information model below are:

$$
\begin{aligned}
u_{a_1^1}^{R_1} &= .333 \times 1 + 0.333 \times 5 + .333 \times 1 = 2.333 \\
u_{a_2^1}^{R_1} &= .333 \times 4 + 0.333 \times 2 + .333 \times 2 = 2.667 \\
u_{a_3^1}^{R_1} &= .333 \times 2 + 0.333 \times 4 + .333 \times 0 = 2
\end{aligned}
$$

Thus, R_1 should conclude that R_2 would think that if R_1 is an agent, then it will pursue the observation from point P2. But since R_1 could also be a bush, with the probability 0.5, R_2 would assign the probability 0.5 to R_1 observing from p2, and 0.5 to it standing still.

This expectation as to R_1 actions results in the following calculation of R_2's expected utilities in R_2's model in the left branch in Figure 4:

$$
\begin{aligned}
u_{a_1^2}^{(R_1,1),R_2} &= 0.5 \times 4 + 0.5 \times 0 = 2 \\
u_{a_2^2}^{(R_1,1),R_2} &= 0.5 \times 3 + 0.5 \times 3 = 3 \\
u_{a_3^2}^{(R_1,1),R_2} &= 0.5 \times 4 + 0.5 \times 0 = 2
\end{aligned}
$$

The calculation of R_2's expected utilities if R_2's model is as in the left branch in Figure 4 is:

$$
\begin{aligned}
u_{a_1^2}^{(R_1,1),R_2} &= 0.5 \times 6 + 0.5 \times 3 = 4.5 \\
u_{a_2^2}^{(R_1,1),R_2} &= 0.5 \times 3 + 0.5 \times 3 = 3 \\
u_{a_3^2}^{(R_1,1),R_2} &= 0.5 \times 4 + 0.5 \times 0 = 2
\end{aligned}
$$

Thus, since R_1 assigns 0.5 probability to the alternative models of R_2, its overall expectation is that R_2 will observe from P1 with probability 0.5, and from P2 with probability 0.5.

[11] Here we again simplify the no-information models into the uniform probability distributions. For the full-blown solution see [12].

Given R_1's decision-making situation described by the top payoff matrix in Figure 4, the expected utilities of its own options are:

$$u_{a_1^1}^{R_1} = 3$$
$$u_{a_2^1}^{R_1} = 3$$
$$u_{a_3^1}^{R_1} = 3$$

In this example, therefore, R_1 would conclude that, given its current state of knowledge, it does not matter what it does, since in all cases the expected payoff is the same.[12]

The two examples above were intended to illustrate the rational choice of an agent action in a multiagent environment. The rational choice of communicative action is addressed in the following section/

4 RATIONAL COMMUNICATIVE BEHAVIOR

(Authors note: This section is based, in part, on [13].) We treat decisions about communication just like decisions about any other actions, and employ decision-theoretic techniques to select the action with the highest expected utility. As in the case of any action, the expected utility is determined by the desirability of the expected outcome. However, unlike a physical action, that changes the physical state of the world, a communicative action changes the state of knowledge of the agents involved. The framework we propose is aimed at calculating the changes in the state of knowledge and quantifying the benefits of communicative actions that bring them about.

The advantage of RMM is that, for the purpose of decision-making, it is able to succinctly represent the agent's state of knowledge, including the agent's preferences, abilities and beliefs about the physical world, as well as the agent's beliefs about the other agents, their preferences and abilities, their beliefs about the world and about other agents, their beliefs about others' beliefs, and so on. The need for considering the nestedness of the agents' beliefs for communication has been widely recognized in the related literature before [2, 3, 5, 6, 14, 20, 23, 26, 27].

We identify the *decision-theoretic (DT) pragmatics* of a communicative act with the transformation of the state of knowledge about the decision-making situation the act brings about. We model DT pragmatics using RMM representation to investigate the utility of the communicative act.[13] The transformation in the agent's decision-making situation, as represented by RMM's recursive model structure,

[12]In this case, all of the alternative actions of R_1 maximize its expected utility, and each of them constitutes a rational choice.

[13]Since our formalism is probabilistic, it naturally handles cases when the meaning of a message is itself uncertain.

may change the expected utilities of alternative actions. It is natural to identify the change of the expected utility brought about by a communicative action as the expected utility of this action itself. By evaluating the alternative communicative acts in this way, the agent can select and send the highest utility message—the message that causes the greatest gain in the expected utility of the agent's action.[14]

DT pragmatics of a communicative act differs in a subtle way from its pragmatic meaning, usually defined [5, 14, 27] as the change of the state of knowledge brought about by the act. to understand the difference imagine two agents engaged together in assembling a bicycle from parts scattered about a garage. A communicative act "The front wheel is in the southwest corner of the garage", uttered by one of the agents, has the pragmatic meaning of changing the other agent's beliefs about the location of the front wheel, if it did not know the location before. This act also changes the decision-making situation the agents are in: The other agent is now in the better position to get the front wheel and complete the bicycle assembly, and the time saved could be of benefit to both agents. The above communicative act, therefore, is endowed with both decision-theoretic pragmatics, as well as pragmatic meaning. But a communicative act "The temperature in the center of Alpha-Centauri is 4357 degrees C", with its pragmatic meaning of changing the hearer's state of knowledge, does nothing to the decision-making situation the agents are facing. Therefore, the second communicative act does not change the decision-making situation the agents are facing, its expected utility is null, and its DT pragmatics is the identity transformation.[15]

In what follows, we denote the probability distribution describing the expected actions of agents other then an agent R_i before communication as $p\{-R_i\}$, and we will call it the prior conjecture. Now, since a communicative act, M, can influence the actions of the other agents, the probability distribution over the others' actions expected after this communicative act can be different. We will denote it as $p^M_{\{-R_i\}}$ and call it the projected conjecture. Further, we will denote the rational act of the speaker agent, R_i, resulting from the prior conjecture as X, with its utility $U_{p_{\{-R_i\}}}(X)$, and the rational act resulting from the projected conjecture as Y, with its utility $U_{p^M_{\{-R_i\}}}(Y)$.[16] We can now define the expected utility of a message itself.

Definition: The expected utility of the message M is the difference between the payoff the agent expects before and after sending the message:

[14]The notion of the utility of a message we define here differs from the notion of the value of information considered in decision theory [22, 26]. The latter expresses the value of information to its recipient. We, on the other hand, consider the value of a message to its sender, since, of course, it is the sender that makes the decision of if, and what, to communicate. The two notions coincide in two special cases, when the preferences of the agents perfectly coincide, and when a speaker requests information from the hearer by asking a question.

[15]Note that the same message may have nontrivial DT pragmatics, as well as a considerable value, in a different decision-making situation.

[16]Y may be the same as X when a communicative act does not change the speaker's intended rational action.

(1) $U(M) = U_{p^M_{\{-R_i\}}}(Y) - U_{p_{\{-R_i\}}}(X).$

We broadly classify messages into types, depending on how a message impacts the recursive model structure of an agent. Some of the types we consider have close correspondents in the speech act theory [2, 28], and in various kinds of performatives considered in KQML [19]. At the beginning of our considerations below we will assume that the agents share a communication language. We will be able to relax this assumption later and take the full advantage of RMM's probabilistic framework.

4.1 Modeling Messages

Modeling messages contain information about the probabilities the agents use to represent information about the properties of their environment. Thus, the modeling messages update the hearer's and the speaker's model of the multiagent world. The close correspondents of these type of messages in speech act theory are the inform, assert and tell acts.

Consider the example of the interaction introduced in Figure 1, and the recursive model structure depicting R_1's state of knowledge, depicted on the left in Figure 5. In case of this structure, two possible models that R_1 has of R_2's decision-making situation reflect R_1's uncertainty as to R_2's being able to see point P2.[17] The bottom-up solution of this structure shows that $R2$'s intended actions can be described by intentional probability distribution $p^{R_1}_{R_2} = [0, 0.1, 0.9]$, which is R_1's prior conjecture about R_2's actions. The best choice for R_1 is then to pursue its option a^1_2, that is to move toward point P2 and make an observation from there, with its expected utility of 2, as shown before.

Let us consider what would happen in this scenario if R_1 were to send a message M_1 stating "There is an observation point P2, twice as high as P1, behind the trees". If, for the time being, we assume that M_1 is encoded in a communication language known to R_2 and that communication channels never lose messages, then R_1 can be sure that R_2 will know about the point P2 as a result of the message having been sent. Thus, the decision-theoretic pragmatics of M_1 is as depicted in Figure 5.

The projected structure, on the right in Figure 5, can be easily solved, showing that R_1 would be sure that R_2 would observe from point P2, taking action a^2_2: $p^{M_1;R_1}_{R_2} = [0, 1, 0]$. The best alternative for R_1 now is to make an observation from P1, but the expected payoff has increased to 5. Thus, by sending the message M_1 to R_2, R_1 was able to increase the expected utility it gets from the interaction from 2 to 5. As defined in Equation 4, the utility of sending the message M_1 is $U(M_1) = 5 - 2 = 3$.

The above analysis assumes that R_2 is guaranteed to receive and properly decode M_1. However, it may be that the language used by R_1 is unknown to R_2,

[17] We simplified both no-information models to be equivalent to uniform probability distributions

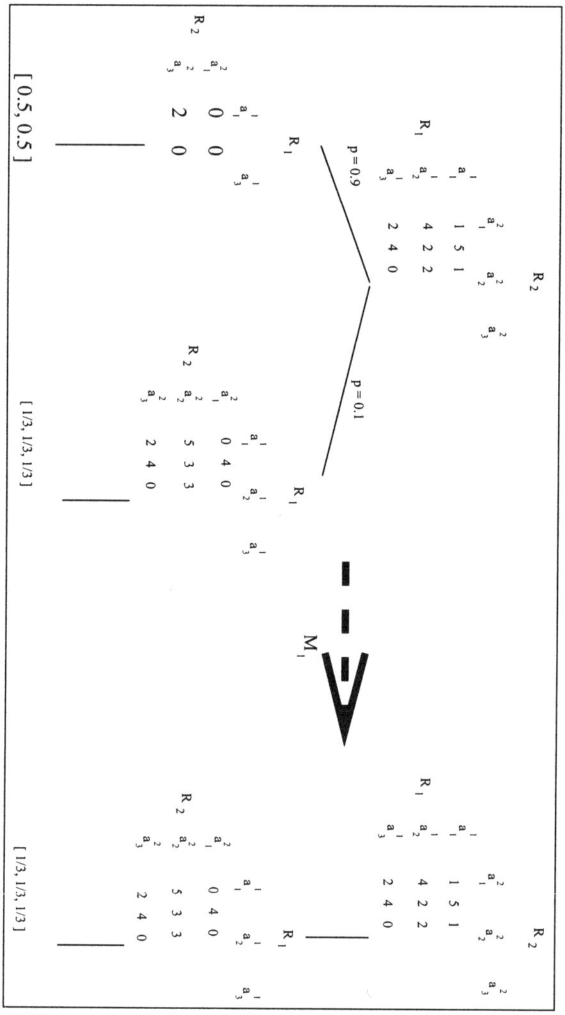

Figure 5. Decision-Theoretic Pragmatics of Message M_1.

or that the communication channel used is unreliable. We can represent these two factors by assigning a probability, p_c ($0 \leq p_c \leq 1$), to R_2's properly receiving and understanding M_1. From R_1's perspective: with probability p_c, M_1 will be received, in which case R_2 would observe from P2, described by the intentional probability of $p_{R_2,}^{M_1;R_1} = [0, 1, 0]$ as just derived; and with probability $(1 - p_c)$, M_1 will not be received, so the probability distribution describing R2's intended actions is the same as with no communication at all: $p_{R_2}^{R_1} = [0, 0.1, 0.9]$. Combining these we get:

$$p_c[0,1,0] + (1-p_c)[0,0.1,0.9] = [0, 0.1 + 0.9p_c, 0.9 - 0.9p_c].$$

The expected utilities of R_1's alternatives can now be computed as:

$$\begin{aligned} u^{R_1}_{a^1_1} &= 1.4 + 3.6p_c \\ u^{R_1}_{a^1_2} &= 2 \\ u^{R_1}_{a^1_3} &= 0.4 + 3.6p_c \end{aligned}$$

From the above we can see that, if $p_c > 1/6$, the value of the message M_1 depends on p_c as follows: $U(M_1) = 3.6p_c - 0.6$. If $p_c < 1/6$, R_1 would prefer to choose a^1_2 and observe from P2, with its payoff of 2. This is the same as without communication, and M_1's expected utility would then be zero.

4.2 *Intentional Messages*

An intentional message corresponds to an agent committing to a choice of action, and informing other agents about it, that is, it contains information about the probabilistic conjectures that the agents can use to describe others expected actions. It corresponds to the *promise* act in the speech act theory. For the purpose of the present discussion we assume that the truthfulness of these messages is guaranteed (see [10] for cases involving lying). Thus, a hearer can use this message to predict exactly what the speaker will do. In modeling the hearer, therefore, the speaker can truncate the projected recursive structure, because it knows that the hearer's conjecture of the speaker's actions correspond exactly to the content of the intentional message.

Let us take again the interaction in Figure 1, and suppose that R_1 considers transmitting an intentional message M_2 to R_2, containing a promise "I will observe from the point P2". The pragmatic meaning of M_2 can be modeled as depicted in Figure 6.

If R_2 is not aware of the point P2 and receives M_2, it models R_1 as pursuing its a^1_3 option, i.e., doing something other then a^1_1,[18] and so, for R_2, the options a^2_1 and a^2_3 are equally good and equally likely. If R_2 can see the point P2 and receives M_2, its options a^2_1 and a^2_3 are also equally good. Thus, the new overall intentional probability distribution over R_2's options is $p^{M_2;R_1}_{R_2} = [0.5, 0, 0.5]$. R_1 has committed itself to a^1_2, but now R_1 computes the expected utility of this action as 3. According to Equation 4, the utility of sending message M_2 is $U(M_2) = 3 - 2 = 1$.

[18] In human communication, this message would likely be interpreted as a conjunction of two messages. First one saying that there is a point P2 out there (modeling), and the second saying that R_1 will do something about it (intentional). In this discussion we treat M_2 as a purely intentional message.

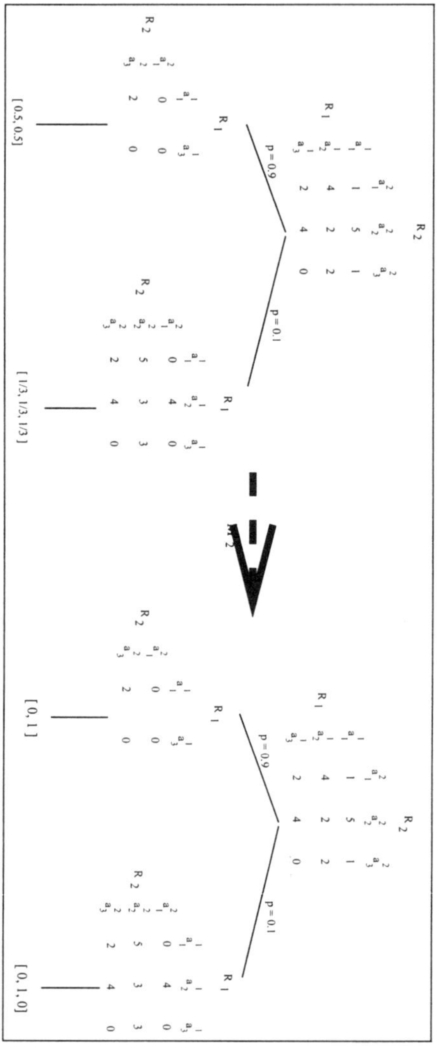

Figure 6. DT pragmatics of Message M_2.

The above analysis assumes that R_2 is guaranteed to receive and properly decode M_2. If the reliability of the communication used is characterized by the probability p_c, as before, the intentional probabilities R_1 ascribes to R_2 will be:

$$p_c[0.5, 0, 0.5] + (1 - p_c)[0, 0.1, 0.9] = [0.5p_c, 0.1 - 0.1p_c, 0.9 - 0.4p_c].$$

The expected utilities of R_1's alternatives can now be computed as:

$$u^{R_1}_{a^1_1} = 1.4 - 0.4p_c$$
$$u^{R_1}_{a^1_2} = 2 + p_c$$
$$u^{R_1}_{a^1_3} = 0.4 + 0.6p_c$$

The expected utility of sending M_2 is thus equal to the probability p_c.

4.3 Modeling vs. Intentional Messages

Note that, assuming reliable communication, the intentional message M_2 from the preceding section was less valuable than the modeling message considered in Section 4.1. That is an intuitive result; in the situation in which an agent is quite unlikely (with only a 10% chance) to be aware of an important feature of the environment, we would also tend to favor the message telling the agent about this feature.

But let us look at a different situation, considered as the second example before, depicted in Figure 3, and the recursive model structure depicting R_1's state of knowledge, now depicted on the left in Figure 7.[19] Here, R_1 knows that R_2 can see both observation points. A modeling message that tells R_2 about any of the observation points is bound to be useless. Now, however, an intentional message may be more appropriate. Say that R_1 considers sending an intentional message M_3: "I will observe from P1". R_1 would model this message's DT pragmatics as the transformation in Figure 7, in a way similar to DT pragmatics of M_2 before.

It is easy to see that R_1 would conclude that, after receiving M_3, R_2 would chose to make an observation from point P2, resulting in the expected utility of observing from P1 for R_1 as 5. The value of this intentional message is in this case equal to 2.

Interestingly, there is a modeling message that R_1 could also consider in this case. It is M_4: "I am not a bush". The pragmatic meaning of M_4 is simple: R_2 would have just one correct model of R_1's decision-making situation. The expected utility of this message turns out to be 0.5 (we invite the reader to check this result), thus, as we would expect, M_4 is viable and beneficial in this case, although not optimal.

The modeling and intentional messages are not the only ones that agents can exchange, of course. Other kinds of communicative acts, including acknowledgments, questions, imperatives, threats and proposals, are discussed in [13].

[19] We again replaced the former $No - Info^1$ models with models denoting uniform probability distributions, to simplify the analysis.

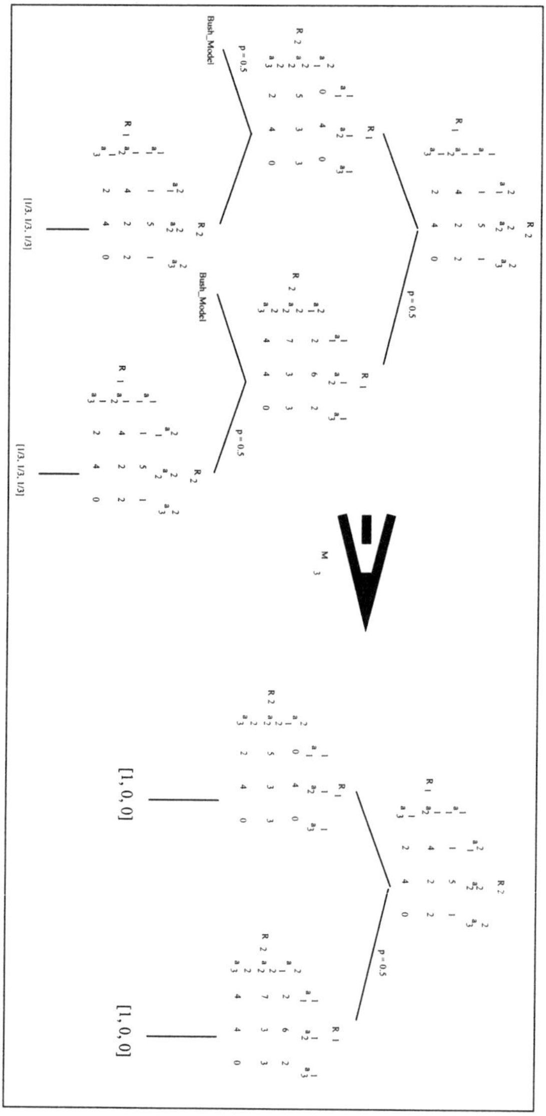

Figure 7. DT Pragmatics of Message M_3.

In summary, our approach to rational communicative behavior was aimed at the agent's ability to compute the expected utility of various communicative acts, and to execute the best one. The expected utilities of messages we have considered seem to coincide with our human assessment of what messages are most appropriate for a limited set of situated scenarios. We intend to undertake a more exhaustive testing of this correlation.

We have not said much about the requirement of language and background knowledge commonality, so frequently required in other treatments (see, for example [5] for interesting discussion) as a prerequisite for effective communication. We see these issues as included in our notion of the speaker's model of the decision-theoretic pragmatics, and thus directly entering into the utility considerations as the reliability of the communication used, p_c. Thus, we do not have to assume that the agents know that they share the same communication language, or that no ambiguities of interpretation will arise. For example, the speaker may think it is possible that the hearer may not understand the language used. The DT pragmatics of any statement made in this language simply includes the possibility (a branch in the recursive model structure) that the hearer succeeded in decoding the message, as well as the possibility (with the probability $1 - p_c$) that the message was not understood. As we show, the utility of the message uttered increases with the probability p_c, as would be expected.

5 BAYESIAN UPDATING WITHIN RMM

In this section, we outline some of our current work addressing the, rather sensitive, issue of "where do these probabilities come from?". In other words, we are interested in the process of revising and updating the probabilities the agents use to represent the uncertainty as to the other agents' preferences, abilities and states of knowledge.

One way of revising these probabilities is, of course, the result of communication among the agents. For example, the probabilities of the two branches in the structure in Figure 2 could be the result of agent R_2 directly communicating to R_1 the reliability of its sensors. Even without communication, however, these probabilities should be updated as the interaction progresses. For example, an agent that enters an interaction without knowing who is the predator and who is the prey, should be able to figure it out with an increasing degree of confidence after observing that some agents are chasing after another one.[20]

A very attractive and well understood method of belief update is Bayesian belief revision [22, 26]. The main advantage of Bayesian revision is that it allows the agent to maintain the rationality of its actions in a dynamic environment by incorporating the information from past observed interactions one piece at a time. The RMM formalism makes it natural to use Bayesian updating to revise the probabilities of alternative models of the other agents; the result of an updated hierarchy of models in RMM is another such hierarchy with the probabilities updated to reflect the newly acquired information.

The simplest case of an update is when a new information is in the form of actual observed behavior of the agents involved, O. It is then straightforward to

[20] As before, we will neglect the possibility bluffing behavior here.

use Bayes formula to update the prior probability assigned to any of the alternative models of the other agent, $p(Mod)$:

$$(2) \qquad p(Mod|O) = p(O|Mod)p(Mod)/p(O),$$

where $p(Mod|O)$ is the new updated probability assigned to a model Mod, and $p(O)$ is computed from the normalizing condition that the probability of all of the alternative models add to unity. The probability $p(O|Mod)$ is the probability with which the model Mod predicts the observed behavior O.

In the simplest case, this probability is the same as the conjecture as to the other agent action, arrived as in the examples discussed before. In a more complex case, the prediction of the models, like the ones in Figure 2, will yield a prediction of behavior on a higher level of abstraction. For example, the model could predict that an agent will observe from the observation point P1, but the observation made is, say, that the agent has moved one meter to the north. In this case, the probability of the observation given model has to include the various ways that the abstract actions can be implemented.

Finally, apart from updating the probabilities of models of the other agent(s), residing on the first level of structures like the ones in Figure 2 and Figure 4, the probabilities of the models residing deeper in the model structure can also be updated. However, while the models of the other agents, on the first level, can be updated based on the other agents' observed behavior, the probabilities of the models that the others may use have to be updated based on the observations that the other agents are expected to acquire. For further details of this approach the reader is referred to [17].

6 CONCLUSION

We have presented an overview of a framework for rational decision-making in multiagent domains. The central parts of it are the method to represent the information that comprises an agent's decision-making situation, and a dynamic programming solution of this representation that yields the rational choice of action in a multiagent environment. We argued that this rational choice constitutes *coordination*, i.e., the choice of action taken in anticipation of the other agents' actions.

We then went on to describe an approach to rational choice of communicative behavior. Here, we have defined the notion of decision-theoretic pragmatics of a communicative act. It is similar to the more traditional notion of pragmatic meaning, but it bring out the decision-theoretic repercussions of the act. Based on that, the definition of the expected utility of a communicative act followed in a natural fashion. Applying the decision-theoretic definition of rationality, we postulated that rational communicative behavior amounts to choosing to execute the communicative act so as to maximize the resulting expected utility. We further showed how the possible unreliability of the communication channel, as well as the possibility that the communication language used is not understood, can be quantified and included in the expected utility calculation.

Finally, we briefly outlined our current ideas on Bayesian belief update in uncertain multiagent environment that uses our framework. Since models within our framework yield a probabilistic conjecture as to the other agents' expected action, it is straightforward to use Bayes formula to update the probability of each model based on how well its prediction matched the actual observed behavior. Using Bayesian updating, the agents can identify the proper models of the other agents and, further, can identify the models the other agents use to model them, as well as others, and so on.

We have applied our framework in a number of domains, with the references and some on-line experiments accessible on http://dali.uta.edu. Probably the most interesting result is the relatively high quality of coordination achieved between our RMM agents and human subjects in simple coordination games. These preliminary results seem to point to the viability of modeling humans as rational utility maximizers during interactions we have experimented with, and parallels the reliance on rationality as a modeling tool in game theory, economics and social sciences.

Piotr Gmytrasiewicz (University of Texas at Arlington, USA)

REFERENCES

[1] Robert J. Aumann and Adam Brandenburger. Epistemic conditions for Nash equilibrium. *Accepted for Publication in Econometrica*, 1995.

[2] J. L. Austin. *How to do Things with Words*. Clarendon Press, 1962.

[3] Afzal Ballim and Yoric Wilks. *Artificial Believers*. Earlbaum Associates, Inc., 1991.

[4] H. Chernoff and L. E. Moses. *Elementary Decision Theory*. John Wiley, New York, 1959.

[5] Herbert H. Clark. *Arenas of Language Use*. The University of Chicago Press, Chicago, 1992.

[6] P. R. Cohen and H. J. Levesque. Rational interaction as the basis for communication. In P. R. Cohen, J. Morgan, and M. E. Pollack, editors, *Intentions in Communication*. MIT Press, 1990.

[7] Thomas H. Cormen, Charles E. Leiserson, and Ronald L. Rivest. *Introduction to Algorithms*. The MIT Press, 1990.

[8] D. Dennett. Intentional systems. In D. Dennett, editor, *Brainstorms*. MIT Press, 1986.

[9] Peter Gardenfors and Nils-Eric Sahlin. *Decision, probability, and utility*. Cambridge University Press, 1988.

[10] Piotr J. Gmytrasiewicz and Edmund H. Durfee. Toward a theory of honesty and trust among communicating autonomous agents. *Group Decision and Negotiation*, 2:237–258, 1993.

[11] Piotr J. Gmytrasiewicz and Edmund H. Durfee. Formalization of recursive modeling. In *In Proceedings of the First International Conference on Multiagent Systems, ICMAS'95*, July 1995.

[12] Piotr J. Gmytrasiewicz and Edmund H. Durfee. Rational interaction in multiagent environments: Coordination. *Submitted for publication*, 1996. Available in postscript from http://www-cse.uta.edu/ piotr/piotr.html.

[13] Piotr J. Gmytrasiewicz and Edmund H. Durfee. Rational interaction in multiagent environments: Communication. *Submitted for publication*, 1996. Available in postscript from http://www-cse.uta.edu/ piotr/piotr.html.

[14] H. P. Grice. Meaning. *Philosophical Review*, (LXVI):377–388, 1957.

[15] John C. Harsanyi. Games with incomplete information played by 'bayesian' players. *Management Science*, 14(3):159–182, November 1967.

[16] Joseph B. Kadane and Patrick D. Larkey. Subjective probability and the theory of games. *Management Science*, 28(2):113–120, February 1982.
[17] Tad Kellogg and Piotr Gmytrasiewicz. Bayesian belief update in multiagent environments. In *Submitted to MAAMAW 1997*, January 1997.
[18] Craig A. Knoblock. *Generating Abstraction Hierarchies: An automated approach to reducing search in planning.* Kluwer Academic Publishers, 1993.
[19] Y. Labrou and T. Finin. A semantics approach for KQML – a general purpose communication language for software agents. In *Proceedings of the Third International Conference on Information and Knowledge Management*, November 1994.
[20] R. Mayerson. Incentive constraints and optimal communication systems. In *Proceedings of the Conference on Theoretical Aspects of Reasoning about Knowladge*, pages 179–193. Morgan Kaufman, March 1988.
[21] Ajit Narayanan. *On Being a Machine.* Ellis Horwood, 1988.
[22] Judea Pearl. *Probabilistic Reasoning in Intelligent Systems: Networks of Plausible Inference.* Morgan Kaufman, 1988.
[23] C. R. Perrault. An application of default logic to speech act theory. In P. R. Cohen, J. Morgan, and M. E. Pollack, editors, *Intentions in Communication.* MIT Press, 1990.
[24] Eric Rasmusen. *Games and Information.* B. Blackwel, 1989.
[25] Jeffrey S. Rosenschein and Gilag Zlotkin. *Rules of Encounter.* MIT Press, 1994.
[26] S. Russell and P. Norvig. *Artificial Intelligence: A Modern Approach.* Prentice Hall, 1994.
[27] Stephen Schiffer. *Meaning.* Clarendon Press, 1972.
[28] Terry Winograd and Fernando Flores. *Understanding Computers and Cognition: A New Foundation for Design.* Ablex Publishing, 1986.

M. P. SINGH

KNOW-HOW

In the knowledge lies the power.
Frederick Hayes-Roth

In the know-how lies the power.
Carl Hewitt

The study of knowledge is crucial to the science of rational agency. This fact is well-recognized in artificial intelligence (AI) and related fields. However, most often the form of knowledge that is studied and formalized is the knowledge of (putative) facts. We refer to this form of knowledge as *know-that*. Know-that has proved an extremely successful concept in AI, being the basis of a large number of AI systems, which are therefore termed *knowledge-based*.

There is great need, however, for other notions of knowledge as well. In particular, since rational agency is intimately related to actions, it is important also to consider the form of knowledge that is about actions and procedures. We refer to this form of knowledge as *know-how*. Intuitively, we might think of the distinction between know-how and know-that as reflecting the distinction between rational agents and on the one hand, and disembodied minds, such as knowledge-based expert systems, on the other.

This chapter introduces know-how and allied concepts from a conceptual standpoint. It presents a formalization of two accounts of know-how borrowed from [33]. It also and compares reviews a selected subset of the approaches available in the literature. It seeks to provide the background with which one may understand the details of the different technical approaches.

Historical Remarks. The noted British philosopher Gilbert Ryle is widely regarded as having been the first, at least in modern times, to have argued for the fundamental difference between knowing how and knowing that. Ryle devotes a chapter of his famous 1949 book *The Concept of Mind* to this distinction, arguing among other things for the key difference between (a) stupidity, that is, not knowing how, and (b) ignorance, that is, not knowing that [28]. He argues that the two are fundamentally separate notions, because often an agent may know how to perform certain complex actions, yet not know that he does a certain specific sequence. This distinction is interesting and related to one discovered years later in the study of reactive systems in AI, for example, by Agre & Chapman [1] and others.

M. Wooldridge and A. Rao (eds.), Foundations of Rational Agency, 105–132.

There has been much work in planning right from the early days of AI. To a large extent, Agre & Chapman and others were rebutting the centrality of planning when they proposed reactive architectures. However, the planning literature did not address the logical notion of know-how *per se*, although it considered the mechanics through which it could be realized. In fact, even the more traditional know-that was not always studied formally in AI, although systems that reasoned with it were abundant.

There was, however, considerable work on logics of know-that in the late 1970s and early 1980s. This was based on previous work in philosophical logic. Building on ideas from McCarthy & Hayes [20], Robert Moore developed a formal logic of knowledge that was essentially an S4 modal logic, but captured in terms of its first-order metatheory [22]. With slight modifications, this is the logic we present in section 2.5. Moore and others did not study know-how *per se*, but rather the knowledge required to execute plans. For example, to execute a conditional plan requires knowing whether its condition is true. Know-How has only recently begun to be studied intensively in AI. Indeed, Jaakko Hintikka, who gave the first formalization of knowledge using ideas from modal logic [14], observed that the logic of knowing how had proved difficult to develop [15] (cited by McCarthy & Hayes [20, p. 447]).

We began looking at know-how as a first-class topic of investigation in the late 1980s. Around the same time, Werner worked on a general theory that included abilities, but not at the present level of detail [41]. Independently, Meyer and associates studied capabilities from a perspective that included other concepts captured as modal operators [38]. Some philosophical work on this subject carried out over roughly the same time-frame by Brown [6], Belnap & Perloff [2], Chellas [8], and Segerberg [31].

Although we take the notion of know-how seriously, we confess that in developing formal theories of it, we shall not be supporting all of the associated philosophical positions. In particular, the very definition of know-how has not much to do—pro or con—with the doctrine of strict reactivity, as evinced in the works of Ryle and Agre & Chapman. This is because the notion of know-that, which we also discuss, can be taken as describing the knowledge of an agent in an explicit conscious sense, or in an implicit sense. Indeed, the specific formal theories we describe mostly develop an implicit notion of knowledge. The distinction between explicit and implicit knowledge, however, is not definitional or logical, but related to the computational power available to an agent. In other words, it will be just as acceptable to us that an agent can describe his know-how as that he cannot.

Organization. This chapter provides a conceptual introduction to several different variants of know-how. Although it includes some technical description to give a flavor of how such formalizations bring together ideas from temporal and dynamic logics, the cited works should be read for their technical details.

The rest of this chapter is organized as follows. Section 1 introduces the key concepts and motivates the study of know-how. Section 2 describes our technical framework, including formalizations of the background concepts of time, actions, and know-that. Section 3 presents our definition of know-how. Section 4 discusses other approaches known in the literature, and relates them to the approach of section 3. Section 5 concludes with a discussion and pointers to some future directions.

1 MOTIVATION

Agency is inherently about performing actions. Because of the intimate relationship between agency and actions, the formal study of rational agents has involved the development of a number of folk concepts of which several relate to actions. Two such key concepts are intentions and desires studied by Rao & Georgeff [27] among others. Of the two, intentions have the closer and more direct relationship with actions, and we consider them in more detail. Intentions are generally understood as having a causal relationship with actions—they not only lead an agent to select suitable actions, but also to perform those actions. As a consequence, intentions have another role in rational agency, namely as explanations of actions, which can be used by designers and analyzers to reason about some agent's behavior, or by the agents themselves to reason about the behavior of other agents.

One of the ways in which intentions are applied in rational agency is as specifying the ends an agent has chosen to pursue. These intentions lead to deliberation by the agent, leading him to adopt additional, more specific intentions as means to his original ends. This process can iterate several times, resulting (if successful) in intentions that the agent can act on directly. This view of deliberation is shared by many researchers, including the philosophers Bratman [4] and Brand [3].

The successful use of intentions in theories of rational agency, therefore, relies upon their linkage to actions. For instance, a natural question is to determine under what circumstances an intention may be taken to lead to success. The simple answer is that an intention can lead to success when it is held long enough, is acted upon, and when the agent has the requisite know-how. This, in our mind, is the single most important motivation for the study of know-how, and was the basis for the work reported in [33].

There are obvious connections between intentions and know-how, some negative. For example, intentions do not entail know-how—you cannot always do what you intend to. Similarly, know-how does not entail intentions—you don't always intend what you have the know-how to do. Although intentions are not formally discussed in this chapter, it will be helpful to keep these connections, at least informally, in mind.

1.1 Actions

When talking of actions, it is conventional to define *basic* actions as those that an agent can perform atomically with a single choice. Philosophers have spent considerable energy in attempting to give necessary and sufficient conditions for when an observed event counts as an agent's action. As for other important topics, there is profound disagreement among the philosophers! Some approaches, exemplified by Searle's work, define actions in terms of what he calls intentions-in-action [29]. Roughly, what this means is that the agent should have the intention to do the given action he is in fact doing, and that his intention should play some causal role in the performance of that action. Another interesting theory is the STIT approach, due to Belnap & Perloff, which states that the actions of an agent are what he has seen to [2]. In a similar vein, Brown argues that actions are exercised abilities [6]. Both of the latter approaches are discussed below.

Somewhat in sympathy with these approaches, the theories of most interest to computer science simply assume that the basic actions are given in the model. We follow this approach in our treatment below. We assume that basic actions can be performed through a single choice by an agent. In other words, the basic actions correspond to the atomic abilities of the agent. Because we do not require that the set of basic actions of an agent is unvarying, the agent must choose from among the basic actions available in the given situation. In this way, there is a component of opportunity wired into the actions. This is quite realistic. For example, a robot can move forward in a hall, but not when pushing against a wall. We could alternatively model the attempt to move as an action in its own right, and leave the success of the move as something to be determined *post hoc*. As far as our theory is concerned, this is not a major step—all we require is that there is a set of basic actions.

A natural extension is to high-level actions. High-level actions can be specified indirectly as propositions that an agent can achieve through a combination of lower-level basic actions. Indeed, many actions can be specified naturally only through the corresponding propositions. This idea too has been long been recognized in the philosophy of actions, for example, by von Wright [39] (cited by Segerberg [31, p. 327]). Although basic actions can be performed directly if the agent has the corresponding physical ability, performing complex high-level actions frequently requires not only the physical ability to perform the underlying basic actions, but also the knowledge to select the appropriate actions to perform at each stage of the complex action.

Thus know-how, when applied to high-level actions, inherently includes or supervenes on the notion of know-that. This is an important connection between the two notions. This connection was recognized in the early work on formalizations of knowledge in AI, for example, by Moore, but framed in terms of the knowledge required to perform specific plans of actions, where the plans, which include conditional actions, correspond to high-level actions. The treatment of high-level actions leads to another view of know-how, which is of course related to intentions as well. Some of the approaches described below will exploit this connection.

Intentions and know-how (or ability) and indeed even plans in general have usually been viewed as being directed toward the achievement of specific conditions. It is equally natural, and in many cases better, to consider not only the achievement of conditions but also their *maintenance*. We have developed an approach to maintenance in [34]. When intentions are similarly expanded, we would expect a similar relationship between intentions to maintain and the know-how to maintain, as between intentions to achieve and know-how to achieve. However, this subject has not yet been thoroughly studied in the literature and the bounds of the expected similarity are not known. For this reason, while acknowledging its importance, we discuss maintenance to a lesser degree here.

1.2 Separating versus Combining Ability and Opportunity

We informally describe two main classes of approaches to know-how. Ability refers to the intrinsic capability of an agent to do something reliably (if he knows what to do). Opportunity refers to the specific openings that an agent may have in specific situations to apply his ability. Know-how refers to the knowledge of how to achieve certain conditions, that is, to perform high-level actions. Intuitively, it is ability combined with the know-that to determine what actions to perform.

Ability and Opportunity

It is traditionally common to distinguish between *ability* and *opportunity*. This understanding is quite natural with respect to the natural language meanings of the two terms. With this understanding, ability refers to the reliable performance of an action by an agent, where the reliability is assessed over all possible situations.

Although natural, this account adds some complexity to the formal treatment. This is because to tease apart the definition of ability from the definition of opportunity requires that we consider *counterfactual* conditions of the following form: for an agent to have an ability means that he would succeed in achieving the given condition *if* he has the opportunity and carries out his actions.

Such approaches makes a subtle distinction between what an agent has the ability for and what he can do now. Importantly, the agent may have the ability but not actually succeed, because he lacks the opportunity. This is technically difficult, because to establish the above conditional statement requires modeling the situations in such a manner as to enable moving from the actual situation (where the agent does not have the opportunity) to a counterfactual situation (where the agent has the opportunity). In doing so, we must ensure that acquiring the opportunity has caused the ability neither to emerge nor to be lost. If either of those is the case, then the opportunity is not independent of the ability, and therefore neither concept is really coherent in itself. This complexity makes this class of approaches less tractable conceptually.

Situated Know-How

There is another class of approaches that do not separate the ability from the opportunity. Therefore, these approaches apply to a given specific situation in which an agent may find himself. In this situation, he has certain abilities and certain opportunities, but we consider them together rather than separately. With this understanding, ability refers to the reliable performance of an action by an agent, where the reliability is assessed only in the given situation.

As a result, the problem of identifying the abilities that cannot be exercised is avoided. Conversely, the accuracy of the concepts studied as formalizations of natural language concepts may be reduced. However, this trade-off is acceptable in coming up with formal concepts that may not be perfect realizations of the folk terms, but are nevertheless useful and more technically tractable than the folk concepts they formalize. We believe that this is just an instance of a pattern that one encounters repeatedly in the formalization of the folk concepts underlying rational agency.

1.3 Possible Worlds versus Representational Approaches

In general, there are a number of possible analyses of informal cognitive concepts, such as knowledge. In particular, for knowledge, there is a family of approaches based on modal logics, which is contrasted with the family of approaches based on sentential logics. The key intuitive difference between these two families is that the modal approaches support a number of inferences, including some inferences that are counterintuitive for humans and other resource-bounded agents.

Modal Approaches

The modal approaches are based on the so-called *possible worlds* approaches that consider alternative sets of situations [7]. These approaches postulate an *alternativeness* relation on situations. This relation is used to give a semantics to the modalities of *necessity* ($\Box$) and *possibility* ($\Diamond$). A proposition is necessarily true at a situation if it holds in all situations that are alternatives of the given situation. A proposition is possibly true if it holds in some alternative situation of the given situation. This definition becomes interesting when additional requirements are stated on the alternativeness relation, for instance, whether it is reflexive, symmetric, transitive, and so on.

The variations among these definitions don't concern us here. However, all of the simpler definitions support the inference of *consequential closure:* if p is necessary, then so are all its logical consequences. The argument is quite simple. Suppose p is necessary. Then p is true in all alternative situations. If q is a logical consequence of p, then q is also true in each of those situations. Hence, q is also necessary. Modal approaches that satisfy consequential closure are termed *normal*.

Some of the more sophisticated modal approaches are *non-normal*. In these approaches, the alternativeness relation relates a situation to a set of situations, and necessity is defined in terms of truth within the alternative set. As a consequence, non-normal modal logics avoid consequential closure. However, such models must validate *closure under logical equivalence:* if p is necessary, then so are all propositions logically equivalent to it.

Possible worlds approaches to knowledge were introduced by Hintikka [14] and developed by several others. These approaches model knowledge as a necessity modal operator. In this case, the underlying relation is one of *epistemic alternativeness*—there is a different relation for each agent. Neither consequential closure nor closure under logical equivalence is acceptable in describing the knowledge of a computationally limited agent. However, sometimes knowledge can be understood from the perspective of an objective designer, in which case it is the designer's capacities of reasoning that are postulated.

Sentential Approaches

The *representational* (typically, *sentential*) approaches contrast with the modal approaches in which an agent is said to have an explicit set of representations (typically, sentences in a formal language) that describe his cognitive state. An agent knows a condition (expressed in a particular sentence) if that sentence is among those in the set of sentences describing his cognitive state. The advantage of this approach is that it is explicit about the agent's knowledge. If a logical consequence of a given sentence is not included in the set of sentences that define an agent's cognitive state, then there is no implication that the logical consequence is known. This is certainly more accurate when describing the knowledge states of agents. However, in restricting ourselves to precisely the sentences that are included in the set of sentences, we also prevent all kinds of other inferences that might be viable. In this sense, the sentential approaches preclude all general inferences; this observation limits their usability in reasoning about agents. A representational approach was developed by Konolige [18].

Hybrid Approaches

There are also some hybrid approaches, which seek to use possible worlds approaches for their semantic ease, in conjunction with some representations to characterize how a computationally bounded agent may reason about his knowledge. These approaches prevent the problematic inferences of the possible worlds approaches, but give a semantic basis for the inferences they do support. Two example approaches are those of Fagin & Halpern [13] and Singh & Asher [35]. The latter also considers intentions in the same framework as beliefs. Although promising, these approaches are technically quite complex, and have not drawn as much attention in the literature as perhaps they deserve.

Consequently, the modal approaches are by far the most common ones in the literature. Accordingly, we primarily consider such approaches below.

2 TECHNICAL FRAMEWORK

It is clear and widely agreed that any formal treatment of any of the shades of know-how requires a mathematical framework that includes actions as primitives. Usually there is also the need for a separate notion of time to help capture other associated intuitions. Traditional approaches, which consider commonsense situations, and especially those that are geared to the natural language meanings of the above terms, include some notion of Newtonian time, usually in terms of a date-based language and semantics. This enables them to express facts such as whether an agent can catch the bus by 3:00 PM, where 3:00 PM is defined independently of any specific course of events, given as it were by a Newtonian clock. This is convenient enough in many cases, so we allow the assignment of real date values to different situations, although the rest of our framework involves a branching time model and considers as primary a qualitative ordering among moments.

2.1 Branching-Time Models

In conceptualizing about actions and know-how, it is important to recognize the choices that the agents can exercise as they go about their business. Intuitively, the world can evolve in several different ways, but the agents constrain it to evolve in a way that suits them by performing appropriate actions. To the extent that they can achieve what they want they can be said to have the requisite ability.

The need to represent choices translates into the requirement of representing multiple courses of events in our technical framework so that our formal definitions can exploit that multiplicity. There are a number of ways of capturing this requirement. One way that is intuitively quite direct is to construct *branching models* of time. There is a large variety of these models; at the very least, because of our need to represent and reason about multiple actions, we must allow the branching to take place into the future. For simplicity, we consider models that are linear in the past. This captures the idea that the past can in principle be fully known, but the future is nondeterministic as long as the agents' choices are open. The ignorance that some agent may have about the past is captured by the general mechanism of beliefs.

The proposed formal model is based on a set of *moments* with a strict partial order, which denotes temporal precedence. Each moment is associated with a possible state of the world, which is identified by the atomic conditions or propositions that hold at that moment. A *scenario* at a moment is any maximal set of moments containing the given moment, and all moments in its future along some particular branch. Thus a scenario is a possible course of events, that is, a specific, possible computation of the system. It is useful for capturing many of our intuitions about the choices and abilities of agents to identify one of the scenarios beginning at a moment as the *real* one. This is the scenario on which the world progresses, assuming it was in the state denoted by the given moment. Constraints on what should or will happen can naturally be formulated in terms of the real scenario.

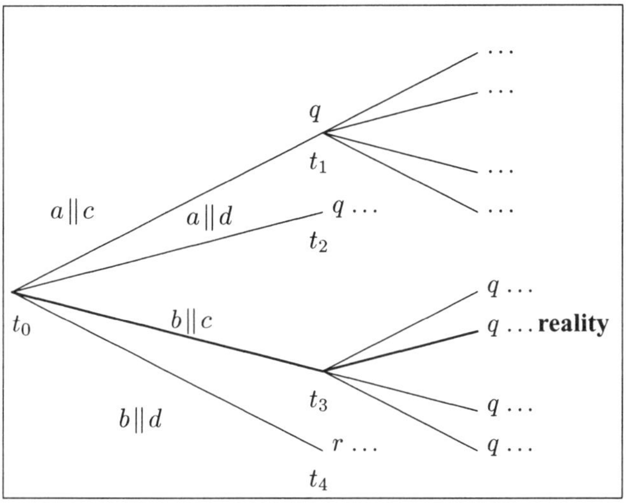

Figure 1. An Example Formal Model

Figure 1 has a schematic picture of the formal model. Each point in the picture is a moment. Each moment is associated with a possible state of the world, which is identified by the atomic conditions or propositions that hold at that moment (atomic propositions are explained in section 2.2). With each moment are also associated the knowledge and intentions of the different agents. A condition p is said to be achieved when a state is attained in which p holds. There is a partial order on moments that denotes temporal precedence. A *scenario* at a moment is any maximal set of moments containing the given moment, and all moments in its future along some particular branch.

Example 1 Figure 1 is labeled with the actions of two agents. Each agent influences the future by acting, but the outcome also depends on other events. For example, in Figure 1, the first agent can constrain the future to some extent by choosing to do action a or action b. If he does action a, then the world progresses along one of the top two branches out of t_0; if he does action b, then it progresses along one of the bottom two branches.

A lot of good research has been carried out on temporal and dynamic logics and models of time and action. We encourage the reader to peruse at least the following works: Emerson [12], Kozen & Tiurzyn [19], van Benthem [36, 37], and Prior [25, 26].

2.2 The Formal Language

Given especially the branching-time models described above, it is convenient to adopt as our formal language one that includes not only traditional propositional logic, but also certain operators borrowed from temporal and dynamic logics. In doing so, we can emphasize the intellectual heritage of the present approaches on research into logics of program, developed for and applied on problems in computing at large. Consequently, our language includes a capacity for expressing conditions, actions, and branching futures. Time is intimately related to actions.

We use a qualitative temporal language, $\mathcal{L}$, based on CTL* [12]. Our language captures the essential properties of actions and time that are of interest in specifying rational agents. Formally, $\mathcal{L}$ is the minimal set closed under the rules given below. Here $\mathcal{L}_s$ is the set of "scenario-formulas," which is used as an auxiliary definition. Φ is a set of atomic propositional symbols, $\mathcal{A}$ is a set of agent symbols, $\mathcal{B}$ is a set of basic action symbols, and $\mathcal{X}$ is a set of variables. We give intuitive meanings of the constructs of our formal language after the following syntactic definitions.

SYN-1. $\psi \in \Phi$ implies that $\psi \in \mathcal{L}$

SYN-2. $p, q \in \mathcal{L}$ and $x \in \mathcal{A}$ implies that $p \wedge q$, $\neg p$, $\mathsf{P}p$, $(\bigvee a : p)$, $(x\mathsf{K_t}p)$, $(x\mathsf{K_h}p), (x\mathsf{K_m}p) \in \mathcal{L}$

SYN-3. $\mathcal{L} \subseteq \mathcal{L}_s$

SYN-4. $p, q \in \mathcal{L}_s$, $x \in \mathcal{A}$, and $a \in \mathcal{B}$ implies that $p \wedge q$, $\neg p$, $p\mathsf{U}q$, $x[a]p$, $x\langle a\rangle p$ $\in \mathcal{L}_s$

SYN-5. $p \in \mathcal{L}_s$ implies that $\mathsf{A}p, \mathsf{R}p \in \mathcal{L}$

SYN-6. $p \in (\mathcal{L}_s - \mathcal{L})$ and $a \in \mathcal{X}$ implies that $(\bigvee a : p) \in \mathcal{L}_s$

2.3 Informal Description

The formulas in $\mathcal{L}$ refer to moments in the model. Each moment has a state corresponding to a possible snapshot of the system. The formulas in $\mathcal{L}_s$ refer to scenarios in the model, that is, to specific computations of the system. Note that $\mathcal{L} \subseteq \mathcal{L}_s$. However, our formal semantics, given in section 2.4, ensures that the formulas in $\mathcal{L}$ are given a unique meaning even if interpreted as being in $\mathcal{L}_s$.

Recall that the semantics of a formal language is given by stating rules through which the interpretation of syntactically acceptable formulae can be determined. This is carried out in the context of some *model*, that is, a description of the world where the formal language is being applied. In logic, the term *model* is used with a specific technical meaning. A model is not just a description of reality, but one that is fine-tuned with respect to the given logical language. Thus our formal model should capture the structure exhibited in Figure 1.

The boolean operators are standard. We introduce two abbreviations. For any $p \in \Phi$: $\mathsf{false} \stackrel{\text{def}}{=} (p \wedge \neg p)$ and $\mathsf{true} \stackrel{\text{def}}{=} \neg \mathsf{false}$.

The temporal and action formulas explicitly consider the evolution of the system's state—the scenario-formulas along a specific scenario and the other formulas along all or some of the possible scenarios. $p\mathsf{U}q$ is true at a moment t on a scenario, iff q holds at a future moment on the given scenario and p holds on all moments between t and the selected occurrence of q. $\mathsf{F}p$ means that p holds sometimes in the future on the given scenario and abbreviates $\mathsf{true}\mathsf{U}p$. $\mathsf{G}p$ means that p always holds in the future on the given scenario; it abbreviates $\neg\mathsf{F}\neg p$. $\mathsf{P}q$ means that q held in a past moment (we assume a linear past). The branching-time operator, A, denotes "in *all* scenarios at the present moment." Here "the present moment" refers to the moment at which a given formula is evaluated. A useful abbreviation is E, which denotes "in *some* scenario at the present moment." In other words, $\mathsf{E}p \Leftrightarrow \neg\mathsf{A}\neg p$.

Example 2 In Figure 1, $\mathsf{EF}r$ and $\mathsf{AF}(q \vee r)$ hold at t_0, since r holds on some moment on some scenario at t_0 and q holds on some moment on each scenario.

The *reality* operator, R, denotes "in the *real* scenario at the present moment." R helps tie together intuitions about what may and what will happen.

Example 3 In Figure 1, $\mathsf{RF}q$ holds at t_0, since q holds on some moment on the real scenario identified at t_0.

$\mathcal{L}$ also contains operators on actions. These are adapted and generalized from dynamic logic [19], in which the action operators behave essentially like state-formulas. Our operators can capture the traditional operators. For an action symbol a, an agent symbol x, and a formula p, $x[a]p$ holds on a given scenario S and a moment t on it, iff, if x performs a on S starting at t, then p holds at some moment while a is being performed. The formula $x\langle a\rangle p$ holds on a given scenario S and a moment t on it, iff, x performs a on S starting at t and p holds at some moment while a is being performed. These definitions require p to hold at any moment in the (left-open and right-closed) period in which the given action is being performed. These definitions generalize naturally to variable length actions, although we restrict our attention in this paper to unitlength actions over discrete time. Under these assumptions, in each of $[\,]$ and $\langle\,\rangle$, p holds at the moment where the action ends. Thus, $x[a]p \Leftrightarrow \neg x\langle a\rangle\neg p$, that is, $[\,]$ and $\langle\,\rangle$ are duals.

Example 4 In Figure 1, $\mathsf{E}\langle b\rangle r$ and $\mathsf{A}[a]q$ hold at t_0, since r holds at the end of b on one scenario, and q holds at the end of a on each scenario. Similarly, $\mathsf{A}[d](q \vee r)$ also holds at t_0. Also, $\mathsf{A}[e]\mathsf{true}$ holds at t_0, because action e does not occur at t_0.

The construct $(\bigvee a : p)$ means that for some action p becomes true. The action symbol a typically occurs in p and is replaced by the specific action which makes p true. The construct $(\bigwedge a : p)$ abbreviates $\neg(\bigvee a : \neg p)$. This means that for all actions p becomes true.

Example 5 In Figure 1, $(\bigvee e : \mathsf{E}x\langle e\rangle\mathsf{true} \wedge \mathsf{A}x[e]q)$ holds at t_0. This means there is an action, namely, a, such that x performs it on some scenario starting at t_0 and on all scenarios on which it is performed, it results in q being true. In other words, some action is possible that always leads to q. This paradigm is used in our formalization of know-how.

The formula $x\mathsf{K_t}p$ means that the agent x *knows that* p. The other important construct is $x\mathsf{K_h}p$. $x\mathsf{K_h}p$ is interpreted to mean that agent x *knows how* to achieve p. The formal definition of these operators is the subject of this paper.

2.4 The Formal Model

Let $M = \langle \mathbf{T}, <, [\![\,]\!], \mathbf{R}, \mathbf{K}\rangle$ be a formal model. $\mathbf{T}$ is the set of moments. Each moment is associated with a possible state of the system—this includes the physical state as identified by the atomic propositions that hold there, as well as the states of the agents described through their beliefs and intentions. The binary relation $<$ is a partial order over $\mathbf{T}$, and is interpreted as the temporal order among the moments of $\mathbf{T}$. Therefore, $<$ must be transitive and asymmetric; it typically branches into the future; we assume it is linear in the past. We further assume that $<$ is discrete and finitely branching. $[\![\,]\!]$ gives the denotation of the various atomic propositions and of the action symbols. For an atomic proposition, p, $[\![p]\!]$ is the set of moments where p is interpreted as holding; for an action a and an agent x, $[\![a]\!]^x$ is the set of periods over which a is performed by x. These periods are notated as $[S; t, t']$ such that a begins at t and ends at t', where $t, t' \in S$.

$\mathbf{R}$ picks out at each moment the *real* scenario at that moment. This is the notion of relativized reality alluded to above, and which is highlighted by a bold line in Figure 1.

For $p \in \mathcal{L}$, $M \models_t p$ expresses "M satisfies p at t." For $p \in \mathcal{L}_s$, $M \models_{S,t} p$ expresses "M satisfies p at moment t on scenario S" (we require $t \in S$). We say p is *satisfiable* iff for some M and t, $M \models_t p$. The satisfaction conditions for the temporal operators are adapted from those given by Emerson [12]. For simplicity, we assume that each action symbol is quantified over at most once in any formula. Below, $p|_b^a$ is the formula resulting from the substitution of all occurrences of a in p by b. We also assume that agent symbols are mapped to unique agents throughout the model. Formally, we have:

SEM-1. $M \models_t \psi$ iff $t \in [\![\psi]\!]$, where $\psi \in \Phi$

SEM-2. $M \models_t p \wedge q$ iff $M \models_t p$ and $M \models_t q$

SEM-3. $M \models_t \neg p$ iff $M \not\models_t p$

SEM-4. $M \models_t \mathsf{A}p$ iff $(\forall S : S \in \mathbf{S}_t \Rightarrow M \models_{S,t} p)$

SEM-5. $M \models_t \mathsf{R}p$ iff $M \models_{\mathbf{R}(t),t} p$

SEM-6. $M \models_t \mathsf{P}p$ iff $(\exists t' : t' < t$ and $M \models_{t'} p)$

SEM-7. $M \models_t (\bigvee a : p)$ iff $(\exists b : b \in \mathcal{B}$ and $M \models_t p|^a_b)$, where $p \in \mathcal{L}$

SEM-8. $M \models_{S,t} (\bigvee a : p)$ iff $(\exists b : b \in \mathcal{B}$ and $M \models_{S,t} p|^a_b)$, where $p \in (\mathcal{L}_s - \mathcal{L})$

SEM-9. $M \models_{S,t} p\mathsf{U}q$ iff $(\exists t' : t \leq t'$ and $M \models_{S,t'} q$ and $(\forall t'' : t \leq t'' \leq t' \Rightarrow M \models_{S,t''} p))$

SEM-10. $M \models_{S,t} x[a]p$ iff $(\forall t' \in S : [S;t,t'] \in [\![a]\!]^x$ implies that $(\exists t'' : t < t'' \leq t'$ and $M \models_{S,t''} p))$

SEM-11. $M \models_{S,t} x\langle a\rangle p$ iff $(\exists t' \in S : [S;t,t'] \in [\![a]\!]^x$ and $(\exists t'' : t < t'' \leq t'$ and $M \models_{S,t''} p))$

SEM-12. $M \models_{S,t} p \wedge q$ iff $M \models_{S,t} p$ and $M \models_{S,t} q$

SEM-13. $M \models_{S,t} \neg p$ iff $M \not\models_{S,t} p$

SEM-14. $M \models_{S,t} p$ iff $M \models_t p$, where $p \in \mathcal{L}$

The above definitions do not include the postulates for know-that and know-how on purpose. We introduce them after further technical motivation in the sections below.

2.5 *Know-That*

We discuss know-that as part of the technical framework, because logics of know-that are standard, but provide a key basis for the study of know-how. As explained in section 1.3, the basic idea of know-that or knowledge as captured in most common formalisms is that the knowledge of an agent helps the agent discriminate among possible states of the world.

K assigns to each agent at each moment the moments that the agent implicitly considers as equivalent to the given moment. This is used in the formal semantics for know-that in the traditional manner. For simplicity, we assume that **K** is an equivalence relation, resulting in K_t being an S5 modal logic operator [7], which grants both positive and negative introspection.

SEM-15. $M \models_t x\mathsf{K}_t p$ iff $(\forall t' : (t,t') \in \mathbf{K}(x) \Rightarrow M \models_{t'} p)$

3 FORMALIZATION

We now use the above technical framework to present some of the common approaches to know-how. In some cases, we modify the details of the approaches a little to facilitate the exposition.

We propose that an agent, x, knows how to achieve p, if he is able to bring about p through his actions, that is, to force p to occur. The agent's beliefs or knowledge must be explicitly considered, since these influence his decision. For example, if an agent is able to dial all possible combinations of a safe, then he is able to open that safe: for, surely, the correct combination is among those that he can dial. On the other hand, for an agent to really know how to open a safe, he must not only have the basic skills to dial different combinations on it, but also know which combination to dial. (Let's assume, for simplicity, that trying a wrong combination precludes the success of any future attempts.)

3.1 Trees

To formalize know-how, we define the auxiliary notion of a *tree* of actions. A tree consists of an action, called its *radix*, and a set of subtrees. The idea is that the agent does the radix action initially and then picks out one of the available subtrees to pursue further. In other words, a tree of actions for an agent is a projection to the agent's actions of a fragment of **T**. Thus a tree includes *some* of the possible actions of the given agent, chosen to force a given condition. Intuitively, a tree encodes the selection function that the agent may use in choosing his actions at each moment. A tree should be bushy enough to cover all the cases.

Let Υ be the set of trees. $\emptyset$ is the empty tree. Then Υ is defined as follows.

T1. $\emptyset \in \Upsilon$

T2. $a \in \mathcal{B}$ implies that $a \in \Upsilon$

T3. $\{\tau_1, \ldots, \tau_m\} \subseteq \Upsilon$, $\tau_1, \ldots, \tau_m$ have different radices, and $a \in \mathcal{B}$ implies that $\langle a; \tau_1, \ldots, \tau_m \rangle \in \Upsilon$

Sometimes it is convenient to just write a as a shorthand for the tree $\langle a; \emptyset \rangle$. Now we extend the formal language with an auxiliary construct.

SYN-7. $\tau \in \Upsilon$, $x \in \mathcal{A}$, and $p \in \mathcal{L}$ implies that $x[\![\tau]\!]p \in \mathcal{L}$

$x[\![\tau]\!]p$ denotes that agent x knows how to achieve p relative to tree τ. As usual, the agent symbol can be omitted when it is obvious from the context. To simplify the notation, we extend $\bigvee$ to apply to a given range of trees. Since distinct trees in each such range have distinct radix actions, the extension of $\bigvee$ from actions to trees is not a major step.

SEM-16. $M \models_t [\![\emptyset]\!]p$ iff $M \models_t \mathsf{K_t}p$

SEM-17. $M \models_t [\![a]\!]p$ iff $M \models_t \mathsf{K_t}(\mathsf{E}\langle a\rangle\mathsf{true} \wedge \mathsf{A}[a]\mathsf{K_t}p)$

SEM-18. $M \models_t [\![\langle a; \tau_1, \ldots, \tau_m\rangle]\!]p$ iff
$M \models_t \mathsf{K_t}(\mathsf{E}\langle a\rangle\mathsf{true} \wedge \mathsf{A}[a](\bigvee_{1 \leq i \leq m} \tau_i : ([\![\tau_i]\!]p)))$

The denotation of a tree, that is, its *know-how denotation*, is implicit in this definition. We need to make the corresponding denotation explicit when we consider maintenance.

3.2 Plain Know-How

Thus an agent knows how to achieve p by following the empty tree, that is, by doing nothing, if he knows that p already holds. As a consequence of his knowledge, the agent will undertake no specific action to achieve p. The nontrivial base case is when the agent knows how to achieve p by doing a single action: this would be the last action that the agent performs to achieve p. In this case, the agent has to know that he will know p immediately after the given action.

It is important to require knowledge in the state in which the agent finally achieves the given condition, because it helps limit the actions selected by the agent. If p holds, but the agent does not know this, then he might select still more actions in order to achieve p.

Lastly, an agent knows how to achieve p by following a nested tree if he knows that he must choose the radix of this tree first and, when it is done, that he would know how to achieve p by following one of its subtrees. Thus know-how presupposes knowledge to choose the next action and confidence that one would know what to do when that action has been performed.

SEM-19. $M \models_t x\mathsf{K_h}p$ iff $(\exists\tau : M \models_t x[\![\tau]\!]p)$

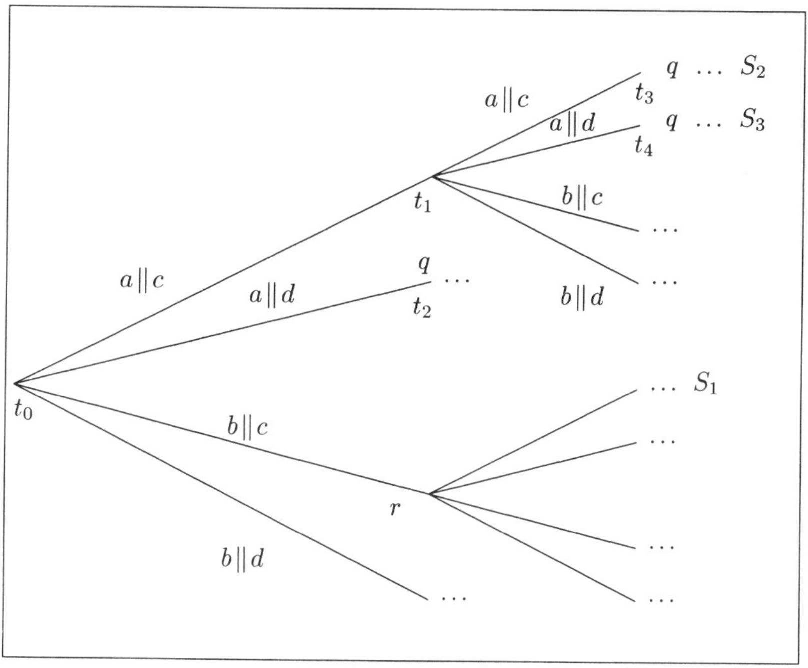

Figure 2. Know-How

Example 6 Consider Figure 2. Let x be the agent whose actions are written first there. Assume for simplicity that each moment is its own unique alternative for x (this is tantamount to assuming that x has perfect knowledge—our formal definitions do not make this assumption). Then, by the above definitions, $x\mathsf{K_t}q$ holds at t_3 and t_4. Also, $x\mathsf{K_h}q$ holds at t_1 (using a tree with the single action a) and at t_2 (using the empty tree). As a result, at moment t_0, x knows that if he performs a, then he will know how to achieve q at each moment where a ends. In other words, we can define a tree, $\langle a; a, \emptyset\rangle$, such that x can achieve q by properly executing that tree. Therefore, x knows how to achieve q at t_0.

Now we present a recursive characterization of know-how. This characterization, which is remarkably simple, forms the basis of the mu-calculus approach developed in [34].

Lemma 1 $\mathsf{K_t}p \vee (\bigvee \alpha : \mathsf{K_t}(\mathbf{E}\langle\alpha\rangle\mathsf{true} \wedge \mathbf{A}[\alpha]\mathsf{K_h}p)) \Leftrightarrow \mathsf{K_h}p$

3.3 Reliable Know-How

The above treatment of know-how captures it essentially as any other modal operator, albeit one that combines temporal and dynamic aspects. The truth and falsity of this operator are determined at a specified moment. The traditional, natural language use of the term know-how, however, includes a greater sense of reliability. In effect, reliability requires looking not only at the given moment, but also at other moments. Once the finer notion has been formalized, its reliable version proves fairly natural. To simplify our presentation, let us assume that 0 is the unique initial moment in $\mathbf{T}$. We also add an operator $\mathsf{K_{rh}}$ (meaning reliably knows-how) to the formal language. Then we can simply state that

SEM-20. $M \models_t x\mathsf{K_{rh}}p$ iff $M \models_0 x\mathsf{K_{rh}}p$, where $0 < t$

SEM-21. $M \models_0 x\mathsf{K_{rh}}p$ iff $M \models_0 \mathsf{AG}x\mathsf{K_h}p$

This states that reliable know-how is obtained if the agent has the restricted know-how in every possible state. Alternative versions of reliable know-how can be readily formulated. In particular, those that select some relevant moments to a given moment would correspond more to natural language, but would also be technically more complex.

4 ALTERNATIVE APPROACHES

We now consider some leading approaches from the computer science and philosophy literatures.

4.1 Actions as Exercised Abilities

Brown distinguishes between ability and opportunity. He formalizes his approach in a modal logic with operators for necessity $\Box$ and possibility $\Diamond$ [6], which builds on his previous work on ability [5]. Brown captures ability and opportunity as nested applications of these operators.

A strength of Brown's approach is its intuitive treatment of the interplay between action and ability. He shows how each can be defined in terms of the other, and how they share various logical inferences, and how they differ. Brown presents a number of interesting axioms and inference rules for his modalities, and studies which of them must be validated by different interpretations of those modalities. For example, $\Box p \Rightarrow p$ is satisfied when $\Box p$ is interpreted as "the agent does p," and not as "the agent is able to do p."

A key notion is that an agent has so acted as to bring about the truth of a given condition. The main intuition is that ability is about the reliable performance of actions. Conversely, actions are exercised abilities (p. 101). Only the reliable consequences of one's actions are counted as actions (p. 96). It also appears that Brown counts all the reliable consequences of an action as actions, which may be too strong, when talking about (intentional) action in general.

Tentatively considering ability as the possibility operator of modal logic, Brown argues that his modal logic be *non-normal* (p. 98), as described in section 1.3. This means that it need not support the inference that an agent who is able to achieve $A \vee B$ is able to achieve A or achieve B. This would clearly be undesirable. Brown's proposed interpretation of ability is stronger than mere modal possibility. Yet the same reason applies for giving it a non-normal semantics.

In Brown's formulation, a *relevance relation* is postulated that relates a possible world to subsets of possible worlds that are somehow "relevant" to it. Each subset is called a *cluster*. The agent is said to be able to achieve p iff there is a relevant cluster such that each world within it satisfies p. Intuitively, each cluster corresponds to the possible outcomes of an action. This is intuitively similar to our definition, and indeed most other definitions of know-how, in that the agent selects an action, such that in each resulting state, the given condition holds.

However, this approach is a purely modal approach, with no reference to any epistemic or temporal aspect. Thus, the agent's knowledge is not taken into account. This agrees with the common definition of ability. Our remark is not meant as a criticism of Brown, but to highlight what is nevertheless an important point of difference with know-how. Even actions are not explicitly modeled. Consequently, although models and a semantics are given and have a direct connection with the logical inferences under discussion, the models are not obviously related to our intuitions about actions and ability. Brown does not offer any reasonable intuitive interpretations of the relevant clusters. Are they actions, action sequences, routines, or regular programs? Let's assume that they are composite actions of some sort.

Brown describes two interesting properties of the relevant clusters. First, he requires that they are *weakly-centered*, meaning that the given world is always a member of each relevant cluster. This effectively means that we are looking at the case where the given action is in fact performed. In other words, the action can take the world in question along its real scenario.

Second, Brown states that the relevant clusters are *closed under pairwise intersection*. Roughly, this means that the "parallel composition" of two actions is also an action. Or, more strongly (in the presence of weak-centering), real actions can be composed to yield another real action. If the agents perform one basic action at a time, the composition can be effected in terms of either interleaving the component actions, or by having one action be a subsequence of the other.

4.2 STIT: Seeing To It That

STIT refers to the *seeing to it that* approaches developed initially by Belnap & Perloff [2], and refined and explained by Chellas [8], from whose exposition we benefited a lot. Perloff compares the STIT approach to leading philosophical approaches in [24]. The STIT approaches seek to characterize the notion of ability in which an agent sees to it that a certain condition is obtained. This presupposes continual actions by the agent leading up to success in achieving the given condition. Informally, an agent sees to it that p if p is not already true, is not inevitable, and he can select and perform certain actions leading up to the truth of p. The STIT approaches are also naturally expressed in branching-time models.

Intuitively, STIT is about the actions that have just been performed. In fact, we find the progressive misleading, and believe a better gloss for STIT would be *has just seen to it that*. This gives its formal logic some characteristics different from the logics of ability or opportunity. Indeed, the concept is better understood as a form of high-level action.

Just like in the approach of section 3 above, Belnap & Perloff consider histories with linear past and branching future (pp. 189–192). The moments are ordered qualitatively, as described above. Belnap & Perloff also assume, as we did above, that each agent can act in different ways, but the future depends on the combination of the actions of the agents and events in the environment. The choices of each agent partition the set of future branches (like $\mathbf{S}_t$ in our framework). Intuitively, each choice-set corresponds to the result of performing some (sequence of) actions.

With this setup, Belnap & Perloff state (p. 191) that an agent x STITs p at moment m iff there is a past moment m_0, such that

- x had a choice set at m_0 such that at every branch in the choice set, p holds at the moments that are alternatives to m, that is, x had a choice that guaranteed p
- x had a choice in which p was not guaranteed.

The definition as stated has a bug in it. We must also ensure that the given moment m itself lies on one of the branches in the choice set being used. The version given by Chellas, however, fixes the bug.

Chellas, in his approach–termed the *imperative approach*, considers linear histories (like scenarios), but relates them intuitively to branching time, so the effect is practically indistinguishable. However, he assumes that a metric time is given with which states in the histories can be identified. Chellas has the notion of an *instigative alternative (IA)* of a history at a time. A (linear) history h' is an IA to h at t if h' is under the control of, or responsive to, the actions of the agent. In this way, actions are defined indirectly via the IAs. The agent's high-level actions are defined in terms of what holds on all of the IAs at the given history and time.

Chellas assumes *historical relevance* of the IAs meaning that the IAs of a history agree with it up to the given time. He also assumes *reflexivity* meaning that a history is an IA to itself.

When relating the IAs to actions, it is not clear if the IAs are the actions the agent is instigating or may instigate. We would expect a set of set of IAs, as in Belnap & Perloff's approach, not a single set. In conjunction with reflexivity, this suggests that the IAs are in fact the chosen IAs that the agent is pursuing on the given history as well.

Neither Chellas nor Belnap & Perloff mention knowledge explicitly, although their intuitive descriptions seem to call for it. An agent could not see to it that something without knowing what he was doing.

The STIT approaches are geared more toward the natural language uses of the term *seeing to it that* than toward the technical definition *per se*. A point where this focus of the STIT approaches is reflected is in their attempt at capturing the felicity of natural language statements involving an agent seeing to it that something obtain. For example, they require that the given condition does not already hold and is not inevitable (independent of the agent's actions). Although these restrictions are appropriate when you announce that a given agent can see to it that something happens, they are not necessarily appropriate as intrinsic components of the concept itself. We believe that these are extrinsic properties that are based on the pragmatics of communication, rather than the semantics of the underlying concept. Indeed, these properties can be thought of as specific Gricean inferences on the *report* of what an agent can see to.

4.3 Strategic Know-How

The approach to know-how described in section 3 considers the actions of the agents directly, although organized into trees. A natural extension is to consider higher level compositions of the actions, which result in a more realistic treatment of know-how [33]. This extension uses *strategies*, which describe at a high level the actions that an agent may perform. Strategies have long been studied in AI and cognitive science. Mention of them goes back to Kochen & Galanter [17] (cited in [21, p. 17]), McCarthy & Hayes [20], and Brand [3].

Strategies do not add any special capability to the agents. They simply help us, designers and analyzers, better organize the skills and capabilities that agents have anyway. Hierarchical or partial plans of agents, thus, turn out to be good examples of strategies. The formal notion of strategies here is based on regular programs, as studied in dynamic logic [19], with an enhancement to allow high-level actions instead of atomic programs, and restricting the language to only allow deterministic programs. The first column of Table 1 shows the syntax. Intuitively, the strategy $\mathbf{do}(q)$ denotes an abstract action, namely, the action of achieving q. It could be realized by any sequence of basic actions that yields q. The remaining constructs are standard.

Y	$\downarrow_t Y$	$\uparrow_t Y$
skip	**skip**	**skip**
$\mathbf{do}(q)$	if $M \models_t \neg q$ then $\mathbf{do}(q)$ else **skip**	**skip**
Y_1;Y_2	if $\downarrow_t Y_1 \neq$ **skip** then $\downarrow_t Y_1$ else $\downarrow_t Y_2$	if $\downarrow_t Y_1 \neq$ **skip** then $(\uparrow_t Y_1)$;Y_2 else $\uparrow_t Y_2$
if q **then** Y_1 **else** Y_2	if $M \models_t q$ then $\downarrow_t Y_1$ else $\downarrow_t Y_2$	if $M \models_t q$ then $\uparrow_t Y_1$ else Y_2
while q **do** Y_1	if $M \models_t \neg q$ then **skip** else $\downarrow_t Y_1$	if $M \models_t \neg q$ then **skip** else if $\downarrow_t Y_1 \neq$ **skip** then $(\uparrow_t Y_1)$;Y else $\uparrow_t Y_2$

Table 1. Strategies: Syntax and Definitions of Current and Rest

It is useful to define two functions, *current* $\downarrow$ and *rest* $\uparrow$, on strategies. These functions depend on the moment at which they are evaluated. Let Y be a strategy. $\downarrow_t Y$ denotes the part of Y up for execution at moment t, and $\uparrow_t Y$ the part of Y that would remain after $\downarrow_t Y$ has been done. Assume that strategies are normalized with respect to the following constraints: (a) **skip;**$Y = Y$ and (b) Y**;skip** $= Y$. Then the $\downarrow_t Y$, which can be either **skip** or $\mathbf{do}(q)$. This helps unravel a strategy for acting on.

Strategies as Abstract Actions

The strategic definition of know-how builds on the definition given previously. To this end, we define $[\![\tau]\!]^x_Y$ as the *know-how denotation* of a tree, τ, relative to a strategy, Y, for an agent, x. $[\![\tau]\!]^x_Y$ is the set of periods on which the given agent knows how to achieve Y by following τ. Precisely those periods are included on which the agent has the requisite knowledge to force the success of the given strategy. The know-how denotation needs to be defined only for the base case of $\downarrow_t Y$. Formally, we have the following cases in the definition of $[\![\tau]\!]^x_Y$.

The agent knows how to satisfy the empty strategy, **skip**, by doing nothing, that is, by following the empty tree.

The agent may know how to satisfy the strategy $\mathbf{do}(q)$ in one of three ways: (a)

by doing nothing, if he knows that q holds; (b) by following a single action tree, if he knows that it will force q; or, (c) by following a general tree, if doing the radix of that tree will result in a state in which he knows how to satisfy $\mathbf{do}(q)$ by following one of its subtrees. Thus we have:

$[S;t,t'] \in [\{\tau\}]^x_{\mathbf{do}(q)}$ iff

1. $\tau = \emptyset$ and $t = t'$ and $M \models_t x\mathsf{K_t}q$
2. $\tau = a$ and $M \models_t [(\tau)]q$ and $M \models_{t'} x\mathsf{K_t}q$ and $(\exists t_1 : t < t' \leq t_1$ and $[S;t,t_1] \in [\![a]\!]$ and $(\forall t_2 : t \leq t_2 < t'$ implies $M \not\models_{t_2} q))$
3. $\tau = \langle a;\tau_1,\ldots,\tau_m\rangle$ and $M \models_{t'} x\mathsf{K_t}q$ and $M \models_t [(\tau)]q$ and $(\exists t_1,t_2,i :$ $[S;t,t_1] \in [\![a]\!]$ and $1 \leq i \leq m$ and $[S;t_1,t_2] \in [\![\tau_i]\!]_{\mathbf{do}(q)}$ and $t_1 \leq t' \leq t_2)$ and $(\forall t_3 : t \leq t_3 < t'$ implies $M \not\models_{t_3} q)$

Intuitively, $[\{\tau\}]^x_{\mathbf{do}(q)}$ corresponds to the denotation of the abstract action performed by agent x of achieving q by exercising his know-how. Based on the above, we extend the formal language by allowing the operators $\langle\rangle$ and $[]$ to apply on strategies. Now we give the semantic conditions for the new operators. We must quantify over trees with which $\mathbf{do}(q)$ can be performed, because those trees are equally legitimate as ways to perform $\mathbf{do}(q)$.

SEM-22. $M \models_{S,t} x\langle\mathbf{do}(q)\rangle p$ iff $(\exists\tau,t' \in S : [S;t,t'] \in [\{\tau\}]^x_{\mathbf{do}(q)}$ and $M \models_{S,t'} p)$

This means that $\mathbf{do}(q)$ can be knowingly and forcibly performed on the given scenario and p holds at the moment at which it ends.

SEM-23. $M \models_{S,t} x[\mathbf{do}(q)]p$ iff $(\forall\tau,t' \in S : [S;t,t'] \in [\{\tau\}]^x_{\mathbf{do}(q)} \Rightarrow M \models_{S,t'} p)$

This means that if the abstract action $\mathbf{do}(q)$ is knowingly and forcibly performed on the given scenario, then at the moment at which it is over, condition p holds.

The notion of know-how relative to a strategy can now be formalized to explicitly reflect the idea that strategies are abstractions over basic actions. An agent knows how to achieve p by following the empty strategy, **skip**, if he knows that p. The justification for this is the same as the one for the case of the empty tree.

For a general strategy, not only must the agent know how to perform the relevant substrategies of a given strategy, he must know what they are when he has to perform them. We introduce two new operators to capture the agent's knowledge of the $\downarrow$ and $\uparrow$ of a strategy. The formula $x\lfloor Y\rfloor Y'$ means that for the agent x to follow Y at the given moment, he must begin by following Y', which is either **skip** or $\mathbf{do}(q)$; $x\lceil Y\rceil Y''$ means that he must continue with Y''.

Thus, $x\lfloor Y\rfloor Y'$ holds only if $Y' = \downarrow_t Y$. However, since the agents' beliefs may be incomplete, $x\lfloor Y\rfloor Y'$ may be false for all Y'. Assuming $x\lfloor Y\rfloor Y'$, $x\lceil Y\rceil Y''$ means that for the agent x to follow Y at the given moment, he must follow Y'' after he has followed Y'. As above, $x\lceil Y\rceil Y''$ holds only if $Y'' =\uparrow_t Y$. We include only some sample definitions, and refer the reader to [33] for additional details.

SEM-24. $M \models_t x\lfloor \mathbf{do}(q)\rfloor \mathbf{skip}$ iff $M \models_t x\mathsf{K}_{\mathsf{t}}q$

SEM-25. $M \models_t x\lfloor \mathbf{if}\ r\ \mathbf{then}\ Y_1\ \mathbf{else}\ Y_2\rfloor Y'$ iff $M \models_t (x\mathsf{K}_{\mathsf{t}}r \wedge x\lfloor Y_1\rfloor Y') \vee (x\mathsf{K}_{\mathsf{t}}\neg r \wedge x\lfloor Y_2\rfloor Y')$

SEM-26. $M \models_t x\lceil \mathbf{do}(q)\rceil \mathbf{skip}$

Strategic Know-How Defined

An agent, x, knows how to achieve a proposition p by following a strategy Y, if there is a strategy Y' such that (a) $x\lfloor Y\rfloor Y'$ holds; (b) he knows how to perform Y'; and, (c) he knows that, in each of the states where Y' is completed, he would know how to achieve p relative to $\uparrow_t Y$. Since Y' is always of one of the forms, **skip** or $\mathbf{do}(q)$, Y is progressively unraveled into a sequence of substrategies of those forms. Formally, we have

SEM-27. $M \models_t x[\langle \mathbf{skip}\rangle]p$ iff $M \models_t x\mathsf{K}_{\mathsf{t}}p$

SEM-28. $M \models_t x[\langle Y\rangle]p$ iff $M \models_t x\mathsf{K}_{\mathsf{t}}(\mathbf{E}x\langle\downarrow_t Y\rangle\mathsf{true} \wedge \mathbf{A}x[\downarrow_t Y]x[\langle\uparrow_t Y\rangle]p)$ and $M \models_t x\lfloor Y\rfloor \downarrow_t Y$

The above definition requires an agent to know what substrategy he must perform only when he has to begin acting on it. The knowledge prerequisites for executing different strategies can be read off from the above semantic definitions. For example, a conditional or iterative strategy can be executed only if the truth-value of the relevant condition is known.

SEM-29. $M \models_t x\mathsf{K}_{\mathsf{hs}}p$ iff $(\exists Y : M \models_t x[\langle Y\rangle]p)$

4.4 Capabilities

van der Hoek *et al.* develop a theory of ability and opportunity [38]. This theory too is based on dynamic logic. However, they separate the actions of dynamic logic from the ability to perform them. van der Hoek *et al.* also use a deterministic variant of dynamic logic. In their notation, $\mathrm{do}_i(\alpha)$ is the *event* corresponding to the agent i doing action α. Here do applies to actions, not to propositions. (We will elide the agent symbol below.) As in dynamic logic, $\langle \mathrm{do}(\alpha)\rangle p$ means that the agent does α and p holds at the end. This is taken to mean that all prerequisites for performing α are satisfied, and the agent performs it. That is, the agent has the opportunity to perform α. $\mathbf{A}(\alpha)$ is a separate operator that denotes that the agent can perform α. This is a primitive, and not formally defined.

van der Hoek *et al.* define $\mathbf{can}(\alpha, p)$ to mean the agent knows that the agent does α resulting in p and has the ability to do α (p. 5). Conversely, $\mathbf{cannot}(\alpha, p)$

means that the agent knows he cannot perform α resulting in p or lacks the ability to do α.

van der Hoek *et al.* define action transformations as ways to manipulate one action (description) into another. They state a number of rules that preserve equivalence of the actions under transformation. The simplest example is that skip; α is equivalent to α. A more complex example involves unraveling a while statement by one loop, but we won't get into the details here. van der Hoek *et al.* then show that if α is equivalent to α', then (a) $[\text{do}(\alpha)]p \Leftrightarrow [\text{do}(\alpha')]p$, (b) $\phi \Leftrightarrow \phi|^{\alpha}_{\alpha'}$, and (c) $\mathbf{A}(\alpha) \Leftrightarrow \mathbf{A}(\alpha')$. As a result, the equivalence of actions satisfy the expected kinds of results. In other words, the definitions are well-formed model-theoretically.

In studying the ability to perform actions, this work generalizes over Moore's analysis of knowing how to perform a plan [22]. Although the idea of separating abilities is interesting, we find the specific definitions a little awkward. For example, $\mathbf{can}(\alpha, p)$ entails not only that the agent can do α, but in fact does α. Clearly, there can be lots of things that an agent can do that he does not. Know-how, in our view, does not entail performance.

4.5 Bringing It About

Segerberg developed a theory of *bringing it about*, which deals with how an agent brings about a particular condition. He proposes a logic of achievements, also in the framework of dynamic logic [19]. Segerberg bases his conceptual account on the notion of *routines*, which roughly are scripts of actions that agents might follow in order to bring things about, that is, to perform high-level actions [30, 31].

Segerberg defines an operator δ, which takes a condition and yields an action, namely, the action of bringing about that condition. Segerberg uses actions of the form δq as the primitive actions in his variant of dynamic logic—he has no other atomic programs. In this respect, Segerberg's work is similar to strategic know-how. The δ operator is intuitively quite close to strategies of the form $\mathbf{do}(q)$. Segerberg defines the denotations of propositions quite in the manner of section 2.4 above. However, he defines the denotation of δp for a proposition p as the set of periods that can result from some program all of whose periods result in the given proposition p.

Segerberg's main intuitions about $[\![\delta p]\!]$ are that it is (a) *reliable* meaning that any of the periods in it will satisfy p, and (b) *maximal* if all periods corresponding to the different executions of a program at a state satisfy p, then all of those periods must be included (p. 329). Segerberg's definition is also similar to STIT in requiring choices to be made that are guaranteed to succeed. However, Segerberg's definitions are forward-looking and do not have the negative condition that is a part of STIT. In this way, Segerberg's definitions are closer to strategic know-how.

However, there are some important dissimilarities from strategic know-how as well. First, Segerberg acknowledges the importance of considering only periods that are "optimal" in some sense, such as being minimal in satisfying the given program. This is in fact done in [33]. However, to keep his approach simple,

Segerberg does not make any assumption of minimality. Optimality of this sort is important in considering executions of strategies and in relating know-how with intentions, because it tells us just how far the current substrategy of a strategy will be executed before the rest of it kicks in.

Second, Segerberg does not consider basic actions at all in his framework, only actions that are derived from propositions. While his results are appealing in terms of their analysis of high-level actions, they lack a connection to the physical actions with which an agent may actually bring something about. In other words, it is not obvious where the semantics is grounded.

Third, Segerberg does not consider the knowledge of agents. Thus the effects of agents' knowledge on their choices cannot be considered. Such choices arise in Segerberg's logic as tests on conditions and in the present approach in conditional and iterative strategies.

The reader might consult Elgesem's paper for a critical review of Segerberg's research program [11].

4.6 Maintenance

Most of the work on know-how and related concepts of interest here has focused on the achievement of different conditions. Sometimes it is important not only to achieve conditions, but to maintain the conditions that hold. Maintenance in this style has not been intensively studied in the literature, but it has recently begun to draw some attention [9, 34].

The following discussion follows the presentation in [34]. Although it bears some resemblance to achievement, maintenance is not easily derived from achievement. For example, simple kinds of duality results between achievement and maintenance do not hold. An agent knows how to maintain a condition if he can continually and knowingly force it to be true, that is, if he can always perform an action that would counteract the potentially harmful actions of other agents. This entails that not only must the actions of other agents not cause any immediate damage, but the given agent should also ensure that they do not lead to a state where he will not be able to control the situation. A key difference with knowing how to achieve some condition is that achievement necessarily requires a bounded number of steps, whereas maintenance does not.

As the base case, we require that the agent know that the given condition holds in the present state. Further, to know how to maintain p, the agent must be able to respond to all eventualities that might cause p to become false. The agent must choose his action such that no combination of the other agents' actions can violate p. Not only must the agent's chosen action maintain p, it should also maintain his ability to maintain p further.

Following the style of the definition for knowing how to achieve, we state that an agent maintains p over an empty tree if he knows that p holds currently. He maintains p over a single action, a, if he knows that he can perform a in the given state and p holds where a begins and where it ends. An agent maintains p over a general tree if he maintains it over its initial action and then over some applicable subtree.

We define $[\![\tau]\!]_{t,p}$, the *maintenance denotation* of a tree τ, as the set of periods beginning at t over which p is maintained by τ. These are the periods over which the agent can knowingly select the right actions. $[\![\tau]\!]_{t,p} = \{\,\}$ means that p cannot be maintained using τ.

- $[\![\emptyset]\!]_{t,p} \stackrel{\text{def}}{=} (\text{if } M \models_t \mathsf{K_t}p, \text{ then } \{[S;t,t]\} \text{ else } \{\,\})$
- $[\![a]\!]_{t,p} \stackrel{\text{def}}{=} \{[S;t,t'] : M \models_t \mathsf{K_t}p \text{ and } (\forall t_k : (t,t_k) \in \mathbf{K}(x) \Rightarrow (\exists S_k, t'_k : [S_k;t_k,t'_k] \in [\![a]\!] \text{ and } [S;t,t'] \in [\![a]\!] \text{ and } (\forall S_k, t'_k : [S_k;t_k,t'_k] \in [\![a]\!] \Rightarrow \text{ and } M \models_{t'_k} \mathsf{K_t}p)))\}$
- $[\![\langle a;\tau_1,\ldots,\tau_m\rangle]\!]_{t,p} \stackrel{\text{def}}{=} \{[S;t,t''] : (\forall t_k : (t,t_k) \in \mathbf{K}(x) \Rightarrow (\exists S_k, t'_k : [S_k;t_k,t'_k] \in [\![a]\!]_{t_k,p} \text{ and } (\forall S_k, t'_k : [S_k;t_k,t'_k] \in [\![a]\!]_{t_k,p} \Rightarrow (\exists t''_k, j : [S_k;t'_k,t''_k] \in [\![\tau_j]\!]_{t'_k,p})))) \text{ and } (\exists t', t'', j : [S;t,t'] \in [\![a]\!]_{t,p} \text{ and } [S;t',t''] \in [\![\tau_j]\!]_{t',p})\}$

In other words, the agent maintains p over $[S;t,t'']$ iff the agent knows at t that he will maintain p over a, that is, till t', and then maintain p till t'' using some subtree.

An agent maintains p to depth i if there is a tree of depth i over which he maintains p. An agent maintains p if he can maintain it to all depths.

SEM-30. $M \models_t \mathsf{K_m}^i p$ iff $(\exists \tau : \text{depth}(\tau) = i \text{ and } [\![\tau]\!]_{t,p} \neq \{\,\})$

SEM-31. $M \models_t \mathsf{K_m} p$ iff $(\forall i : M \models_t \mathsf{K_m}^i p)$

Now we present a recursive characterization of maintenance. This characterization resembles the one given in section 3.2, and is also used in [34], where we develop an approach based on the μ-calculus for computing know-how and maintenance.

Lemma 2 $\mathsf{K_t}p \wedge (\bigvee \alpha : \mathsf{K_t}(\mathbf{E}\langle\alpha\rangle\mathsf{true} \wedge \mathbf{A}[\alpha]\mathsf{K_m}p)) \Leftrightarrow \mathsf{K_m}p$

5 CONCLUSIONS

We discussed a number of variants of the broad concept of know-how that has been studied in the literature on theoretical aspects of rational agency. These variants fill an essential need in the theories of agency that relate the intentions and knowledge of agents with their actions.

To summarize briefly, our initial approach, Brown, Chellas, Belnap & Perloff do not use dynamic logic, whereas our latter approach (strategic know-how), van der Hoek *et al.*, and Segerberg do use dynamic logic. Branching time is explicit in some and implicit in the other approaches, but is a key unifying theme. While there remain important differences, it is remarkable that the different approaches, although developed independently, share many important intuitions. We take this as a promising sign that this subarea of rational agency is maturing.

There are important problems that require additional study. One problem is to sensitize the know-how to the real-time aspects of decision-making in practical settings, both in terms of being able to achieve the desired conditions in bounded time, and to determine the appropriate actions with bounded reasoning. A step in this direction would be develop computational techniques for know-how that are related to planning.

Another challenge is to give a probabilistic account of know-how, which can give a more realistic treatment of the notion of reliability. We believe that such an account will preserve many of the intuitions of the qualitative approaches discussed above.

Another set of issues is opened up when we turn our attention to multiagent settings. If the agents can cooperate with each other, they can together achieve more than any of them can individually. There has been some work on this problem, for example, [32], but additional research is needed to relate the know-how of agents with the structures of the organizations in which they exist.

The foregoing should have made it clear that there are considerable overlaps and similarities between approaches to rational agency in computer science and philosophy. There are also some important differences. Unfortunately, the relationships are not always as well understood as they ought to be.

After all is said and done, have we understood know-how as that term is commonly used, for example, in the quotation by Hewitt given at the beginning of this paper? In its entirety, we believe, not. However, good progress has been made in this small community of computer scientists and philosophers. We encourage the reader to participate in the program of research described above. Its challenges remain important, and provide a fertile ground on which to explore the key concepts of both philosophy and computer science.

ACKNOWLEDGMENTS

This paper being expository in nature is based on previous work by us and others. This work is supported by the NCSU College of Engineering, the National Science Foundation under grants IRI-9529179 and IRI-9624425, and IBM corporation.

Munindar Singh (North Carolina State University, USA)

REFERENCES

[1] Philip Agre and David Chapman. Pengi: An implementation of a theory of activity. In *Proceedings of the National Conference on Artificial Intelligence (AAAI)*, pages 268–272, 1987.

[2] Nuel Belnap and Michael Perloff. Seeing to it that: A canonical form for agentives. *Theoria*, 54(3):175–199, 1988.

[3] Myles Brand. *Intending and Acting*. MIT Press, Cambridge, MA, 1984.

[4] Michael E. Bratman. *Intention, Plans, and Practical Reason.* Harvard University Press, Cambridge, MA, 1987.

[5] Mark A. Brown. On the logic of ability. *Journal of Philosophical Logic*, 17:1–26, 1988.

[6] Mark A. Brown. Action and ability. *Journal of Philosophical Logic*, 19:95–114, 1990.

[7] Brian F. Chellas. *Modal Logic.* Cambridge University Press, New York, 1980.

[8] Brian F. Chellas. Time and modality in the logic of agency. *Studia Logica*, 51(3/4):485–517, 1992.

[9] Giuseppe De Giacomo and Xiao Jun Chen. Reasoning about nondeterministic and concurrent actions: A process algebra approach. In *Proceedings of the National Conference on Artificial Intelligence*, pages 658–663, 1996.

[10] Yves Demazeau and Jean-Pierre Müller, editors. *Decentralized Artificial Intelligence, Volume 2.* Elsevier/North-Holland, Amsterdam, 1991.

[11] Dag Elgesem. Intentions, actions and routines: A problem in Krister Segerberg's theory of actions. *Synthese*, 85:153–177, 1990.

[12] E. A. Emerson. Temporal and modal logic. In J. van Leeuwen, editor, *Handbook of Theoretical Computer Science*, volume B, pages 995–1072. North-Holland, Amsterdam, 1990.

[13] Ronald Fagin and Joseph Y. Halpern. Belief, awareness, and limited reasoning. *Artificial Intelligence*, 34:39–76, 1988.

[14] Jaakko Hintikka. *Knowledge and Belief: An Introduction to the Logic of the Two Notions.* Cornell University Press, Ithaca, 1962.

[15] Jaakko Hintikka. Alternative constructions in terms of the basic epistemological attitudes. In *[23]*. 1972.

[16] Michael N. Huhns and Munindar P. Singh, editors. *Readings in Agents.* Morgan Kaufmann, San Francisco, 1997.

[17] Manfred Kochen and Eugene Galanter. The acquisition and utilization of information in problem solving and thinking. *Information and Control*, 1:267–288, 1958.

[18] Kurt Konolige. *A Deduction Model of Belief.* Morgan Kaufmann, 1986.

[19] Dexter Kozen and Jerzy Tiurzyn. Logics of program. In J. van Leeuwen, editor, *Handbook of Theoretical Computer Science*, volume B, pages 789–840. North-Holland, Amsterdam, 1990.

[20] John McCarthy and Patrick J. Hayes. Some philosophical problems from the standpoint of artificial intelligence. In *Machine Intelligence 4*. American Elsevier, 1969. Page numbers from the version reprinted in [40].

[21] George A. Miller, Eugene Galanter, and Karl Pribram. *Plans and the Structure of Behavior.* Henry Holt, New York, 1960.

[22] Robert C. Moore. A formal theory of knowledge and action. In Jerry R. Hobbs and Robert C. Moore, editors, *Formal Theories of the Commonsense World*, pages 319–358. Ablex, Norwood, NJ, 1984.

[23] Raymond E. Olson and Anthony M. Paul, editors. *Contemporary Philosophy in Scandinavia.* Johns Hopkins Press, Baltimore, 1972.

[24] Michael Perloff. *Stit* and the language of agency. *Synthese*, 86(3):379–408, 1991.

[25] Arthur N. Prior. *Time and Modality.* Clarendon Press, Oxford, 1957.

[26] Arthur N. Prior. *Past, Present and Future.* Clarendon Press, Oxford, 1967.

[27] Anand S. Rao and Michael P. Georgeff. Modeling rational agents within a BDI-architecture. In *Proceedings of the International Conference on Principles of Knowledge Representation and Reasoning*, pages 473–484, 1991. Reprinted in [16].

[28] Gilbert Ryle. *The Concept of Mind.* Hutchinson's University Library, London, 1949.

[29] John R. Searle. *Intentionality: An Essay in the Philosophy of Mind.* Cambridge University Press, Cambridge, UK, 1983.

[30] Krister Segerberg. Routines. *Synthese*, 65:185–210, 1985.

[31] Krister Segerberg. Bringing it about. *Journal of Philosophical Logic*, 18:327–347, 1989.

[32] Munindar P. Singh. Group ability and structure. In *[10]*, pages 127–145. 1991.

[33] Munindar P. Singh. *Multiagent Systems: A Theoretical Framework for Intentions, Know-How, and Communications.* Springer-Verlag, Heidelberg, 1994.

[34] Munindar P. Singh. Applying the mu-calculus in planning and reasoning about action. *Journal of Logic and Computation*, 1998. In press.

[35] Munindar P. Singh and Nicholas M. Asher. A logic of intentions and beliefs. *Journal of Philosophical Logic*, 22(5):513–544, October 1993.

[36] J. F. A. K. van Benthem. Temporal logic. In D. Gabbay, C. Hogger, and J. Robinson, editors, *Handbook of Logic in Artificial Intelligence and Logic Programming*, volume 3. Oxford University Press, Oxford, 1990.

[37] Johan F. A. K. van Benthem. *The Logic of Time: A Model-Theoretic Investigation into the Varieties of Temporal Ontology and Temporal Discourse*, volume 152 of *Synthese Library*. Kluwer, Dordrecht, Holland, 2nd edition, 1991.

[38] Wiebe van der Hoek, Bernd van Linder, and John-Jules Ch. Meyer. A logic of capabilities. TR IR-330, Vrije Universiteit, Amsterdam, 1993.

[39] Georg Henrik von Wright. *Norm and Action*. Routledge & Kegan Paul, London, 1963.

[40] Bonnie L. Webber and Nils J. Nilsson, editors. *Readings in Artificial Intelligence*. Morgan Kaufmann, 1981.

[41] Eric Werner. A unified view of information, intention and ability. In *[10]*, pages 109–125, 1991.

W. VAN DER HOEK, B. VAN LINDER AND J-J.CH. MEYER

AN INTEGRATED MODAL APPROACH TO RATIONAL AGENTS

ABSTRACT: In this paper we give an overview of work we have done to provide a framework in which many aspects of rational agency are integrated. The various attitudes of a rational agent, viz. the informational as well as the motivational ones, are modelled in the framework by means of a variety of modal operators that are interpreted by means of possible worlds, as usual in modal logic. A main point here is that we incorporate all these modal operators into one model, so that in principle the various modal operators can be mixed to describe an agent's complex attitudes.

1 INTRODUCTION

We present a single formal framework for describing crucial aspects of rational or intelligent agents. Before proposing such a formal framework, let us try and distinguish some of its necessary features in a more informal way. An essential feature of being an *agent* is that it has access to a variety of *actions*; by using the adjective *rational*, we mean that the agent is able to *reason*. Obviously there is an interesting interaction between these two features: in our view, a vital feature that links the two mentioned above is that of *information*.

Information is a key issue when describing an agent's existence and behaviour in its environment, the world. On the one hand, information about this world is received by the agent, and it can use this, together with its specific reasoning mechanism, to draw certain conclusions (about the world). Based on these conclusions, it may decide to undertake some specific actions. These actions, on the other hand, may change the world itself, and thereby the information that can be obtained about it.

We may distinguish two aspects or levels regarding information: the *declarative* or *static* level of information, at which the agent may state that he knows, believes or doubts a certain piece of information. But there is also an *operational* or *dynamic* aspect to information: an agent may perform actions that provide it with (new) information about the world: it may, for instance, perform observations, or communicate with other agents.

Thus, by performing particular actions, the agent may acquire more information on which it can base its reasoning. Then, how does this reasoning lead it to the performance of new actions? We think (and many with us) that a rational agent is *directed* towards certain actions more than other ones. And, by using the idea that an action is aimed at changing the world, we even adhere to the principle that the agent considers certain states of the world more preferable than other states. In

M. Wooldridge and A. Rao (eds.), Foundations of Rational Agency, 133–167.

order to approximate such states, the agent must be able to reason about complex actions, and *choose* some actions as candidates for a plan. Again, here we can distinguish a declarative or static level, at which it can for instance be expressed that the agent is *committed* (in the sense of being in a state of 'committedness') to some action, and an operational or dynamic level at which the agent can perform an action like committing itself to do something.

But a rational agent should also be aware of the *limits* and *constraints* of certain actions. These limits may stem from the world itself: an agent does not always have the *opportunity* to perform an action. But other constraints are the consequence of the agent's very own design: it may simply not be equipped with certain *abilities*.

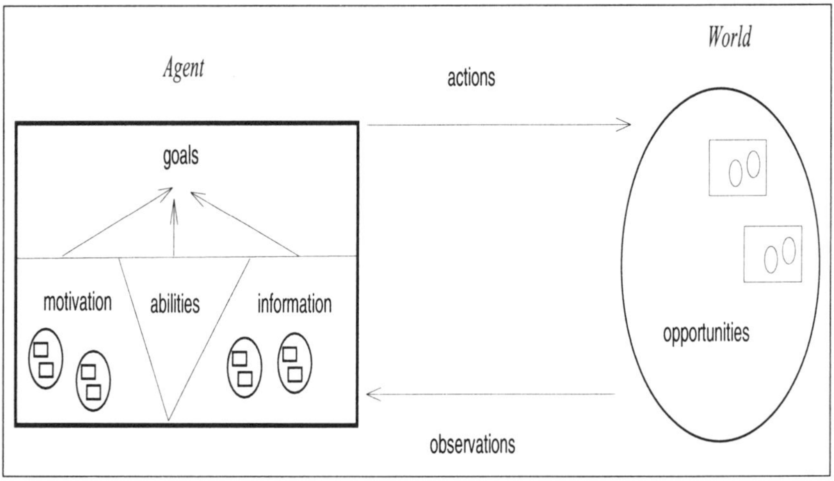

Figure 1. The agent and the world

Figure 1 depicts our framework of agents in a nutshell. Depending on the agent's information, abilities and motivation, it can formulate certain goals. This generally leads to performing certain actions, which have their impact on the world. Since the agent itself is part of the world, some of its actions may also change its own state, like the act of 'committing' that takes it to a state of 'being committed'. Thus, one of the agents in the world of agent a is a itself. The figure also shows samples of the world in the agent's informational and motivational state: in the former state they represent the possible ways the world looks like on the basis of the agent's information, in the latter state they denote possible desired states of the world.

To express all these aspects of agency we will employ modal logic with Kripke-style possible world semantics. In fact, we will use a blend of dynamic, epistemic and doxastic logic, which we extend with certain elements (operators and actions) to suit our purpose, viz. giving a proper description of the informational and motivational attitudes of agents.

This chapter is organized as follows. In Section 2 we introduce the basic framework. Here we shall discuss the language that we will use as well as general considerations about its semantics. In Section 3 we will concentrate on the instantiation of the agent's informational attitudes, while in Section 4 we treat the motivational attitudes of an agent. We conclude the paper with some remarks about related work and an indication of further topics in Section 5.

2 THE FRAMEWORK

In this section we present a core language that is rich enough to reason about some of the agent's attitudes mentioned above–other operators will be defined on top of this– and indicate formal models for this language. Doing so, we will try not to lose ourselves in technical details (these can be found elsewhere, [13, 24, 23, 25]), but rather provide the reader with an intuitive grasp for the ideas underlying our formal definitions.

2.1 Language

The language $\mathcal{L}$ that we use to formalise these notions is based on a fixed set of propositional atoms, and the connectives $\wedge, \vee, \rightarrow, \neg$ to build formulas $\varphi, \psi, \ldots$ with their usual meaning; the canonical tautology $\top$ is defined to be $p \vee \neg p$ for p some arbitrary propositional atom, and the canonical contradiction $\bot$ is defined to be $\neg\top$. We denote the pure propositional language with $\mathcal{L}_0$. We extend this core language to deal with actions and epistemic and motivational attitudes.

Let i be a variable over a set of agents $\{1, \ldots, n\}$. Actions in the set Ac are either atomic actions ($\mathrm{At} = \{a, b, \ldots\}$) or composed ($\alpha, \beta, \ldots$) by means of confirmation of formulas ($\texttt{confirm}\,\varphi$), sequencing ($\alpha; \beta$), conditioning ($\texttt{if}\ \varphi\ \texttt{then}\ \alpha\ \texttt{else}\ \beta$) and repetition ($\texttt{while}\ \varphi\ \texttt{do}\ \alpha$). (Actions that are either atomic or $\texttt{confirm}$ actions will be called *semi-atomic* in this paper.) These actions α can then be used to build new formulas to express the possible *result* of the execution of α by agent i (the formula $\langle do_i(\alpha)\rangle\varphi$ denotes that φ is a result of i's execution of α), the *opportunity* for i to perform α ($\langle do_i(\alpha)\rangle\top$) and i's *capability* of performing the action α ($\mathbf{A}_i\alpha$). The formula $[do_i(\alpha)]\varphi$ is shorthand for $\neg\langle do_i(\alpha)\rangle\neg\varphi$, thus expressing that all possible results of performance of α by i imply φ, thereby being non-committal about the agent's opportunity to perform α. We shall refer to the language built out of the constructs mentioned in this paragraph as the *core language*, denoted as $\mathcal{L}_c$. So $\mathcal{L}_c \subseteq \mathcal{L}$.

In order to successfully complete an action, both the opportunity and the ability to perform the action are necessary (we call this combination the *practical possibility* to do the action). Although these notions are interconnected, they are surely not identical: the abilities of agents comprise mental and physical powers, moral capacities, and physical possibility, whereas the opportunity to perform actions is best described by the notion of circumstantial possibility (cf. [20]). To formalise the knowledge of agents on their practical (im)possibilities, we introduce the so-called Can-predicate and Cannot-predicate. These are binary predicates, pertaining to a pair consisting of an action and a proposition, and denoting that an agent knows that performing the action constitutes a practical (im)possibility to bring about the proposition. We consider practical possibility to consist of two parts, viz. correctness and feasibility: action α is *correct* with respect to φ iff $\langle \mathrm{do}_i(\alpha)\rangle\varphi$ holds and α is *feasible* iff $\mathbf{A}_i\alpha$ holds.

Definition 1 The Can-predicate and the Cannot-predicate are, for all agents i, actions α and formulae φ, defined as follows in terms of practical possibilities.

- $\mathbf{PracPoss}_i(\alpha,\varphi) \triangleq \langle \mathrm{do}_i(\alpha)\rangle\varphi \wedge \mathbf{A}_i\alpha$
- $\mathbf{Can}_i(\alpha,\varphi) \triangleq \mathbf{K}_i\mathbf{PracPoss}_i(\alpha,\varphi)$
- $\mathbf{Cannot}_i(\alpha,\varphi) \triangleq \mathbf{K}_i\neg\mathbf{PracPoss}_i(\alpha,\varphi)$

Thus the Can-predicate and the Cannot-predicate express the agent's knowledge about its practical possibilities and impossibilities, respectively. Therefore these predicates are important for the agent's planning of actions.

In this paper we only consider *deterministic* actions, i.e. actions that have *at most one* successor state. In this case, the diamond-formula $\langle do_i(\alpha)\rangle\varphi$ is stronger than the box-formula$[do_i(\alpha)]\varphi$. To be more precise, we have the equivalence expressed by $\langle do_i(\alpha)\rangle\varphi \leftrightarrow ([do_i(\alpha)]\varphi \wedge \langle do_i(\alpha)\rangle\top)$. The diamond-formula expresses that agent i has the opportunity to perform α, and that φ is one of its effects, whereas the box-formula only asserts that agent i would end up in a situation in which φ holds if i would perform α, and as such it says nothing about the opportunity for i to do α.

Roughly speaking, the language $\mathcal{L}$ allows for *objective* formulas, i.e., formulas about an actual state of affairs, and we have *operators* to make assertions about the agents' (informational and motivational) attitudes, capacities and dynamics. Such operators can be *practitional*, when their argument is an action (like in $\mathbf{A}_i\alpha$ saying that i is able of doing α, or $\mathbf{Com}_i\alpha$ which expresses that i is committed to α). But operators can also be *assertional*, when their argument is a formula (like in $\mathbf{Goal}_i\varphi$—agent i has φ as a goal). In such a case, one could also speak about just a *modal* operator. In this dichotomy, $\langle do_i(\alpha)\rangle\varphi$ (and, for that matter, α, $[do_i(\alpha)]\varphi$) lives in both worlds: it expresses the practitional fact that i has the opportunity to do α, and the assertional fact that φ is one of its possible effects. However, it is also important to note that the informational and motivational attitudes of the agents not only can be expressed at a *declarative* level (agent i believes φ, is committed to α), but are also considered at an *operational* level (agent i updates its beliefs with φ, or commits itself to α).

Whereas the compositional behaviours of results, abilities and opportunities are a main issue in our research, we feel that in order to dress up agents with cognitive features, one also has to come up with specific instances of atomic actions. In Section 3 we study some typical 'informational' actions. To be more precise, for every propositional formula φ and agent j, we consider the actions $\texttt{obs}\,\varphi$, (agents can do observations), $\texttt{inform}(\varphi, j, \mathbf{t})$ (an agent may inform another agent j about φ, 'tagged' with an additional $\mathbf{t}$, see Section 3) and $\texttt{try\_jump}\,\varphi$ (an agent may apply one of its defaults in order to conclude φ). The three actions just mentioned provide the agent with mechanisms that trigger it to change its mind (he can observe or hear something, or have its own 'rules of thumb') but they don't necessarily specify *how* the agent exactly changes its mind. Under some general conditions, informative actions typically have the effect of changing the *epistemic state* of an agent. In the literature, a popular triple of such actions is *expansions, contractions* and *revisions*. In Section 3 we will study three corresponding informative actions (we use the generic term 'updates' for them) for each agent.

The operational aspects of the informational attitudes of agents are addressed by the atomic actions mentioned above. For the declarative aspects of information, we use modal operators $\mathbf{B}_i^k$, $\mathbf{B}_i^o$, $\mathbf{B}_i^c$ and $\mathbf{B}_i^d$. $\mathbf{B}_i^k\varphi$ means that agent i *knows* φ. Here, knowledge must be understood as the information that the agent is born or designed with, like propositional tautologies or domain-specific facts. As such, it comprises to the agent available, true, but fixed information: it determines the agent's basic perspective throughout its existence. A characteristic property of the other $\mathbf{B}_i$ operators is that they model information that is believed to be true by the agent and that is subject to change. In particular in planning it may be necessary for agents to acquire additional information about their world from whatever source possible. The $\mathbf{B}_i^o$ and $\mathbf{B}_i^c$ operator correspond to external sources the agent can use to gain information: the agent can *observe* a state of affairs, or he may *communicate* about it, respectively. The operator $\mathbf{B}_i^d$ represents those beliefs that the agent adopts *by default*. We assume that observational beliefs are always true, whereas this is not a prerequisite for the communicational and default beliefs. In Section 3 we will link these attitudes to the corresponding informative actions mentioned above.

Section 4 adds 'motivational' actions to the atomic kernel of actions. For any action α defined so far, we add the action $\texttt{commit\_to}\,\alpha$ to the set of actions. In some situations, agents cannot fulfil their promises. Therefore, we also consider $\texttt{uncommit}\,\alpha$ as a basic action. Our language is then completed with operators that deal with motivational attitudes. At the *assertion* level, we consider *wishes* and *goals*, represented as $\mathbf{W}_i$ and $\mathbf{Goal}_i$, respectively. At the *practition* level we define *commitments*, represented as $\mathbf{Com}_i$.

2.2 Semantics

Recalling that the full language described above is denoted with $\mathcal{L}$, we will now briefly indicate how to formally interpret formulas of $\mathcal{L}$. We use the following kind of Kripke models $M = \langle W, \pi, \mathtt{D}, \mathtt{I}, \mathtt{M}\rangle$ where W is a non-empty set of worlds. In order to determine whether a formula $\varphi \in \mathcal{L}$ is true in w (if so, we write $(M, w) \models \varphi$, where (M, w) is called a *state*), π, $\mathtt{P}$, $\mathtt{I}$ and $\mathtt{M}$ encode how to interpret the propositional atoms and the Dynamic, Informational and Motivational operators, respectively. For the propositional part, we stipulate $\mathcal{M}, s \models p$ iff $\pi(s)(p) = \mathit{true}$, and the logical connectives are interpreted in $\mathcal{M}, s$ as expected. We are going to extend the definition of $\mathcal{M}, s \models \varphi$ to arbitrary φ; if $\mathcal{M}$ is clear from context, we abbreviate $\{s \mid \mathcal{M}, s \models \varphi\}$ to $[\![\varphi]\!]$.

Let us agree to call a modal operator X to be a necessity operator for a relation R (or a function f) if the truth of $X\varphi$ at state M, w is defined as the truth of φ in all states s for which Rws (or $s \in f(w)$, respectively). Then, the $\mathtt{D}$-part of a model consists of two functions: a result function $\mathtt{r}$ such that, for each atomic action a and agent i, $\mathtt{r}(i, a)(\mathcal{M}, w)$ yields all resulting states (M', w') of i doing a in w. We will explain below why we generalize the function $\mathtt{r}$ from a world transformation (as it is usually defined in dynamic logic, cf. [10]) to a transformation on states (M, w). $\mathtt{D}$ also contains a capability function $\mathtt{c}$ where $\mathtt{c}(i, a)(w)$ is true exactly for those atomic actions a that i is capable of in w. The function $\mathtt{r}$ is extended to arbitrary actions in a way standard for dynamic logic (cf. [10]) and then $[do_i(\alpha)]$ is interpreted as the necessity operator for this extended $\mathtt{r}^*$. In a similar fashion, $\mathtt{c}$ is extended for arbitrary actions. Then we have:
$\mathcal{M}, s \models [do_i(\alpha)]\varphi$ iff $\mathcal{M}, t \models \varphi$ for all $\mathcal{M}, t \in \mathtt{r}^*(\mathcal{M}, s)$
$\mathcal{M}, s \models \mathbf{A}_i\varphi$ iff $\mathtt{c}^*(\mathcal{M}, s) = \mathit{true}$

Definition 2 We now give the extensions of $\mathtt{r}$ and $\mathtt{c}$, respectively. Recall that we assume that all actions are *deterministic*. For convenience, we introduce the special state $\mathcal{E} = (\mathcal{M}, \epsilon)$, where ϵ is a world where all impossible actions lead to: from there, no other actions can be performed anymore.

$$
\begin{array}{ll}
\mathtt{r}^*(i, a)(\mathcal{M}, s) & = \mathtt{r}(i, a)(\mathcal{M}, s) \\
\mathtt{r}^*(i, \mathtt{confirm}\,\varphi)(\mathcal{M}, s) & = (\mathcal{M}, s) \text{ if } \mathcal{M}, s \models \varphi \text{ and } \mathcal{E} \text{ otherwise} \\
\mathtt{r}^*(i, \alpha_1; \alpha_2)(\mathcal{M}, s) & = \mathtt{r}^*(i, \alpha_2)(\mathtt{r}^*(i, \alpha_1)(\mathcal{M}, s)) \\
\mathtt{r}^*(i, \mathtt{if}\,\varphi\,\mathtt{then}\,\alpha_1 & = \mathtt{r}^*(i, \alpha_1)(\mathcal{M}, s) \text{ if } \mathcal{M}, s \models \varphi \text{ and} \\
\quad \mathtt{else}\,\alpha_2\,\mathtt{fi})(\mathcal{M}, s) & \quad \mathtt{r}^*(i, \alpha_2)(\mathcal{M}, s) \text{ otherwise} \\
\mathtt{r}^*(i, \mathtt{while}\,\varphi & = (\mathcal{M}', s') \text{ iff } \exists k \in \mathbb{N} \exists \mathcal{M}_0, s_0 \ldots \exists \mathcal{M}_k, s_k \\
\quad \mathtt{do}\,\alpha_1\,\mathtt{od})(\mathcal{M}, s) & \quad [\mathcal{M}_0, s_0 = \mathcal{M}, s \,\&\, \mathcal{M}_k, s_k = \mathcal{M}', s' \,\&\, \forall j < k \\
 & \quad [\mathcal{M}_{j+1}, s_{j+1} = \mathtt{r}^*(i, \mathtt{confirm}\,\varphi; \alpha_1)(\mathcal{M}_j, s_j)] \\
 & \qquad \& \, \mathcal{M}', s' \models \neg\varphi] \\
\text{where } \mathtt{r}^*(i, \alpha)(\mathcal{E}) & = \mathcal{E} \\
\text{and} & \\
\mathtt{c}^*(i, a)(\mathcal{M}, s) & = \mathtt{c}(i, a)(s) \\
\mathtt{c}^*(i, \mathtt{confirm}\,\varphi)(\mathcal{M}, s) & = \mathit{true} \text{ if } \mathcal{M}, s \models \varphi \text{ and } \mathit{false} \text{ otherwise} \\
\mathtt{c}^*(i, \alpha_1; \alpha_2)(\mathcal{M}, s) & = \mathtt{c}^*(i, \alpha_1)(\mathcal{M}, s) \,\&\, \mathtt{c}^*(i, \alpha_2)(\mathtt{r}^*(i, \alpha_1)(\mathcal{M}, s)) \\
\mathtt{c}^*(i, \mathtt{if}\,\varphi\,\mathtt{then}\,\alpha_1 & = \mathtt{c}^*(i, \mathtt{confirm}\,\varphi; \alpha_1)(\mathcal{M}, s) \text{ or} \\
\quad \mathtt{else}\,\alpha_2\,\mathtt{fi})(\mathcal{M}, s) & \quad \mathtt{c}^*(i, \mathtt{confirm}\,\neg\varphi; \alpha_2)(\mathcal{M}, s) \\
\mathtt{c}^*(i, \mathtt{while}\,\varphi & = \mathit{true} \text{ if } \exists k \in \mathbb{N}[\mathtt{c}^*(i, (\mathtt{confirm}\,\varphi; \alpha_1)^k; \\
\quad \mathtt{do}\,\alpha_1\,\mathtt{od})(\mathcal{M}, s) & \qquad \mathtt{confirm}\,\neg\varphi)(\mathcal{M}, s) = \mathit{true}] \\
 & \quad \text{and } \mathit{false} \text{ otherwise}
\end{array}
$$

where $\mathtt{c}^*(i,\alpha)(\mathcal{E}) \qquad = \textit{true}$

Note that the clauses above only indicate how composed actions are interpreted. One main feature of our approach is the possibility to define actions that may effect the semantic units: actions may change the set of worlds that an agent considers epistemically possible, desirable, and also the sequence of actions it is planning to perform next. This will be exploited further in the following sections. With regard to the abilities of agents, the motivation for the choices made in Definition 2 is the following. The definition of $\mathtt{c}(i, \mathtt{confirm}\,\varphi)$ expresses that an agent is able to get confirmation for a formula φ iff φ holds. An agent is capable of performing a sequential composition $\alpha_1; \alpha_2$ iff it is capable of performing α_1 (now), and it is capable of executing α_2 after it has performed α_1. An agent is capable of performing a conditional composition, if either it is able to get confirmation for the condition and thereafter perform the then-part, or it is able to confirm the negation of the condition and perform the else-part afterwards. An agent is capable of performing a repetitive composition $\mathtt{while}\,\varphi\,\mathtt{do}\,\alpha\,\mathtt{od}$ iff it is able to perform the action $(\mathtt{confirm}\,\varphi; \alpha_1)^k; \mathtt{confirm}\,\neg\varphi$ for some natural number k, i.e. it is able to perform the k-th unwinding of the while-loop. Also note that in the ';-clause' for $\mathtt{c}^*$ we have chosen to model *optimistic* agents: as a consequence, we have that any agent finds itself capable of performing $\alpha; \beta$, whenever it lacks the opportunity to perform α (for a discussion on this choice, cf. [22]).

To explain the content of the informational layer $\mathtt{I}$ of the model, recall that, concerning the declarative part, we had four informational operators $\mathbf{B}_i^k, \mathbf{B}_i^o, \mathbf{B}_i^c$ and $\mathbf{B}_i^d$. Let x be a variable over $\{k, o, c, d\}$, then for each x, $\mathtt{I}$ contains a function B^x such that, for each state $w \in W$, and agent i, $\mathrm{B}^x(i, w)) \subseteq W$ denotes all the worlds that are x-believed to be possible by agent i, in world w. The interpretation of the $\mathbf{B}_i^x$ now becomes:

$\mathcal{M}, s \models \mathbf{B}_i^x\varphi$ iff $\forall t \in \mathrm{B}^x(i, s), \mathcal{M}, t \models \varphi$

This truth condition for the operators $\mathbf{B}_i^x$ ensures that the notions of belief satisfy at least the properties of the logic **K45** (cf. [7]). To ensure that both $\mathbf{B}_i^k$ and $\mathbf{B}_i^o$ behave knowledge-like (or **S5**-like), and to relate the different kinds of belief, we require some additional properties on our models, which will be made explicit in Section 3.

Now, since we allow for actions that change the beliefs of an agent it is clear why actions must be considered to transform states into states, since an informative action, like an update of an agent i's beliefs at level x in a state (M, w) typically does not effect objective formulas in w, nor any of the functions $\mathrm{B}^x(j, \cdot)(i \neq j)$, but it will effect the function $\mathrm{B}^x(i, \cdot)$: as a result of an update, agent i may consider other states as epistemically indistinguishable. There is both a mathematical and a conceptual difference between the result function $\mathtt{r}$ and the information functions B^x. Firstly, $\mathtt{r}$ may pick states from a model $\mathcal{M}'$ different from the model $\mathcal{M}$ it is applied to. Secondly, the function $\mathtt{r}$ models a state *transition* caused by certain actions, whereas the informational functions yield states that are *compatible* with agent i's knowledge, observations, communication or default beliefs, respectively.

In Section 3 we will see the need for further ingredients of the `I`-part of our models as we'll go along.

Dealing with motivational attitudes of agents is the part where we depart significantly from standard modal approaches. We leave the instantiation of the `M`-part of the model to Section 4.

Without going into technical details at this point, it will be clear right now that our framework definitely embodies a modal-logical flavour. However, it is well-known that normal modal operators have certain properties that are occasionally considered undesirable for the common-sense notions that they are intended to formalise. For example, although the formal notions of knowledge and belief are closed under logical consequence, this property will in general not hold for human knowledge and belief (although it will, for instance, hold for the information that is recorded in a database, and it can also be defended for the knowledge and belief of an artificial agent, in some restricted situations). While using modal operators for epistemic notions may lead to *over-idealisation* of such a notion, when formalising motivational attitudes the undesired properties induced by closure under logical consequence become even more problematic. For agents do, in general, not desire all the logical consequences of their wishes, nor do they consider the logically inevitable to be among their goals. To facilitate discussion of some of those properties, let us collect them in the following seven 'Logical Omniscience properties', which are all valid properties for normal modal operators.

Definition 3 Let $\varphi, \psi \in \mathcal{L}$ be formulae, and let $\mathbf{X}$ be some operator.

- $\models \mathbf{X}\varphi \wedge \mathbf{X}(\varphi \to \psi) \to \mathbf{X}\psi$ *LO*1
- $\models \varphi \Rightarrow \models \mathbf{X}\varphi$ *LO*2
- $\models \varphi \to \psi \Rightarrow \models \mathbf{X}\varphi \to \mathbf{X}\psi$ *LO*3
- $\models \varphi \leftrightarrow \psi \Rightarrow \models \mathbf{X}\varphi \leftrightarrow \mathbf{X}\psi$ *LO*4
- $\models (\mathbf{X}\varphi \wedge \mathbf{X}\psi) \to \mathbf{X}(\varphi \wedge \psi)$ *LO*5
- $\models \mathbf{X}\varphi \to \mathbf{X}(\varphi \vee \psi)$ *LO*6
- $\models \neg(\mathbf{X}\varphi \wedge \mathbf{X}\neg\varphi)$ *LO*7

3 GATHERING INFORMATION

In this section, we first describe various notions of belief in our framework, and then we discuss the notion of belief change. In particular, we distinguish ways that describe how to perform such a belief change (Section 3.2) from the effects of a belief change as a result of a particular action (Section 3.3).

3.1 Notions of Belief

The logics of knowledge and belief are by now well established and understood ([7, 27]). To cater for the desired properties, we dress up our models with the following requirements. For the function B^k we require that $w \in \mathrm{B}^k(i, w)$ and, for all $u, v \in W, u \in \mathrm{B}^k(i,v) \Leftrightarrow \mathrm{B}^k(i,u) = \mathrm{B}^k(i,v)$. This ensures that the relation $\mathrm{R}_i^k xy$ defined as $y \in \mathrm{B}^k(i,x)$ is an equivalence relation, and hence, that knowledge satisfies the **S5** properties. We put the same constraints on B^o, such that observational beliefs have the same properties as knowledge. Finally, we require that the B^x-functions are such that for all i and $s, s' \in W$:

- $\mathrm{B}^d(i,s) \neq \emptyset$
- $\mathrm{B}^d(i,s) \subseteq \mathrm{B}^c(i,s) \subseteq \mathrm{B}^o(i,s) \subseteq \mathrm{B}^k(i,s)$
- if $s' \in \mathrm{B}^o(i,s)$ then $\mathrm{B}^c(i,s') = \mathrm{B}^c(i,s)$ and $\mathrm{B}^d(i,s') = \mathrm{B}^d(i,s)$

The properties above ensure the following theorem.

Theorem 1 Any belief operator $\mathbf{X} \in \{\mathbf{B}_i^k, \mathbf{B}_i^o, \mathbf{B}_i^c, \mathbf{B}_i^d\}$ satisfies all the logical omniscience properties $LO1 - LO2$ of definition 3. Moreover, for all $\varphi \in \mathcal{L}$ we have:

1. $\models (\mathbf{B}_i^o\varphi \to \varphi) \wedge (\mathbf{B}_i^k\varphi \to \varphi)$ T
2. $\models \neg(\mathbf{X}\varphi \wedge \mathbf{X}\neg\varphi)$ D
3. $\models \mathbf{X}\varphi \to \mathbf{X}\mathbf{X}\varphi$ 4
4. $\models \neg\mathbf{X}\varphi \to \mathbf{X}\neg\mathbf{X}\varphi$ 5
5. $\models (\mathbf{B}_i^k\varphi \to \mathbf{B}_i^o\varphi) \wedge (\mathbf{B}_i^o\psi \to \mathbf{B}_i^c\psi) \wedge (\mathbf{B}_i^c\chi \to \mathbf{B}_i^d\chi)$

Here, we will not address the problem of logical omniscience for belief operators any further (for a discussion, see [35, 15]). Item 1 of theorem 1 says that knowledge and observational beliefs are veridical: they must be true. Item 2 states a weaker property, saying that one cannot have false beliefs. Items 3 and 4 are known as positive and negative introspection, respectively. Finally, item 5 relates the various notions of beliefs. It may seem a bit counter-intuitive to say, for instance, that knowing φ implies your 'observational-believing' φ, but this only indicates relative strength. There *is* a way to express that i o-believes φ purely on the basis of its observations. Let, for $x = k, o, c, d$, $\mathbf{Ignorant}_i^x\varphi \triangleq \neg\mathbf{B}_i^x\varphi \wedge \neg\mathbf{B}_i^x\neg\varphi$. Then we can also define the following:

Definition 4 *For each $\varphi \in \mathcal{L}$ we define:*

- $\mathbf{Saw}_i\varphi \triangleq \mathbf{Ignorant}_i^k\varphi \wedge \mathbf{B}_i^o\varphi$
- $\mathbf{Heard}_i\varphi \triangleq \mathbf{Ignorant}_i^o\varphi \wedge \mathbf{B}_i^c\varphi$
- $\mathbf{Jumped}_i\varphi \triangleq \mathbf{Ignorant}_i^c\varphi \wedge \mathbf{B}_i^d\varphi$

Note that, for instance, the $\mathbf{Heard}_i$ operator does not formalise hearing *per se*, but rather *believing on the basis of being told*. One easily verifies that the operators defined above satisfy logical omniscience property $LO1$. However, they generally don't satisfy all properties of Logical Omniscience. For instance, $LO2$ is not valid: for no tautology ψ, one can have $\mathbf{Saw}_i\psi$. A last property we mention (for more, see also [22, Proposition 5.12]) is the following weakening of $LO3$: if $\models \varphi \to \psi$

then $\models \mathbf{B}_i^c\psi$, and hence $\models \varphi \to \psi$ implies $\models \mathbf{Heard}_i\varphi \to (\mathbf{Heard}_i\psi \vee \mathbf{Saw}_i\psi \vee \mathbf{B}_i^k\psi)$ The latter implication expresses that if an agent believes a formula on the basis of its communication, it must believe all its consequences on at least a reliable level: it must have heard or seen them, or else even know them.

3.2 Belief Changes: How?

Concerning the agent's attitude towards *new* information, we like to address two main issues within our framework. The first is: what makes an agent change its beliefs? When is it prepared to give up (part of) its ideas about how the world looks like? The other question is: how do agents update their beliefs? Can we guarantee that they don't change it unnecessarily, for instance, and if so, how do they decide which beliefs to hold on to, and which to give up? We will first tackle the second question, since it eases, technically speaking, our way to answer the first one.

The work of Alchourrón, Gärdenfors and Makinson (AGM, [1]) has become a standard reference in the area of belief revision. The basic building blocks here are a theory T, which is closed under classical consequences, some propositional formula ϕ representing novel information to be incorporated into T, and an operation that performs the change and yields a theory T'. AGM distinguishes the operations *expansion* (where ϕ is added to T), *contraction* (the resulting T' should not imply ϕ anymore) and *revision* (which can be defined as a contraction with $\neg\phi$, followed by an expansion with ϕ). Here, we will use the term *updates* as a generic term.

We will now define our updates, and, having done that, we will relate them to the AGM framework. In fact, we consider knowledge to be static, and thus not subject to any updates. Thus, our notion of knowledge comprises the properties the agent is designed or born with. In fact, since the *implementation* of the basic update operations are the same for o, c and d beliefs (not the *condition* under which they are performed, see below), in the following we will omit the superscript x, and just study updates of beliefs. Thus, we assume, for each formula $\varphi \in \mathcal{L}_0$—we will explain this restriction to $\mathcal{L}_0$ shortly—to have the atomic actions $\texttt{expand}\ \varphi, \texttt{contract}\ \varphi$ and $\texttt{revise}\ \varphi$. We think the following properties must be shared by all of them:

Definition 5 We distinguish the following properties of actions α, where $\chi \in \mathcal{L}$.

- $\models \langle \mathrm{do}_i(\alpha)\rangle \top$ *realizability*
- $\models \langle \mathrm{do}_i(\alpha)\rangle\chi \to [\mathrm{do}_i(\alpha)]\chi$ *determinism*
- $\models \langle \mathrm{do}_i(\alpha;\alpha)\rangle\chi \leftrightarrow \langle \mathrm{do}_i(\alpha)\rangle\chi$ *idempotence*

Realizability of an action implies that agents have the opportunity to perform the action regardless of circumstances; determinism of an action means that performing the action results in a unique state of affairs, and idempotence of an action implies that performing the action an arbitrary number of times has the same effect as performing the action just once. We say that $\Delta \in \{\texttt{expand}, \texttt{contract}, \texttt{revise}\}$ satisfies any of the properties of Definition 5 if the action $\Delta\varphi$ satisfies that property for all $\varphi \in \mathcal{L}_0$.

Let us now see how we can model expansions. Informally, a belief expansion is an action that leads to a state of affairs in which some formula is included in the set of beliefs of an agent. In our framework uncertainties of agents are formalized through the different doxastic alternatives that the agent has: if an agent believes neither φ nor $\neg\varphi$ then it considers both doxastic alternatives supporting φ and doxastic alternatives supporting $\neg\varphi$ possible. Expanding the beliefs of the agent with φ may then be implemented by declaring all alternatives supporting $\neg\varphi$ to be 'doxastically impossible', i.e., on the ground of its beliefs the agent no longer considers these alternatives to be possible. Hence the *expansion* of the belief set of an agent can be modelled through a *restriction* of its set of doxastic alternatives.

Definition 6 Recall that we only consider two doxastic accessibility functions B^k,and B respectively, in the informational layer $\mathtt{I}$ now. Let some model $\mathcal{M} = \langle W, \pi, \mathtt{D}, \mathtt{I}, \mathtt{M}\rangle$ with $s \in W$, and $\varphi \in \mathcal{L}_0$ be given. We define:

$$\begin{aligned}
&\mathtt{r}(i, \mathtt{expand}\,\varphi)(\mathcal{M}, s) = \mathcal{M}', s \text{ where}\\
&\quad \mathcal{M}' = \langle W, \pi, \mathtt{D}, \mathtt{I}', \mathtt{M}\rangle \text{ with}\\
&\quad\quad \mathrm{B}'(i', s') = \mathrm{B}(i', s') \text{ if } i' \neq i \text{ or } s' \notin \mathrm{B}^k(i, s)\\
&\quad\quad \mathrm{B}'(i, s') = \mathrm{B}(i, s') \cap [\![\varphi]\!] \text{ if } s' \in \mathrm{B}^k(i, s)
\end{aligned}$$

Definition 6 provides for an intuitively acceptable formalization of belief expansions as can be seen in the following proposition.

Proposition 7 *For all* $\varphi, \psi \in \mathcal{L}_0$ *we have:*

- $\models [\mathrm{do}_i(\mathtt{expand}\,\varphi)]\mathbf{B}_i\varphi$
- $\models \mathbf{B}_i\psi \rightarrow [\mathrm{do}_i(\mathtt{expand}\,\varphi)]\mathbf{B}_i\psi$
- $\models \mathbf{B}_i\varphi \rightarrow (\mathbf{B}_i\psi \leftrightarrow [\mathrm{do}_i(\mathtt{expand}\,\varphi)]\mathbf{B}_i\psi)$

The first clause of Proposition 7 states that an expansion with some formula results in the formula being believed. The second clause states that beliefs are persistent under expansions. In this clause the restriction to *propositional* formulae ψ is in general necessary. For consider a situation in which an agent does not believe φ and by negative introspection believes that it does not believe φ. After expanding its beliefs with φ, the agent believes φ and, assuming that the resulting belief set is not the absurd one, it no longer believes that it does not believe φ. Hence not all beliefs of the agent persist in situations like these. Note that the first two clauses combined indicate that our definition of belief, in the context of allowing absurd belief sets, is a good one when dealing with expansions. For an expansion with some formula φ in a situation in which $\neg\varphi$ is already believed, results in the agent believing both φ and $\neg\varphi$ and hence having inconsistent beliefs. The third clause states that in situations where some formula is already believed, nothing is changed as the result of an expansion with that formula. This latter property is suggested by the *criterion of informational economy* [8], which states that since information is in general not gratuitous, unnecessary losses of information are to be avoided.

A belief contraction is the change of belief through which in general some formula that is believed beforehand is no longer believed afterwards. As such, apparent beliefs that an agent has are turned into doubts as the result of a contraction. In terms of our framework, this change of belief may be implemented by *extending* the set of doxastic alternatives of an agent in order to encompass at least one state not satisfying the formula that is to be contracted. Consider for example the situation of an agent i that believes p, i.e., p holds in all its doxastic alternatives. When contracting p from the belief set of the agent, some $\neg p$-worlds are added to the set of doxastic alternatives of the agent. In order to end up with well-defined Kripke models, these worlds that are to be added, need to be in the set of epistemic alternatives of s. For in our Kripke models, the set of doxastic alternatives for a given agent in a given state is contained in its set of epistemic alternatives in that state. Thus the worlds that are to be added to the set of doxastic alternatives of the agent are elements of the set of epistemic alternatives not supporting p.

The problem with defining contractions in this way, is that it is not straightforward to decide which worlds need to be added. From the basic idea that knowledge — acting as the principles of agents — provides some sort of lower bound of the belief set of an agent, it is clear that in the case of a contraction with φ some states need to be added that are elements of the set of epistemic alternatives of the agent and do not support φ, but it is not clear exactly *which* elements of this set need to be chosen. The approach that we propose to solve this problem is based on the use of so called *selection functions*. These are functions $\sigma : \mathcal{A} \times W \times \mathcal{L}_0 \to \wp(W)$ that (whenever possible) select a subset of the set of epistemic alternatives in such a way that the resulting $\texttt{contract}$ action behaves rationally. Without giving the full details here (see [25], let us here mention the two requirements that $\sigma(i, s, \varphi) \subseteq \mathrm{B}^k(i, s)$ and that $\sigma(i, s, \varphi) = \sigma(i, s'\varphi)$, whenever $s' \in \mathrm{B}^k(i, s)$)), we claim that such a function can be defined in such a way that the following definition guarantees Proposition 9, stating such rationality properties. The reader may verify that the two requirements mentioned above guarantee the items 3 and 9 of Proposition 9, respectively).

The definition of $\texttt{r}^*$ for the $\texttt{contract}$ action is based on the use of selection functions: a contraction is performed by adding to the set of doxastic alternatives of the agent exactly those worlds that are picked out by the selection function.

Definition 8 Let some model $\mathcal{M} = \langle W, \pi, \mathrm{D}, \mathrm{I}, \mathrm{M} \rangle$ with $s \in W$ and $\varphi \in \mathcal{L}_0$ be given. Furthermore, let σ be an arbitrary but fixed selection function for $\mathcal{M}$. We define:

$$\begin{array}{l} \texttt{r}^*(i, \texttt{contract}\,\varphi)(\mathcal{M}, s) = \mathcal{M}', s \text{ where} \\ \quad \mathcal{M}' = \langle W, \pi, \mathrm{D}, \mathrm{I}', \mathrm{M} \rangle \text{ with} \\ \qquad \mathrm{B}'(i', s') = \mathrm{B}(i', s') \text{ if } i' \neq i \text{ or } s' \notin \mathrm{B}^k(i, s) \\ \qquad \mathrm{B}'(i, s') = \mathrm{B}(i, s') \cup \sigma(i, s, \varphi) \text{ for all } s' \in \mathrm{B}^k(i, s) \end{array}$$

Proposition 9 *For all $\varphi, \psi, \vartheta \in \mathcal{L}_0, \chi \in \mathcal{L}$ we have:*

- $\models [\mathrm{do}_i(\texttt{contract}\,\varphi)]\mathbf{B}_i\psi \to \mathbf{B}_i\psi$

- $\models \neg\mathbf{B}_i\varphi \rightarrow ([\text{do}_i(\texttt{contract}\,\varphi)]\mathbf{B}_i\psi \leftrightarrow \mathbf{B}_i\psi)$
- $\models \neg\mathbf{K}_i\varphi \rightarrow [\text{do}_i(\texttt{contract}\,\varphi)]\neg\mathbf{B}_i\varphi$
- $\models \mathbf{B}_i\varphi \rightarrow (\mathbf{B}_i\psi \rightarrow [\text{do}_i(\texttt{contract}\,\varphi;\texttt{expand}\,\varphi)]\mathbf{B}_i\psi)$
- $\models ([\text{do}_i(\texttt{contract}\,\varphi)]\mathbf{B}_i\vartheta \wedge [\text{do}_i(\texttt{contract}\,\psi)]\mathbf{B}_i\vartheta) \rightarrow$ $[\text{do}_i(\texttt{contract}\,\varphi \wedge \psi)]\mathbf{B}_i\vartheta$
- $\models [\text{do}_i(\texttt{contract}\,\varphi \wedge \psi)]\neg\mathbf{B}_i\varphi \rightarrow$ $([\text{do}_i(\texttt{contract}\,\varphi \wedge \psi)]\mathbf{B}_i\vartheta \rightarrow [\text{do}_i(\texttt{contract}\,\varphi)]\mathbf{B}_i\vartheta)$
- $\models \neg\mathbf{B}_i\varphi \rightarrow ([\text{do}_i(\texttt{contract}\,\varphi)]\chi \leftrightarrow \chi)$
- $\models \mathbf{B}_i\varphi \rightarrow (\chi \leftrightarrow [\text{do}_i(\texttt{contract}\,\varphi;\texttt{expand}\,\varphi)]\chi)$
- $\models \mathbf{K}_i(\varphi \leftrightarrow \psi) \rightarrow ([\text{do}_i(\texttt{contract}\,\varphi)]\chi \leftrightarrow [\text{do}_i(\texttt{contract}\,\psi)]\chi)$
- `contract` *satisfies realizability, determinism and idempotence*

The first clause of Proposition 9 states that after a contraction an agent believes at most the formulae that it believed before the contraction. The second clause states that in situations in which an agent does not believe φ, nothing changes as the result of contracting φ. Again this property reflects the criterion of informational economy. The third clause states that a contraction with a contractable formula, i.e., a formula not belonging to the agent's knowledge, results in the agent not believing the contracted formula. The fourth clause states that whenever an agent believes a formula, all beliefs in its original belief set are recovered after a contraction with that formula followed by an expansion with the same formula. The fifth clause formalizes the idea that all formulae that are believed both after a contraction with φ and after a contraction with ψ, are believed after a contraction with $\varphi \wedge \psi$. Clause 6 states that if a contraction with $\varphi \wedge \psi$ results in φ not being believed, then in order to contract φ no more formulae need to be removed than those that were removed in order to contract $\varphi \wedge \psi$. This clause is related to the property of *minimal change* for contractions. The seventh clause states that contractions with disbelieved formulae cause no change at all. The next clause states that whenever an agent believes a formula, a contraction with that formula followed by an expansion with the same formula reduces to the void action and therefore causes no change. Clause 9 states that for formulae that an agent knows to be equivalent, a contraction with one formula causes exactly the same change as a contraction with the other formula. By the last clause, contractions obey the properties given in Definition 5.

Having defined actions that model expansions and contractions, we now turn to defining actions that model revisions. A revision is a change of belief through which some formula is added to the beliefs of an agent, while preserving consistency. Our definition of actions that model revisions is based on the *Levi identity* [21]. Levi suggested that revisions can be defined in terms of contractions and expansions: a revision with φ can be defined as a contraction with $\neg\varphi$ followed by an expansion with φ. Given the definitions of contractions and expansions of the previous paragraphs and the fact that the class of actions Ac that we consider is closed under sequential composition, the Levi identity provides for a means to define revisions as the sequential composition of a contraction and an expansion action.

Definition 10 Let some model $\mathcal{M} = \langle W, \pi, \mathtt{D}, \mathtt{I}, \mathtt{M}\rangle$ with $s \in W$ and $\varphi \in \mathcal{L}_0$ be given. We define:

- $\mathtt{r}^*(i, \mathtt{revise}\,\varphi)(\mathcal{M}, s) = \mathtt{r}^*(i, \mathtt{contract}\,\neg\varphi; \mathtt{expand}\,\varphi)(\mathcal{M}, s)$

Proposition 11 *For all* $\varphi, \psi, \vartheta \in \mathcal{L}_0$ *we have:*

- $\models [\mathrm{do}_i(\mathtt{revise}\,\varphi)]\mathbf{B}_i\varphi$
- $\models [\mathrm{do}_i(\mathtt{revise}\,\varphi)]\mathbf{B}_i\vartheta \to [\mathrm{do}_i(\mathtt{expand}\,\varphi)]\mathbf{B}_i\vartheta$
- $\models \neg\mathbf{B}_i\neg\varphi \to ([\mathrm{do}_i(\mathtt{expand}\,\varphi)]\mathbf{B}_i\vartheta \leftrightarrow [\mathrm{do}_i(\mathtt{revise}\,\varphi)]\mathbf{B}_i\vartheta)$
- $\models \mathbf{K}_i\neg\varphi \leftrightarrow [\mathrm{do}_i(\mathtt{revise}\,\varphi)]\mathbf{B}_i\bot$
- $\models \mathbf{K}_i(\varphi \leftrightarrow \psi) \to ([\mathrm{do}_i(\mathtt{revise}\,\varphi)]\mathbf{B}_i\vartheta \leftrightarrow [\mathrm{do}_i(\mathtt{revise}\,\psi)]\mathbf{B}_i\vartheta)$
- $\models [\mathrm{do}_i(\mathtt{revise}\,\varphi \wedge \psi)]\mathbf{B}_i\vartheta \to [\mathrm{do}_i(\mathtt{revise}\,\varphi; \mathtt{expand}\,\psi)]\mathbf{B}_i\vartheta$
- $\models \neg[\mathrm{do}_i(\mathtt{revise}\,\varphi)]\mathbf{B}_i\neg\psi \to$
  $([\mathrm{do}_i(\mathtt{revise}\,\varphi; \mathtt{expand}\,\psi)]\mathbf{B}_i\vartheta \to [\mathrm{do}_i(\mathtt{revise}\,\varphi \wedge \psi)]\mathbf{B}_i\vartheta)$

The first clause of Proposition 11 states that agents believe φ as the result of revising their beliefs with φ. The second clause states that a revision with φ results in the agent believing at most the formulae that it would believe after expanding its beliefs with φ, i.e., changing the belief set to incorporate φ consistently (if possible) — this is a revision with φ — results in a subset of the set of beliefs that results from straightforward inserting φ in the belief set — an expansion with φ. The third clause formalizes the idea that expansion is a special kind of revision: in cases where $\neg\varphi$ is not believed, expanding with φ and revising with φ amount to the same action. The left-to-right implication of the fourth clause states that if $\neg\varphi$ is known, i.e., $\neg\varphi$ is among the agent's knowledge, then the revision with φ results in the agent believing $\bot$, i.e., the revision results in the absurd belief set. The right-to-left implication of the fourth clause states that the agent will believe $\bot$ only after it performs a revision with a negation of something that it knows. The fifth clause states that revisions with formulae that are known to be equivalent have identical results. The sixth clause formalizes the idea that the revision with the conjunction $\varphi \wedge \psi$ results in the agent believing at most the formulae that it would believe after a revision with φ followed by an expansion with ψ. The seventh clause states that if a revision with φ does not result in $\neg\psi$ being believed, then after revising with $\varphi \wedge \psi$ the agent believes at least the formulae that it would believe as the result of performing a revision with φ followed by an expansion with ψ. As Gärdenfors remarks, clauses 6 and 7 provide for some sort of *minimal change* condition on revisions.

Let us finally explain how our approach can be related to the AGM postulates for expansion, contraction and revision. In order to do so, let the belief set of agent i at state $\mathcal{M}, s$ be defined as $\mathcal{B}(i, \mathcal{M}, s) = \{\varphi \in \mathcal{L}_0 \mid \mathcal{M}, s \models \mathbf{B}_i\varphi\}$. Moreover, the *contraction* of $\mathcal{B}(i, \mathcal{M}, s)$ with $\varphi \in \mathcal{L}_0$, notation $\mathcal{B}^-_\varphi(i, \mathcal{M}, s)$ is defined as: $\mathcal{B}^-_\varphi(i, \mathcal{M}, s) = \{\psi \in \mathcal{L}_0 \mid \mathcal{M}, s \models [\mathrm{do}_i(\mathtt{contract}\,\varphi)]\mathbf{B}_i\psi\}$ Similar sets

$\mathcal{B}^{+}_{\varphi}(i, \mathcal{M}, s)$ and $\mathcal{B}^{*}_{\varphi}(i, \mathcal{M}, s)$ can be defined as the expansion and the revision at s with φ, respectively. Then, in [25] we show that these sets exactly satisfy the AGM postulates for AGM contraction, expansion and revision, respectively. There is one difference: whereas the AGM postulates assume classical logic as a backup system, in our setup this is provided by the agent's knowledge. For instance, whereas the AGM Success-postulate for contraction states that a formula φ that is not a classical tautology does not follow from a theory T contracted with φ, our variant of that postulate reads

$$\text{If } \mathcal{M}, s \models \neg \mathbf{B}^{k}_{i}\varphi \text{ then } \varphi \notin \mathcal{B}^{-}_{\varphi}(i, \mathcal{M}, s)$$

The ability to change one's mind

In the previous (sub)sections, we dealt with the formalization of the opportunity for, and the result of, the actions that model the belief changes of agents. Here we look at the *ability* of agents to change their beliefs.

For 'mental' actions, like testing (observing) and communicating, the abilities of agents are closely related to their (lack of) information. This observation seems to hold *a fortiori* for the abstract actions that cause agents to change their beliefs. For when testing and communicating, at least some interaction takes place, either with the real world in case of testing, or with other agents when communicating, whereas the changing of beliefs is a strictly mental, agent-internal, activity. Therefore, it seems natural to let the ability of an agent to change its beliefs be determined by its informational state only.

The intuitive idea behind the definitions as we present them, is that the ability to change beliefs can be used to guide the changes that the beliefs of an agent are subjected to. In particular, if an agent is able to change its beliefs in a certain way, then this change of belief should work out as desired, i.e., it should neither result in an absurd belief set nor cause no change at all. Another point of attention is given by the observation that the Levi identity should also be respected for abilities, i.e., an agent is capable of revising its beliefs with a formula φ if and only if it is able to contract its beliefs with $\neg\varphi$ and thereafter perform an expansion with φ.

Definition 12 Let $\mathcal{M}$ be some Kripke model with state s, and let $\varphi \in \mathcal{L}_0$ be arbitrary. We define the capability function c for the `expand`, `contract` and `revise` actions in the following manner:

$$\begin{array}{lll} \texttt{c}(i, \texttt{expand}\,\varphi)(\mathcal{M}, s) = \mathbf{1} & \Leftrightarrow \mathcal{M}, s \models \neg \mathbf{B}_i \neg\varphi \\ \texttt{c}(i, \texttt{contract}\,\varphi)(\mathcal{M}, s) = \mathbf{1} & \Leftrightarrow \mathcal{M}, s \models \neg \mathbf{K}_i \varphi \\ \texttt{c}(i, \texttt{revise}\,\varphi)(\mathcal{M}, s) & = \texttt{c}(i, \texttt{contract}\,\neg\varphi; \texttt{expand}\,\varphi)(\mathcal{M}, s) \end{array}$$

The first clause of Definition 12 states that an agent is able to expand its set of beliefs with a formula if and only if it does not already believe the negation of the formula. The second clause formalizes the idea that an agent is able to remove some formula from its set of beliefs if and only if it does not consider the formula to be one of its principles. The ability for the `revise` action is defined through the Levi identity.

Proposition 13 *For all* $\varphi \in \mathcal{L}_0$ *we have:*

- $\models \mathbf{A}_i\mathtt{expand}\,\varphi \leftrightarrow \mathbf{K}_i\mathbf{A}_i\mathtt{expand}\,\varphi$
- $\models \mathbf{A}_i\mathtt{expand}\,\varphi \rightarrow \langle \mathrm{do}_i(\mathtt{expand}\,\varphi)\rangle\neg\mathbf{B}_i\bot$
- $\models \mathbf{A}_i\mathtt{contract}\,\varphi \leftrightarrow \mathbf{K}_i\mathbf{A}_i\mathtt{contract}\,\varphi$
- $\models \mathbf{A}_i\mathtt{contract}\,\varphi \rightarrow \langle \mathrm{do}_i(\mathtt{contract}\,\varphi)\rangle\neg\mathbf{B}_i\varphi$
- $\models \mathbf{A}_i\mathtt{revise}\,\varphi \leftrightarrow \mathbf{A}_i\mathtt{contract}\,\neg\varphi$
- $\models \mathbf{A}_i\mathtt{revise}\,\varphi \rightarrow \langle \mathrm{do}_i(\mathtt{revise}\,\varphi)\rangle(\mathbf{B}_i\varphi \wedge \neg\mathbf{B}_i\bot)$

The first and third clause of Proposition 13 state that agents know of their ability to expand and contract their beliefs; a consequence of the fifth clause is that agents also know of their ability to revise their beliefs. The second, fourth and sixth clause formalize the idea that belief changes of which the agent is capable, behave as desired, i.e., an expansion does not result in absurd belief sets, a contraction leads to disbelief in the contracted formula, and a revision results in a combination of these.

3.3 Belief Changes: When?

Observations: seeing is believing

Through observations an agent *learns whether* some proposition is true of the state in which it is residing. For artificial agents it seems to be a reasonable assumption to demand that observations are *truthful*. That is, if some observation yields information that φ, then it should indeed be the case that φ. Observations form the most trustworthy way of acquiring information: utterances like 'I've seen it with my own eyes' or 'Seeing is believing' support this claim. The formalisation that we propose is therefore such that observations overrule any beliefs acquired by other means. In situations where an agent does not observationally believe whether φ, its beliefs are revised by applying the $\mathtt{revise}^o$ function to the model and the state under consideration. If the agent already observationally believed whether φ no revision takes place. Note that observations will never conflict with an agent's observational beliefs or its knowledge: since both these notions are veridical, and observations are truthful, it is not possible that an observation that some formula φ holds contradicts observational belief or knowledge that $\neg\varphi$ holds, since this would force both φ and $\neg\varphi$ to be true in one and the same state.

Let, for $x = k, o, c, d$, $\mathbf{Bwh}_i^x\varphi \triangleq \neg\mathbf{Ignorant}_i^x\varphi$.

Definition 14 *For all* $\mathcal{M} \in \mathbf{M}^{\mathrm{I}}$ *with state* s *and all* $\varphi \in \mathcal{L}_0$ *we define:*

$$\mathtt{r}(i, \mathrm{obs}\,\varphi)(\mathcal{M}, s) = \begin{cases} \mathcal{M}, s & \text{if } \mathcal{M}, s \models \mathbf{Bwh}_i^o\varphi \\ \mathtt{revise}^o(i,\varphi)(\mathcal{M}, s), s & \text{otherwise} \end{cases}$$

Now we can define $\texttt{revise}^o$ in the same fashion as the semantic revision procedures of Section 3.2, but we have to do some additional work in order to keep the relations between $\mathbf{B}_i^o$, $\mathbf{B}_i^c$ and $\mathbf{B}_i^d$ in tact. To do so, let us agree on the following notation. Let $\mathcal{M}$ be given. For any formula φ and B^x, let

$$\kappa([\![\varphi]\!]) \cdot \mathrm{B}^x(i,w) = \begin{cases} \mathrm{B}^x(i,w) \cap [\![\varphi]\!] & \text{if } \mathcal{M}, w \models \varphi \\ \mathrm{B}^x(i,w) \cap [\![\neg\varphi]\!] & \text{else} \end{cases}$$

Definition 15 *Let* $\mathcal{M} = \langle W, \pi, \mathtt{D}, \mathtt{I}, \mathtt{M}\rangle$ *be given,* $\mathtt{I} = \{\mathrm{B}^k, \mathrm{B}^o, \mathrm{B}^c, \mathrm{B}^d\}$. *Then* $\texttt{revise}^o(i,\varphi)(\mathcal{M}, s) = \langle W, \pi, \mathtt{D}, \{\mathrm{B}^k, \mathrm{B}^{o\prime}, \mathrm{B}^{c\prime}, \mathrm{B}^{d\prime}\}, \mathtt{M}, \rangle, s$ *where* ($y \in \{c, d\}$, $x \in \{o, c, d\}$):

$$\begin{array}{ll} \mathrm{B}^{x\prime}(j,t) = \mathrm{B}^x(j,t) & \textit{if } j \neq i \textit{ or } t \notin \mathrm{B}^x(i,s) \\ \mathrm{B}^{o\prime}(i,t) = \kappa([\![\varphi]\!]) \cdot \mathrm{B}^o(i,t) & \textit{if } t \in \mathrm{B}^o(i,s) \\ \mathrm{B}^{y\prime}(i,t) = \mathrm{B}^y(i,t) \cap \mathrm{B}^{o\prime}(i,t) & \textit{if } \mathrm{B}^y(i,t) \cap \mathrm{B}^{o\prime}(i,t) \neq \emptyset,\ t \in \mathrm{B}^o(i,s) \\ \mathrm{B}^{c\prime}(i,t) = \mathrm{B}^{o\prime}(i,t) & \textit{if } \mathrm{B}^c(i,t) \cap \mathrm{B}^{o\prime}(i,t) = \emptyset,\ t \in \mathrm{B}^o(i,s) \\ \mathrm{B}^{d\prime}(i,t) = \mathrm{B}^{c\prime}(i,t) & \textit{if } \mathrm{B}^d(i,t) \cap \mathrm{B}^{o\prime}(i,t) = \emptyset,\ t \in \mathrm{B}^o(i,s) \end{array}$$

Intuitively, Definition 15 expresses the following strategy, if agent i tries to revise its observational beliefs, when it is in state s of the model $\mathcal{M}$. First of all, there is no effect whatsoever for other agents (see item 3 of the next Proposition), or for states that are not observational equivalent to s, according to i. The next clause of the definition ensures that an agent has a correct observational belief after revising: if it revises with φ, and φ is true in s, then it only considers as new observational alternatives those in which φ holds; if $\neg\varphi$ is true in s, however, it will only consider the $\neg\varphi$-alternatives, afterwards. The last three clauses of the definition deal with communicational and default alternatives. They stay within the observational cluster, so to speak, but if there are no communicational alternatives left in the new observational cluster, the agent performs a kind of reset: it then equalizes the communicational alternatives with the new observational ones. A similar property holds for the new default-compatible alternatives. The intuitive acceptability of this definition can be seen from the following proposition.

Proposition 16 *For all* $i, j \in \mathcal{A}$, $\varphi, \psi \in \mathcal{L}_0$ *and* $\chi \in \mathcal{L}$ *we have:*

1. $\mathtt{obs}\,\varphi$ *is deterministic, idempotent and realisable*
2. $\models [\mathrm{do}_i(\mathtt{obs}\,\varphi)]\mathbf{Bwh}_i^o\varphi \wedge ((\varphi \rightarrow [\mathrm{do}_i(\mathtt{obs}\,\varphi)]) \wedge (\neg\varphi \rightarrow [\mathrm{do}_i(\mathtt{obs}\,\neg\varphi)]$
3. $\models \mathbf{B}_j^k\psi \leftrightarrow \langle \mathrm{do}_i(\mathtt{obs}\,\varphi)\rangle \mathbf{B}_j^k\psi$
4. $\models \mathbf{B}_j^o\psi \rightarrow \langle \mathrm{do}_i(\mathtt{obs}\,\varphi)\rangle \mathbf{B}_j^o\psi$
5. $\models (\varphi \wedge \langle \mathrm{do}_i(\mathtt{obs}\,\varphi)\rangle \mathbf{B}_i^o\psi) \leftrightarrow (\varphi \wedge \mathbf{B}_i^o(\varphi \rightarrow \psi))$
6. $\models (\neg\varphi \wedge \langle \mathrm{do}_i(\mathtt{obs}\,\varphi)\rangle \mathbf{B}_i^o\psi) \leftrightarrow (\neg\varphi \wedge \mathbf{B}_i^o(\neg\varphi \rightarrow \psi))$
7. $\models \langle \mathrm{do}_i(\mathtt{obs}\,\varphi)\rangle\chi \leftrightarrow \langle \mathrm{do}_i(\mathtt{obs}\,\neg\varphi)\rangle\chi$
8. $\models \varphi \wedge \mathbf{Ignorant}_i^k\varphi \rightarrow \langle \mathrm{do}_i(\mathtt{obs}\,\varphi)\rangle \mathbf{Saw}_i\varphi$
9. $\models \neg\varphi \wedge \mathbf{Ignorant}_i^k\varphi \rightarrow \langle \mathrm{do}_i(\mathtt{obs}\,\varphi)\rangle \mathbf{Saw}_i\neg\varphi$
10. $\models \varphi \wedge (\mathbf{Heard}_i\neg\varphi \vee \mathbf{Jumped}_i\neg\varphi) \rightarrow \langle \mathrm{do}_i(\mathtt{obs}\,\varphi)\rangle \mathbf{Saw}_i\varphi$
11. $\models \varphi \wedge \mathbf{B}_i^c\neg\varphi \rightarrow \langle \mathrm{do}_i(\mathtt{obs}\,\varphi)\rangle((\mathbf{B}_i^c\chi \leftrightarrow \mathbf{B}_i^o\chi) \wedge (\mathbf{B}_i^d\chi \leftrightarrow \mathbf{B}_i^o\chi))$

The properties given in the first item are not uncommon for informative actions: determinism and idempotence are strongly related to the AGM postulates, and are also encountered for the other informative actions. Realisability is typical for observations; it models the idea that our agents are perfect observers which always have the opportunity to make an observation. The second item of Proposition 16 formalises two essential properties of observations, viz. their informativeness and truthfulness. Item 3 states that the knowledge fluents — the propositional formulae known to be true — of all agents remain unaffected under execution of an `observe` action by one of them, and item 4 states an analogous, but slightly weaker property, for their observational beliefs. The fifth and sixth item express the fact that an agent adapts only new beliefs ψ after doing an observation, if it already believed on beforehand that this observation would entail this new belief ψ. In item 7 it is formalised that the $\texttt{obs}\,\varphi$ action models 'observing whether φ': observing whether φ is in all aspects equivalent to observing whether $\neg\varphi$. Items 8 and 9 state that for knowledge-ignorant agents observations actually lead to *learning by seeing*. Item 10 — a special case of item 8 — is intuitively a very nice one: it states that observations are the most credible source of information. Observations overrule other beliefs acquired through communication or adopted by default, i.e. incorrect communicational or default beliefs are *revised* in favour of observational beliefs. The last item of Proposition 16 sheds some more light on the (rigorous) way in which beliefs are revised: observing something that contradicts communicational beliefs leads to a *reset* of both the latter and the default beliefs of the agent, i.e. after such a revision all the beliefs of the agent are at least grounded in its observations. Phrased differently, there no longer is any formula that the agent believes due to it being told or assuming it by default, i.e. after such a revision there is no formula ψ for which either $\mathbf{Heard}_i\psi$ or $\mathbf{Jumped}_i\psi$ holds.

Communication: hearing is believing

The second source of information available to an agent consists of the information communicated by other agents. As we present it here, communication is reduced to its barest form, viz. the transfer of information. That is, we are not dealing with concepts like communication protocols, synchronisation and the like, but instead consider communication as consisting of an agent transferring some of its information to another agent. In our implementation of revising by communication, we made some choices that are in some sense rather arbitrary. On the other hand, we believe our framework demonstrates to be both expressive in that it makes those choices explicit, and flexible in that it can easily be adapted to other choices. In general, agents have the opportunity to send all of their beliefs, and nothing else. However, the sending agent should also provide his information with his source, in order to let the receiving agent rationally decide on how to deal with the new incoming information. For these sources, we take the tags $\mathbf{k}, \mathbf{s}, \mathbf{h}$ and $\mathbf{j}$, denoting that the sending agent's source for the message he is sending is that he knows it, observed it, heard it or concluded it by using some default, respectively. This assumes that each agent knows, or has some awareness about how he comes to his beliefs, for instance whether $\neg\mathbf{B}_i^c \to \mathbf{B}_i^k \neg\mathbf{B}_i^c\varphi$ hold. Such a systematic investigation is beyond the scope of this chapter, but one may read [12]. Thus, we introduce, for each $\varphi \in \mathcal{L}_0$ and $\mathbf{t} \in \{\mathbf{k}, \mathbf{s}, \mathbf{h}, \mathbf{j}\}$, an action $\mathtt{inform}(\varphi, i, \mathbf{t})$, expressing that agent i is informed about φ, and that the status of this fact at the sender's perspective is $\mathbf{t}$.

Depending on the credibility of both the sending agent and the information that it sends, the receiving agent may use this information to revise its beliefs. For reasons of simplicity, we define the credibility of the sending agent as a binary notion, i.e. the agent is either credible or it is not credible, without distinguishing degrees of credibility. The notion of credibility is modelled through the so-called dependence operator, originally proposed by Huang [17]. This ternary operator, pertaining to a pair of agents and a formula, models that there is a relation of trust, dependence or credibility between the agents with respect to the formula: $\mathbf{D}_{i,j}\varphi$ indicates that agent j accepts agent j as an authority on φ, or that j is a teacher of i on the subject φ. The $\mathbf{D}_{i,j}$ operator is interpreted by incorporating a function $\mathrm{D} : \mathcal{A} \times \mathcal{A} \to W \to \wp(\mathcal{L})$ in the $\mathtt{I}$-part of our models: $\mathbf{D}_{i,j}\varphi$ is true in a state s of some model iff φ is in $\mathrm{D}(i,j)(s)$. The credibility of the transferred information is determined by both the credibility that the sending agent attaches to this information and the credibility that the receiving agent attaches to any of i beliefs that contradict this information. That is, if the sending agent itself observationally believes the information that it is sending, then this information overrules any contradicting beliefs that the receiving agent may have. If the sending agent itself was told the information that it is now transferring, the receiving agent will accept this information only if it does not have any contradicting information that it believes at least with the credibility attached to communicational beliefs. Information that the sending agent adopted by default is considered to be too weak to ever justify a revision of the receiving agent's beliefs. These intuitive ideas are formalised in Definition 17. The definition of $\mathtt{revise}^c$ below is in the same spirit of Definition 15 for $\mathtt{revise}^o$, we omit it here for reasons of space (see [22] for details).

Definition 17 *For all* $\mathcal{M} \in \mathbf{M}^{\mathrm{I}}$ *with state* s, $i, j \in \mathcal{A}$, $\mathbf{t} \in \{\mathbf{k}, \mathbf{s}, \mathbf{h}, \mathbf{j}\}$, $\psi \in \mathcal{L}$ *and* $\varphi \in \mathcal{L}_0$ *we define:*

$$\mathcal{M}, s \models \mathbf{D}_{i,j}\psi \Leftrightarrow \psi \in \mathrm{D}(i,j)(s)$$

$$\mathtt{r}(j, \mathtt{inform}(\varphi, i, \mathbf{t}))(\mathcal{M}, s) = \begin{cases} \emptyset & \text{if } \mathbf{t} = \mathbf{k} \text{ and } \mathcal{M}, s \models \neg\mathbf{B}_j^k\varphi, \text{ or} \\ & \text{if } \mathbf{t} = \mathbf{s} \text{ and } \mathcal{M}, s \models \neg\mathbf{Saw}_j\varphi, \text{ or} \\ & \text{if } \mathbf{t} = \mathbf{h} \text{ and } \mathcal{M}, s \models \neg\mathbf{Heard}_j\varphi, \text{ or} \\ & \text{if } \mathbf{t} = \mathbf{j} \text{ and } \mathcal{M}, s \models \neg\mathbf{Jumped}_j\varphi \\ \mathtt{revise}^c(i,\varphi)(\mathcal{M}, s), s & \text{if } \mathbf{t} = \mathbf{j} \text{ and } \mathcal{M}, s \models \mathbf{D}_{i,j}\varphi \wedge \mathbf{Ignorant}_i^c\varphi) \\ & \text{if } \mathbf{t} \in \{\mathbf{k}, \mathbf{s}\} \text{ and } \mathcal{M}, s \models \mathbf{D}_{i,j}\varphi \wedge \mathbf{Ignorant}_i^o\varphi \\ \mathcal{M}, s & \text{otherwise} \end{cases}$$

In the following proposition we summarise some validities describing the behaviour of the communication actions. These validities show that our formalisation indeed corresponds to the intuitive ideas unfolded above.

Proposition 18 *For all* $i, i', j \in \mathcal{A}$, $\varphi, \psi \in \mathcal{L}_0$ *and* $\chi \in \mathcal{L}$, $\mathbf{t} \in \{\mathbf{k}, \mathbf{s}, \mathbf{h}, \mathbf{j}\}$, $\mathbf{t}' \in \{\mathbf{k}, \mathbf{s}, \mathbf{h}\}$ *we have:*

1. $\mathtt{inform}(\varphi, i, \mathbf{t})$ *is deterministic and idempotent*
2. $\models \mathbf{B}^x_{i'}\psi \rightarrow [\mathrm{do}_j(\mathtt{inform}(\varphi, i, \mathbf{t}))]\mathbf{B}^x_{i'}\psi$ *for* $x \in \{k, o\}$
3. $\models \mathbf{Jumped}_j\varphi \leftrightarrow \langle \mathrm{do}_j(\mathtt{inform}(\varphi, i, \mathbf{j}))\rangle\top$
4. $\models \mathbf{B}^d_j\varphi \wedge \neg\mathbf{D}_{i,j}\varphi \rightarrow (\langle \mathrm{do}_j(\mathtt{inform}(\varphi, i, \mathbf{t}))\rangle\chi \leftrightarrow \chi)$
5. $\models \mathbf{D}_{i,j}\varphi \wedge \mathbf{B}^c_j\varphi \rightarrow (\langle \mathrm{do}_j(\mathtt{inform}(\varphi, i, \mathbf{k}))\rangle\mathbf{Bwh}^c_i\varphi \vee$ $\langle \mathrm{do}_j(\mathtt{inform}(\varphi, i, \mathbf{s}))\rangle\mathbf{Bwh}^c_i\varphi \vee \langle \mathrm{do}_j(\mathtt{inform}(\varphi, i, \mathbf{h}))\rangle\mathbf{Bwh}^c_i\varphi)$
6. $\models \mathbf{D}_{i,j}\varphi \wedge \mathbf{Heard}_j\varphi \wedge \mathbf{Ignorant}^c_i\varphi \rightarrow \langle \mathrm{do}_j(\mathtt{inform}(\varphi, i, \mathbf{t}'))\rangle\mathbf{Heard}_i\varphi$
7. $\models \mathbf{D}_{i,j}\varphi \wedge \mathbf{Heard}_j\varphi \wedge \mathbf{Bwh}^c_i\varphi \rightarrow (\langle \mathrm{do}_j(\mathtt{inform}(\varphi, i, \mathbf{h}))\rangle\chi \leftrightarrow \chi)$
8. $\models \mathbf{D}_{i,j}\varphi \wedge \mathbf{Heard}_j\varphi \wedge \mathbf{Ignorant}^c_i\varphi \rightarrow$ $(\langle \mathrm{do}_j(\mathtt{inform}(\varphi, i, \mathbf{t}'))\rangle\mathbf{B}^c_i\psi \leftrightarrow \mathbf{B}^c_i(\varphi \rightarrow \psi))$
9. $\models \mathbf{D}_{i,j}\varphi \wedge \mathbf{Saw}_j\varphi \wedge \mathbf{Ignorant}^o_i\varphi \wedge \mathbf{B}^c_i\neg\varphi$ $\rightarrow (\langle \mathrm{do}_j(\mathtt{inform}(\varphi, i, \mathbf{s}))\rangle\mathbf{B}^c_i\psi \leftrightarrow \mathbf{B}^o_i(\varphi \rightarrow \psi))$
10. $\models \mathbf{D}_{i,j}\varphi \wedge \mathbf{Jumped}_j\varphi \rightarrow (\langle \mathrm{do}_j(\mathtt{inform}(\varphi, i, \mathbf{j}))\rangle\chi \leftrightarrow \chi)$
11. $\models \mathbf{B}^x_i\varphi \rightarrow (\langle \mathrm{do}_i(\mathtt{inform}\varphi, i, \mathbf{t}))\rangle\chi \rightarrow \chi)$ *for* $x \in \{k, o, c, d\}$

The first item of Proposition 18 states that the inform action also obeys the properties of determinism and idempotence that are intuitively related to the AGM postulates for belief revision. The second item states that both the knowledge and the observational belief fluents of all agents persist under execution of an inform action by one of them. Thus communication does at most affect the regions of beliefs that it should affect, viz. the communicational and default belief clusters. Note that, in contrast with the corresponding items of Propositions 16 and 20, the communicational beliefs of the receiving agent do not necessarily persist. The reason for this is given in item 9, where communicational beliefs are genuinely revised (and not just expanded). Item 3 states that agents may transfer, using the tag **j**, all, and nothing but, the beliefs they jumped to. Attempts to use a wrong tag are doomed to fail. Agents are therefore not even allowed to tell white lies; they are utterly honest. Note that item 3 also shows that the inform action is — in contrast with the observe action — generally not realisable. Corresponding properties hold for the other tags. Item 4 formalises that authority is a *conditio sine qua non* for effectual communication: if the receiving agent does not trust the sending agent on the transferred information, it lets the information pass without revising its beliefs (or changing anything else for that matter). Item 5 states that if some trustworthy agent j tells another agent i some formula φ that j either knows, observed or was told, this leads to a state of affairs in which the receiving agent believes whether φ at least with the credibility attached to communicational beliefs; whenever i is beforehand ignorant with regard to φ on the level of communicational beliefs, the receiving agent actually *learns* φ by being told (item 6). Item

7 states that if agent j tells i some formula that j itself was told while i already communicationally believes whether φ, nothing changes, really. Items 8 and 9 deal with the ways in which the receiving agent's beliefs are revised as the result of it acquiring information through communication. If the sending agent j heard the transferred formula φ and the receiving agent i is communicationally ignorant with respect to φ, then an expansion with φ of the communicational beliefs of i takes place (item 8). If j saw φ and i is observationally ignorant with respect to φ while communicationally believing $\neg\varphi$, then i's communicational beliefs consist henceforth of the observational beliefs that are implied by φ (item 9). Item 10 states that default beliefs are not transferable: the credibility of this kind of information is too low to have any effect upon being heard. The last item in particular shows that agents cannot increase the credibility they attach to information by talking to themselves, since this talking to themselves does not change anything. Thus our agents are not susceptible to this kind of auto-suggestion.

Default jumps: jumping is believing

The last possible source of information that we consider in this chapter is the only endogenous one, and consists of the possibility to adopt beliefs by default. In general in default reasoning, plausible yet fallible conclusions are derived on the basis of the presence of certain information and the absence of other information. In the formalisation of default reasoning as proposed by Reiter [33], defaults are formalised as special inference rules $\varphi_1 : \varphi_2/\varphi_3$, which should be interpreted as stating that if φ_1 holds and it is consistent to assume φ_2 then φ_3 may be derived. Here we consider the most basic form of default reasoning, in which no information is required to be present and the information that needs to be absent is strongly related to the conclusion that is to be derived. In terms of Reiter's framework, the kind of default reasoning that we consider here uses only *supernormal* defaults, i.e. defaults of the form $: \varphi/\varphi$; these defaults can be seen as *possible hypotheses* in Poole's system [31]. Here we shall introduce these supernormal defaults as syntactical constructs that are at the agent's disposal. Formally, we include in the $\mathtt{I}$-part of our models a function $\mathrm{N} : \mathcal{A} \to W \to \wp(\mathcal{L})$, such that $\mathrm{N}(i, s)$ yields the set of defaults available to agent i at state s. In the language we include an operator $\mathbf{N}_i$, such that $\mathbf{N}_i\varphi$ expresses that φ is a default of i; $\mathbf{N}_i\varphi$ is true in a state s of a model iff $\varphi \in \mathrm{N}(i, s)$.

Using the $\mathbf{N}_i$-operator to represent defaults, we now come to the formalisation of the attempted jumps to conclusion that constitute (supernormal) default reasoning. Since adopting a belief by default accounts for acquiring information of the lowest credibility, it is obvious that default jumps are effective only for agents that are completely ignorant with respect to the default that is jumped to.

Definition 19 *For all $\mathcal{M} \in \mathbf{M}^{\mathrm{I}}$ with state s, $i \in \mathcal{A}$, and $\varphi \in \mathcal{L}_0$ we define:*

$\mathtt{r}(i, \mathtt{try\_jump}\,\varphi)(\mathcal{M}, s) =$

$$\begin{cases} \emptyset & \textit{if } \mathcal{M}, s \models \neg \mathbf{N}_i\varphi \\ \texttt{revise}^d(i,\varphi)(\mathcal{M},s), s & \textit{if } \mathcal{M}, s \models \mathbf{N}_i\varphi \wedge \mathbf{Ignorant}_i^d\varphi \\ \mathcal{M}, s & \textit{otherwise} \end{cases}$$

Proposition 20 *For all $i, j \in \mathcal{A}$, $\varphi, \psi \in \mathcal{L}_0$ and $\chi \in \mathcal{L}$, we have:*

1. $\texttt{try\_jump}\,\varphi$ *is deterministic, idempotent and d-informative with regard to* φ
2. $\models \mathbf{B}_j^x\psi \to [\mathrm{do}_i(\texttt{try\_jump}\,\varphi)]\mathbf{B}_j^x\psi$ *for* $x \in \{d, c, o, k\}$
3. $\models \mathbf{N}_i\varphi \leftrightarrow \langle \mathrm{do}_i(\texttt{try\_jump}\,\varphi)\rangle\top$
4. $\models \langle \mathrm{do}_i(\texttt{try\_jump}\,\varphi)\rangle\top \leftrightarrow \langle \mathrm{do}_i(\texttt{try\_jump}\,\varphi)\rangle\mathbf{Bwh}_i^d\varphi$
5. $\models \mathbf{N}_i\varphi \wedge \mathbf{Ignorant}_i^d\varphi \to \langle \mathrm{do}_i(\texttt{try\_jump}\,\varphi)\rangle\mathbf{Jumped}_i\varphi$
6. $\models \mathbf{N}_i\varphi \wedge \mathbf{Ignorant}_i^d\varphi \to (\langle \mathrm{do}_i(\texttt{try\_jump}\,\varphi)\rangle\mathbf{B}_i^d\psi \leftrightarrow \mathbf{B}_i^d(\varphi \to \psi))$
7. $\models \mathbf{N}_i\varphi \wedge \mathbf{Bwh}_i^d\varphi \to (\langle \mathrm{do}_i(\texttt{try\_jump}\,\varphi)\rangle\chi \leftrightarrow \chi)$

The first item of Proposition 20 deals again with the properties more or less typical for informative actions. It is obvious that the property of realisability is not validated in general since an agent may attempt to jump to a non-default, and in 3 it is formalised that such attempted jumps are doomed to fail. Item 2 states that all information of all agents persists under the attempted jump to a formula by one of them. Item 4 states that default jumps for which an agent has the opportunity always result in the agent believing by default whether the formula that is jumped to holds. The fifth item formalises the idea that agents that are completely ignorant with respect to some formula φ jump to new default beliefs by applying the $\texttt{try\_jump}\,\varphi$ action. The incorporation of these new beliefs is brought about by expanding the default beliefs of the agent with the default that is adopted (item 6). The last item states that attempted jumps to default conclusions yield information for totally ignorant agents only.

The ability to gather information

As we see it, in the ability of intelligent information agents to execute informative actions two different notions are combined. On the one hand, the abilities of an agent restrict its practical possibility to acquire information: only the actions that are within the agent's capacities can be used as means to extend its information. This corresponds to the idea that agents are not just able to acquire all the information they would like to acquire. On the other hand, we use the agents' abilities to steer the way in which information is acquired. That is, through its abilities an agent is forced to prefer more credible sources of information. We will, for instance, define an agent to be able to try to adopt some formula by default only if it cannot acquire this information through its — more credible — observations.

Since, by nature, observations provide the most credible source of information, the ability to observe is determined strictly by the fact that the agents' information gathering is limited, and does not depend on other means to acquire information. In our opinion it is reasonable to consider this limit to the agent's ability to perform observations as given by the construction of the agent. For instance, human agents are built in such a way that they cannot observe objects at great distances, or objects that are outside the spectrum of human observation, like for instance X-rays. In the case of artificial agents, one could think of two robots, working together to explore a strange, new world. Each robot is equipped with its own, personal set of sensors: one robot is for instance able to observe whether its environment is radioactive, the other whether the planet's atmosphere contains oxygen, but neither robot is able to observe both. Being a construction decision, we assume that the observational capacities of agents are determined beforehand.

In defining the capabilities of agents to inform other agents, we consider the *moral* component of ability to be most relevant. That is, we demand our agents to be sincere in that they are morally unable to lie or gossip. The things that an agent is capable of telling to other agents are exactly the things that it itself believes, be it with the lowest credibility. In this way the information acquisition of an agent that is due to communication is restricted to those formulae that are believed by some authority.

The ability to attempt to jump to a default captures both aspects described above, i.e. both the aspect of restricted information acquisition as well as that of preferring the most credible source of information are visible in the definition of the ability to (attempt to) jump. Concerning the first aspect, an agent is able to jump to a default only if it *knows* it to be a default, i.e. agents have to know their defaults in order to be able to use them. With respect to the second aspect, the capability to jump is defined to depend on the observational capacities of the agent: an agent is able to attempt a jump to a formula only if it knows that it is not able to observe whether the formula holds. In this way it is ensured that agents resort to default jumps only if the possibility of acquiring the information through observations is excluded.

Definition 21 *Let $\mathcal{M}$ be some model. The function* c *is for the informative actions defined as follows, where s is some state in $\mathcal{M}$, i, j are agents and $\varphi \in \mathcal{L}_0$ is some propositional formula.*

$$\begin{aligned} \mathtt{c}(j, \mathtt{inform}(\varphi, i, \mathbf{t}))(\mathcal{M}, s) &= \mathtt{r}(j, \mathtt{inform}(\varphi, i, \mathbf{t}))(\mathcal{M}, s) \neq \emptyset \\ \mathtt{c}(i, \mathtt{try\_jump}\,\varphi)(\mathcal{M}, s) &= \mathtt{c}(i, \mathtt{confirm}\,\mathbf{K}_i(\neg\mathbf{A}_i\mathtt{obs}\,\varphi \wedge \mathbf{N}_i\varphi))(\mathcal{M}, s) \end{aligned}$$

where $\lambda\varphi \in \mathcal{L}_0.(\mathtt{c}(i, \mathtt{obs}\,\varphi)(\mathcal{M}, s))$ is a function in φ such that

$$\mathtt{c}(i, \mathtt{obs}\,\varphi)(\mathcal{M}, s) = \mathtt{c}(i, \mathtt{obs}\,\neg\varphi)(\mathcal{M}, s)$$

Proposition 22 *For all $i, j \in \mathcal{A}$ and $\varphi \in \mathcal{L}_0$ and any tag* $\mathbf{t}$ *we have:*

1. $\models \mathbf{A}_i\mathtt{obs}\,\varphi \leftrightarrow \mathbf{A}_i\mathtt{obs}\,\neg\varphi$
2. $\models \mathbf{A}_j\mathtt{inform}(\varphi, i, \mathbf{t}) \leftrightarrow \langle \mathrm{do}_j(\mathtt{inform}(\varphi, i, \mathbf{t}))\rangle\top$
3. $\models \mathbf{A}_i\mathtt{try\_jump}\,\varphi \rightarrow \langle \mathrm{do}_i(\mathtt{try\_jump}\,\varphi)\rangle\top$
4. $\models \mathbf{A}_i\mathtt{obs}\,\varphi \rightarrow \neg\mathbf{A}_i\mathtt{try\_jump}\,\varphi$

The first item of Proposition 22 again expresses that the observe actions formalise 'observing whether'. Item 2 and 3 state that both the actions modelling communication and those modelling default reasoning, though not realisable *per se*, are *A-realisable*, i.e. having the ability to perform these actions implies having the opportunity to do so. The property of A-realisability is considered unacceptable for mundane actions, but given our view on the notion of ability for the non-mundane, informative actions, this property seems much less controversial. For both the ability and the opportunity to perform informative actions are defined in terms of informational attitudes, and thus the distinction between these notions is less clear than it is for mundane actions. The last item of Proposition 22 expresses that agents are able to attempt to jump to a formula only if it is not within their capacities to observe whether the formula holds.

3.4 Possible extensions

One can think of several obvious yet interesting extensions of the framework presented in this Section, most of which are in fact not too hard to implement. A possible extension was suggested by Castelfranchi [3], and consists of refining the communication part of information acquisition. In this more refined form of communication it should be possible to relate the credibility that an agent attaches to some formula that it has been told, to the number of agents that have been communicating this formula, i.e. the more agents that tell that some formula is true, the more fiercely it is believed. It is possible to formalise this intuitive idea by splitting the communicational belief cluster of an agent into different parts, one for each of the other agents. Whenever an agent j informs an agent i of the truth of φ, i's communicational belief cluster that is associated with j is revised with φ. Over the different communicational belief clusters a kind of *graded* belief modality could be defined (in the spirit of [16]), which formalises the credibility attached to the agent's communicational beliefs. The formula $\mathbf{B}_i^{c,0.5}\varphi$ would then be taken to represent that agent i communicationally believes φ with credibility 0.5, i.e. φ holds in at least half of the communicational belief clusters of i. It is clear that in this way it can indeed be modelled that the credibility attached by i to one of its communicational beliefs depends on the number of agents that have informed i of the truth of this formula. Another possibility to refine communication could be to allow a form of 'demand-driven' communication in addition to the 'supply-driven' one considered here. In this extended form of communication an agent may request information on the truth or falsity of certain proposition from another, trusted, agent. One possible way of formalising this kind of extended communication is given in [23], where an additional modal operator is used to record the agents' requests.

Another very interesting extension of the framework presented in this chapter concerns the incorporation of actions associated with belief *updates* (cf. [19]) in addition to the ones associated with belief revision that we considered here. Whereas belief revisions are changes in information on an unchanging world, updates are information changes that are associated with a change in the state of the world. The interaction between these different information changes might well be worth looking at.

The framework could also be extended to make it a suitable formalisation tool for special agents, like *intelligent information retrieval agent*. These agents assist a user with finding relevant information, particularly in cyberspace, that satisfies one of its information needs. To this end they communicate, either with their user or with other (intelligent information retrieval) agents, go to World Wide Web sites to seek for relevant information, or make assumptions by default. A preliminary formalisation of these intelligent information retrieval agents based on the framework proposed in this chapter was presented by Huibers & Van Linder [18].

4 SETTING GOALS

As explained in the introduction a rational agent does not only possess information processing capabilities; it should also be endowed with motivations and should be able to modify these. In this section we will discuss how these 'motivational attitudes' can be incorporated into our modal framework.

Informally speaking, we will equip an agent with an *agenda*, which contains the action(s) that the agent has committed itself to and is supposed to perform. Formally, this agenda will take the form of a function that yields for a given agent in a given possible world the set of actions that the agent is committed to in that world. The main technical difficulty that has to be solved is the proper representation in the model of the way the agent maintains its agenda while performing actions. Intuitively speaking, the agent will have to drop from its agenda the actions that it has performed already. Although this sounds rather commonsense and clear, the formal implementation of this idea turns out to be rather involved, as we shall see, mainly due to the need for considering the right mix of syntax and semantics of the actions in the agenda.

Formally, we go about as follows. We start with the formalisation of the concept of a *wish* or *desire*. We then continue with that of the notions of selecting wishes which may then be committed to (and next possibly also be uncommitted to). We will again interpret the acts of selecting, committing and uncommitting as model transformations rather than mere world transformations. The act of selecting changes a model by affecting the set of choices, and the act of (un)committing transforms the agent's agenda.

We extend the language as follows. We build the motivational part on top of the core language $\mathcal{L}_c$. (At the moment we have no need for more complicated formulas expressing, for example, that one wishes to believe something, or one is committed to a revision operator, but in principle this could be done if one likes.) The language $\mathcal{L}^{\mathrm{M}}$ is obtained by adding the following clauses:

- if $\varphi \in \mathcal{L}_c$ and $i \in \mathcal{A}$ then $\mathbf{W}_i\varphi \in \mathcal{L}^{\mathrm{M}}$
- if $\varphi \in \mathcal{L}_c$ and $i \in \mathcal{A}$ then $\Diamond_i\varphi \in \mathcal{L}^{\mathrm{M}}$
- if $\varphi \in \mathcal{L}_c$ and $i \in \mathcal{A}$ then $\mathbf{C}_i\varphi \in \mathcal{L}^{\mathrm{M}}$
- if $\alpha \in \mathrm{Ac}$ and $i \in \mathcal{A}$ then $\mathbf{Com}_i\alpha \in \mathcal{L}^{\mathrm{M}}$

We also consider an extended class Ac^{M} of actions that is the smallest superset of At closed under the clauses of the core language and such that

- if $\varphi \in \mathcal{L}_c$ then $\mathtt{select}\,\varphi \in \mathrm{Ac}^{\mathrm{M}}$
- if $\alpha \in \mathrm{Ac}$ then $\mathtt{commit\_to}\,\alpha \in \mathrm{Ac}^{\mathrm{M}}$
- if $\alpha \in \mathrm{Ac}$ then $\mathtt{uncommit}\,\alpha \in \mathrm{Ac}^{\mathrm{M}}$

4.1 Formalising wishes

In our approach we consider wishes to be the most primitive, fundamental motivational attitudes, that is to say, *in ultimo* agents are motivated to fulfil their wishes. Wishes are represented by means of a plain normal modal operator, i.e. wishes are straightforwardly interpreted as a necessity operator over an accessibility relation W.

Thus we instantiate the M-part of our models to cater for wishes as follows:

The M-part of a model $\mathcal{M} = \langle W, \pi, \mathtt{D}, \mathtt{I}, \mathtt{M} \rangle$ contains the functions $\mathrm{W} : \mathcal{A} \to \wp(W \times W)$, which determines the desirability relation of an agent in a state, and $\mathrm{C} : \mathcal{A} \times W \to \wp(\mathcal{L}_c)$ denoting the choices made by an agent in a state, and a function $\mathrm{Agenda} : \mathcal{A} \times W \to \wp(\mathrm{AcSeq})$, which records the commitments of agents, per state. Here AcSeq stands for the set of sequences of semi-atomic actions.

We can now interpret the $\mathbf{W}_i$ operator as usual:
$\mathcal{M}, s \models \mathbf{W}_i\varphi \Leftrightarrow \forall s' \in W((s, s') \in \mathrm{W}(i) \Rightarrow \mathcal{M}, s' \models \varphi)$

4.2 Selecting wishes

In order to transform wishes to goals, an agent has to first select candidate goals from its set of wishes on the basis of the criteria of unfulfilledness and implementability. In more ordinary language this means that an agent can choose a wish if it is (as yet) unfulfilled and implementable, i.e., the agent is able and has the opportunity to fulfil the wish by means of a finite sequence of atomic actions. Unfulfilledness of a formula φ is easily expressed in the language by means of the classical negation $\neg\varphi$. The notion of implementability is somewhat more involved. For this purpose we introduce an implementability operator $\Diamond_i$, which we interpret as follows:

$$\mathcal{M}, s \models \Diamond_i\varphi \Leftrightarrow \exists k \in \mathbb{N}\exists a_1, \ldots, a_k \in \mathrm{At}(\mathcal{M}, s \models \mathbf{PracPoss}_i(a_1; \ldots; a_k, \varphi)).$$

That is to say, φ is implementable by i if i has the practical possibility to perform a finite sequence of atomic actions yielding φ.

Having defined unfulfilledness and implementability, we can now formally introduce the $\mathtt{select}$ action.

Definition 23 *For model $\mathcal{M}$ with state s, $i \in \mathcal{A}$ and $\varphi \in \mathcal{L}_c$ we define:*

$$\texttt{r}(i, \texttt{select}\,\varphi)(\mathcal{M}, s) = \begin{cases} \emptyset & \textit{if}\ \mathcal{M}, s \models \neg \mathbf{W}_i\varphi \\ \texttt{choose}(i, \varphi)(\mathcal{M}, s), s & \textit{if}\ \mathcal{M}, s \models \mathbf{W}_i\varphi \end{cases}$$

where for $\mathcal{M} = \langle W, \pi, \texttt{D}, \texttt{I}, \texttt{M}\rangle$ *with* $\texttt{M} = \langle \mathrm{W}, \mathrm{C}, \mathrm{Agenda}\rangle$ *we define*

$$\texttt{choose}(i, \varphi)(\mathcal{M}, s) = \langle W, \pi, \texttt{D}, \texttt{I}, \texttt{M}'\rangle \textit{ with } \texttt{M}' = \langle \mathrm{W}, \mathrm{C}', \mathrm{Agenda}\rangle \textit{ such that}$$
$$\mathrm{C}'(i', s') = \mathrm{C}(i', s') \textit{ if } i \neq i' \textit{ or } s \neq s'$$
$$\mathrm{C}'(i, s) \ \ = \mathrm{C}(i, s) \cup \{\varphi\}$$

$$\texttt{c}(i, \texttt{select}\,\varphi)(\mathcal{M}, s) = \mathbf{1} \Leftrightarrow \mathcal{M}, s \models \neg\varphi \wedge \Diamond_i\varphi$$

Finally we define the interpretation of the $\mathbf{C}_i$ operator:
$\mathcal{M}, s \models \mathbf{C}_i\varphi \Leftrightarrow \varphi \in \mathrm{C}(i, s)$

It can be shown ([26]) the act of selecting a formula φ causes minimal change in the sense that the formula φ is marked to be chosen, and nothing else of the model is changed. This has as a corollary that, for example, wishes and implementability formulas remain true after selecting if they were so before.

4.3 Goals

Having defined wishes and selections, one might be tempted to straightforwardly define goals to be selected wishes, i.e. $\mathbf{Goal}_i\varphi \triangleq \mathbf{W}_i\varphi \wedge \mathbf{C}_i\varphi$. This definition is however not adequate to formalise the idea of goals being selected, *unfulfilled*, *implementable* wishes. In the selection operator the criteria of unfulfilledness and implementability have not been incorporated yet. So, an easy way to do this is just to add them. Therefore, goals are defined to be those wishes that are unfulfilled, implementable and selected.

Definition 24 *The* **Goal**$_i$ *operator is for* $i \in \mathcal{A}$ *and* $\varphi \in \mathcal{L}_c$ *defined by:*

$$\mathbf{Goal}_i\varphi \triangleq \mathbf{W}_i\varphi \wedge \neg\varphi \wedge \Diamond_i\varphi \wedge \mathbf{C}_i\varphi$$

Below we state a few properties of wishes, selections and goals.

Proposition 25 *For all* $i \in \mathcal{A}$ *and* $\varphi \in \mathcal{L}_c$ *we have:*

1. $\models \mathbf{W}_i\varphi \leftrightarrow \langle \mathrm{do}_i(\texttt{select}\,\varphi)\rangle\top$
2. $\models \langle \mathrm{do}_i(\texttt{select}\,\varphi)\rangle\top \leftrightarrow \langle \mathrm{do}_i(\texttt{select}\,\varphi)\rangle\mathbf{C}_i\varphi$
3. $\models \neg\mathbf{A}_i\texttt{select}\,\varphi \rightarrow [\mathrm{do}_i(\texttt{select}\,\varphi)]\neg\mathbf{Goal}_i\varphi$
4. $\models \mathbf{PracPoss}_i(\texttt{select}\,\varphi, \top) \leftrightarrow \langle \mathrm{do}_i(\texttt{select}\,\varphi)\rangle\mathbf{Goal}_i\varphi$
5. $\models \varphi \Rightarrow \models \neg\mathbf{Goal}_i\varphi$
6. $(\varphi \rightarrow \psi) \rightarrow (\mathbf{Goal}_i\varphi \rightarrow \mathbf{Goal}_i\psi)$ *is not for all* $\varphi, \psi \in \mathcal{L}_c$ *valid*
7. $\mathbf{K}_i(\varphi \rightarrow \psi) \rightarrow (\mathbf{Goal}_i\varphi \rightarrow \mathbf{Goal}_i\psi)$ *is not for all* $\varphi, \psi \in \mathcal{L}_c$ *valid*

The first item of Proposition 25 states that agents have the opportunity to select all, and nothing but, their wishes. The second item formalises the idea that every choice for which an agent has the opportunity results in the selected wish being marked chosen. In the third item it is stated that whenever an agent is unable to select some formula, then selecting this formula will not result in it becoming one of its goals. The related item 4 states that all, and nothing but, practically possible selections result in the chosen formula being a goal. The fifth item states that no logically inevitable formula qualifies as a goal. Hence whenever a formula is valid this does not only not necessarily imply that it is a goal but it even necessarily implies that it is not. The last two items of Proposition 25 state that goals are neither closed under implications nor under known implications.

4.4 Commitments

After having defined goals in our framework, we go on with a formal description of how an agent can commit itself to a goal. To this end we introduce a `commit_to` α action that, when successful, will have as a result that the agent is committed to the action α. In order to accommodate our model to this action, we need some extra formal machinery.

In order to prepare grounds for the formal semantics of the `commit_to` action we introduce the notion of a transition relation in the spirit of Structural Operational Semantics of Plotkin ([30]). This method is widely used in computer science and we can employ it here fruitfully to describe below what happens with the agent's agenda when it performs actions (from that agenda, so to speak).

In our set-up we consider transitions of the form $\langle \alpha, s \rangle \rightharpoonup_{i,a} \langle \alpha', s' \rangle$, where $\alpha, \alpha' \in \mathrm{Ac}, i \in \mathcal{A}, a$ semi-atomic and $s, s' \in W$. Such a transition expresses that if in state s, i has to perform the action α, after performing the (semi-)atomic action a, this leads to a state s' in which a remaining action α' has still to be performed. We use the symbol Λ for the empty action, with as property that $\Lambda; \alpha = \alpha; \Lambda = \alpha$. Furthermore, we use the projection function π_2, which is assumed to yield the second element of a pair.

Transitions are given by the following deductive system, often called a transition system:

Definition 26 *The transition system T is given by the following axioms:*

- $\langle \alpha, s \rangle \rightharpoonup_{i,\alpha} \langle \Lambda, s' \rangle$ *with* $s' = \pi_2(\mathtt{r}(i, \alpha)(\mathcal{M}, s))$ *if* α *is semi-atomic.*
- $\langle \mathtt{if}\ \varphi\ \mathtt{then}\ \alpha_1\ \mathtt{else}\ \alpha_2\ \mathtt{fi}, s \rangle \rightharpoonup_{i,\mathtt{confirm}\ \varphi} \langle \alpha_1, s \rangle$ *if* $s \models \varphi$
- $\langle \mathtt{if}\ \varphi\ \mathtt{then}\ \alpha_1\ \mathtt{else}\ \alpha_2\ \mathtt{fi}, s \rangle \rightharpoonup_{i,\mathtt{confirm}\ \neg\varphi} \langle \alpha_2, s \rangle$ *if* $s \not\models \varphi$
- $\langle \mathtt{while}\ \varphi\ \mathtt{do}\ \alpha\ \mathtt{od}, s \rangle \rightharpoonup_{i,\mathtt{confirm}\ \varphi} \langle \alpha; \mathtt{while}\ \varphi\ \mathtt{do}\ \alpha\ \mathtt{od}, s \rangle$ *if* $s \models \varphi$
- $\langle \mathtt{while}\ \varphi\ \mathtt{do}\ \alpha\ \mathtt{od}, s \rangle \rightharpoonup_{i,\mathtt{confirm}\ \neg\varphi} \langle \Lambda, s \rangle$ *if* $s \not\models \varphi$

  and the following rule:

- $\frac{\langle \alpha_1, s\rangle \rightharpoonup_{i,a} \langle \alpha_1', s'\rangle}{\langle \alpha_1;\alpha_2, s\rangle \rightharpoonup_{i,a} \langle \alpha_1';\alpha_2, s'\rangle}$

Next, we introduce for convenience's sake an Intend predicate, analogously to the Can predicate. This predicate expresses that the agent intends to do an action with some result if it can perform the action with this result and, moreover, he knows that this result is a goal.

Definition 27 *For $\alpha \in \mathrm{Ac}^{\mathrm{M}}, i \in \mathcal{A}$ and $\varphi \in \mathcal{L}_c$ we define:*

$$\mathbf{Intend}_i(\alpha, \varphi) \triangleq \mathbf{Can}_i(\alpha, \varphi) \wedge \mathbf{K}_i \mathbf{Goal}_i \varphi$$

Having established the formal prerequisites, we can now present the formal semantics of the `commit_to` action. Informally, a `commit_to` α is only successful if the agent intends to do α with some result φ. If it is successful, the agent's agenda is updated in the worlds it is in. Moreover, also in all other possible worlds that are related to this world, either by performing actions or by considering epistemic alternatives, the agenda is updated. The general idea is that if the agent performs (part of) an action that is in its agenda, the agenda in the resulting state will contain the remainder of the action that was in its agenda in the original state. Moreover, since we like to model that an agent knows what it is committed to, we also stipulate that epistemic alternatives of states contain the same agenda.

Finally, it is stipulated that an agent is able to commit to an action iff its agenda is empty, and is ready to receive another commitment, so to speak. This models a simple kind of agent which could be called a *single-minded* one. By varying this definition one may model other agents as well. However, for the sake of simplicity—it is very convenient to have to consider only at most one action sequence in the agenda—we have chosen this definition here.

Definition 28 [1] *For all models $\mathcal{M} = \langle W, \pi, \mathrm{D}, \mathrm{I}, \mathrm{M}\rangle$ with state s, for all $i \in \mathcal{A}$ and $\alpha \in \mathrm{Ac}$ we define:*

$\mathtt{r}(i, \mathtt{commit\_to}\,\alpha)(\mathcal{M}, s) = \emptyset$ *if* $\mathcal{M}, s \models \neg\mathbf{Intend}_i(\alpha, \varphi)$ *for all* $\varphi \in \mathrm{C}(i, s)$
$\mathtt{r}(i, \mathtt{commit\_to}\,\alpha)(\mathcal{M}, s) = \mathcal{M}', s$
with $\mathcal{M}' = \langle W, \pi, \mathrm{D}, \mathrm{I}, \mathrm{M}'\rangle$ *and* $\mathrm{M}' = \langle \mathrm{W}, \mathrm{C}, \mathrm{Agenda}'\rangle$
where Agenda' *is minimal such that it is closed under the following:*
for all $s' \in \mathrm{B}^k(i, s)$, $\mathrm{Agenda}'(i, s') = \mathrm{Agenda}(i, s') \cup \{\alpha\}$
and for all $s', s'', s''' \in W, \alpha' \in \mathrm{Agenda}'(i, s')$
such that, for some semi-atomic a,
$\langle \alpha', s'\rangle \rightharpoonup_{i,a} \langle \alpha'', s''\rangle$ *and* $s''' \in \mathrm{B}^k(i, s'')$:
$\mathrm{Agenda}'(i, s''') = \mathrm{Agenda}(i, s''') \cup \{\alpha''\}$
otherwise

$\mathtt{c}(i, \mathtt{commit\_to}\,\alpha)(\mathcal{M}, s) = \mathbf{1}$ *iff* $\mathrm{Agenda}(i, s) = \emptyset$

[1] In fact, in order for this definition to be well-defined regarding the agenda function some restrictions have to be put on the models regarding the function $\mathtt{r}_0$ and the interpretation of the `confirm` actions. For ease of presentation these are omitted here. Details can be found in [26].

Again one can show ([26]) that the `commit_to` action is minimal in the sense that only the agenda of agent i is updated and that only the agenda in the states that are affected in the sense described above are updated.

Next we define an operator $\mathbf{Com}_i$ that indicates that the agent i is committed to an action. We define the interpretation of this operator such that, in any state epistemically equivalent with the state it is in, the agent is committed to all actions that are (semantically equivalent to) 'initial parts' of the actions written in its agenda. In order to capture the notion of semantical equivalence of actions we use our transition systems again, and define the notion of a *computation run*.

Definition 29 $\mathrm{CR}^{\mathrm{C}}_{\mathcal{M}}(i, \alpha, s) \ni a_1; a_2; \ldots; a_n$ *iff* $\langle \alpha, s\rangle \rightharpoonup_{i,a_1} \langle \alpha_1, s_1\rangle \rightharpoonup_{i,a_2} \langle \alpha_2, s_2\rangle \rightharpoonup_{i,a_3} \cdots \rightharpoonup_{i,a_n} \langle \alpha_n, s_n\rangle$ *for some* $\alpha_1, \ldots, \alpha_n \in \mathrm{Ac}, s_1, \ldots, s_n \in W$, *such that* $\alpha_n = \Lambda$.

Note that due to the fact that our actions are deterministic, the set $\mathrm{CR}^{\mathrm{C}}_{\mathcal{M}}(i, \alpha, s)$ contains *at most one* element. This should be kept in mind while considering the definitions below. Actions that have the same computation run (with respect to a certain starting state) are considered to be semantically equivalent (in that state). So now we are able to give our interpretation of the $\mathbf{Com}_i$ operator.

$\mathcal{M}, s \models \mathbf{Com}_i\alpha \Leftrightarrow$
$\forall s' \in \mathrm{B}^k(i, s) \exists \alpha_1 \in \mathrm{CR}^{\mathrm{C}}_{\mathcal{M}}(i, \alpha, s') \exists \alpha_2 \in \mathrm{Agenda}(i, s') \exists \alpha_2' \in \mathrm{CR}^{\mathrm{C}}_{\mathcal{M}}(i, \alpha_2, s')$ $(\mathrm{Prefix}(\alpha_1, \alpha_2'))$
where Prefix stands for the prefix relation on computation runs (which are sequences of actions).

Finally, we note that to let an agent be really rational, it should also be capable in certain situations to abandon its commitments, for instance, when the goal is achieved or is not implementable any more. In the definition below this is put as follows: an `uncommit` α action is only successful if the agent i was committed to the action α, and the agent is able to uncommit iff it does no longer intend to do α for any purpose φ.

Definition 30 *For all models* $\mathcal{M} = \langle W, \pi, \mathtt{D}, \mathtt{I}, \mathtt{M}\rangle$ *with state* s, *for all* $i \in \mathcal{A}$ *and* $\alpha \in \mathrm{Ac}$ *we define*[2]*:*

$$\begin{aligned}
&\mathtt{r}(i, \mathtt{uncommit}\,\alpha)(\mathcal{M}, s) = \emptyset \text{ if } \mathcal{M}, s \models \neg\mathbf{Com}_i\alpha\\
&\mathtt{r}(i, \mathtt{uncommit}\,\alpha)(\mathcal{M}, s) = \mathcal{M}', s\\
&\quad \textit{with } \mathcal{M}' = \langle W, \pi, \mathtt{D}, \mathtt{I}, \mathtt{M}'\rangle \textit{ and } \mathtt{M}' = \langle \mathrm{W}, \mathrm{C}, \mathrm{Agenda}'\rangle\\
&\quad \textit{where for all } s' \in \mathrm{B}^k(i, s),\\
&\qquad \mathrm{Agenda}'(i, s') = \mathrm{Agenda}(i, s') \setminus\\
&\qquad\qquad \{\beta \mid \mathrm{Prefix}(\mathrm{CR}^{\mathrm{C}}_{\mathcal{M}}(i, \alpha, s'), \mathrm{CR}^{\mathrm{C}}_{\mathcal{M}}(i, \beta, s'))\}\\
&\quad \textit{and for all } s', s'', s''' \in W \textit{ with } \alpha' \in \mathrm{Agenda}'(i, s') \textit{ and such that,}\\
&\quad \textit{for some semi-atomic } a,\ \langle \alpha', s'\rangle \rightharpoonup_{i,a} \langle \alpha'', s''\rangle \textit{ and } s''' \in \mathrm{B}^k(i, s''),
\end{aligned}$$

[2] Actually, in order to let this notion be well-defined, we need certain minimality conditions which we omit here for simplicity's sake (for a more rigorous treatment, see [26])

$$\text{Agenda}'(i, s'') = \text{Agenda}(i, s'')\setminus \{\beta \mid \text{Prefix}(\text{CR}^{\text{C}}_{\mathcal{M}}(i, \alpha'', s''), \text{CR}^{\text{C}}_{\mathcal{M}}(i, \beta, s''))\}$$

otherwise

$$\texttt{c}(i, \texttt{uncommit}\,\alpha)(\mathcal{M}, s) = \mathbf{1} \textit{ iff } \mathcal{M}, s \models \neg\mathbf{Intend}_i(\alpha, \varphi) \textit{ for all } \varphi \in \text{C}(i, s)$$

The complication in this definition, as compared to that of the $\texttt{commit\_to}$ operator, is due to the fact that 'committedness' is closed under taking prefixes of (computation runs of) actions, so that in order to successfully uncommit to an action α also all actions that have α as a prefix (with respect to computation runs) should be removed from the agent's agenda.

Some properties of the operators treated in this section are given in the following proposition.

Proposition 31 *For all* $i \in \mathcal{A}$*,* $\alpha, \beta \in \text{Ac}$ *and* $\varphi \in \mathcal{L}_c$ *we have:*

1. $\models \mathbf{Intend}_i(\alpha, \varphi) \rightarrow \langle \text{do}_i(\texttt{commit\_to}\,\alpha)\rangle\top$
2. $\models \langle \text{do}_i(\texttt{commit\_to}\,\alpha)\rangle\top \leftrightarrow \langle \text{do}_i(\texttt{commit\_to}\,\alpha)\rangle\mathbf{Com}_i\alpha$
3. $\models \mathbf{Com}_i\alpha \rightarrow \neg\mathbf{A}_i\texttt{commit\_to}\,\beta$
4. $\models [\text{do}_i(\texttt{commit\_to}\,\alpha)]\neg\mathbf{A}_i\texttt{commit\_to}\,\beta$
5. $\models \mathbf{Com}_i\alpha \leftrightarrow \langle \text{do}_i(\texttt{uncommit}\,\alpha)\rangle\neg\mathbf{Com}_i\alpha$
6. $\models \mathbf{Intend}_i(\alpha, \varphi) \rightarrow \neg\mathbf{A}_i\texttt{uncommit}\,\alpha$
7. $\models (\mathbf{C}_i\varphi \leftrightarrow \mathbf{K}_i\mathbf{C}_i\varphi) \rightarrow (\mathbf{A}_i\texttt{uncommit}\,\alpha \leftrightarrow \mathbf{K}_i\mathbf{A}_i\texttt{uncommit}\,\alpha)$
8. $\models \mathbf{Com}_i\alpha \wedge \neg\mathbf{Can}_i(\alpha, \top) \rightarrow \mathbf{Can}_i(\texttt{uncommit}\,\alpha, \neg\mathbf{Com}_i\alpha)$

In the third item it is stated that being committed prevents an agent from having the ability to (re)commit. The fourth item states that the act of committing is ability-destructive with respect to future $\texttt{commit}$ actions, i.e. by performing a commitment an agent loses its ability to make any other commitments. Item 5 states that being committed is a necessary and sufficient condition for having the opportunity to uncommit; as mentioned above, agents have the opportunity to undo all of their commitments. In item 6 it is stated that agents are (morally) unable to undo commitments to actions that are still known to be correct and feasible to achieve some goal. In item 7 it is formalised that agents know of their abilities to uncommit to some action. The last item states that whenever an agent is committed to an action that is no longer known to be practically possible, it knows that it can undo this impossible commitment.

Finally, the following proposition states some intuitive properties of the $\mathbf{Com}_i$ operator with respect to complex actions.

Proposition 32 *For all* $i \in \mathcal{A}$*,* $\alpha, \alpha_1, \alpha_2 \in \text{Ac}$ *and all* $\varphi \in \mathcal{L}_c$ *we have:*

1. $\models \mathbf{Com}_i\alpha \rightarrow \mathbf{K}_i\mathbf{Com}_i\alpha$
2. $\models \mathbf{Com}_i(\alpha_1; \alpha_2) \rightarrow \mathbf{Com}_i\alpha_1 \wedge \mathbf{K}_i[\text{do}_i(\alpha_1)]\mathbf{Com}_i\alpha_2$
3. $\models \mathbf{Com}_i\texttt{if}\,\varphi\,\texttt{then}\,\alpha_1\,\texttt{else}\,\alpha_2\,\texttt{fi} \wedge \mathbf{K}_i\varphi \rightarrow \mathbf{Com}_i(\texttt{confirm}\,\varphi; \alpha_1)$
4. $\models \mathbf{Com}_i\texttt{if}\,\varphi\,\texttt{then}\,\alpha_1\,\texttt{else}\,\alpha_2\,\texttt{fi} \wedge \mathbf{K}_i\neg\varphi \rightarrow \mathbf{Com}_i(\texttt{confirm}\,\neg\varphi; \alpha_2)$

5. $\models \mathbf{Com}_i \texttt{while}\,\varphi\,\texttt{do}\,\alpha\,\texttt{od} \wedge \mathbf{K}_i\varphi \rightarrow$
$\mathbf{Com}_i((\texttt{confirm}\,\varphi;\alpha);\texttt{while}\,\varphi\,\texttt{do}\,\alpha\,\texttt{od})$

The first item of Proposition 32 states that commitments are known. The second item states that a commitment to a sequential composition $\alpha_1;\alpha_2$ of actions implies a commitment to the initial part α_1, and that the agent knows that after execution of this initial part α_1 it will be committed to the remainder α_2. The third and fourth item formalise the rationality of agents with regard to their commitments to conditionally composed actions. The last item concerns the unfolding of a while-loop: if an agent is committed to a while-loop while knowing the condition of the loop to be true, then the agent is also committed to the then-part of the while-loop.

5 CONCLUSION

All the characterizing features of agents that emerge in our formal model have popped up in the literature before.

Dynamic logic has been studied extensively in computer science (see Goldblatt's [9]), *epistemic* and *doxastic* logic have been of interest among philosophers since Hintikka's [11], and have since then been formalized and proven useful for computers science and artificial intelligence (see the volumes [7, 27]). It was Moore ([28, 29]) who realized that dynamic and epistemic logic can be perfectly combined into one modal framework for actions and knowledge. Our contribution here is that we formalized and developed the ideas of Moore: we were able to come up with an axiomatization for a basic multi-agent system for actions and knowledge ([13]). We also introduces several epistemic attitudes([25]), thus taking into account the source of the information the knowledge (the variants are called 'beliefs') stems from.

We have also come across several accounts of *abilities* in the literature. The approach of Brown [2] and Elgesem [6] is essentially *endogenous* in the sense that actions are not referred to explicitly. They propose modal operators, ranging over formulae, to formalise ability: $\mathbf{A}_i\varphi$ is then to be read as 'agent i is able to bring about circumstances in which φ is true'. The only formal system that we know of in which it is hinted at an exogenous approach towards ability is the one proposed by Penther. And we agree with her that it is the *compositional behaviour* of ability that should be at the focus of attention, here.

There is also an extensive literature on *motivational attitudes*. Probably the most influential account of motivational attitudes is due to Cohen & Levesque [4]. Starting from the primitive notions of implicit goals and beliefs, Cohen & Levesque define so-called persistent goals, which are goals which agents give up only when they think they are either satisfied or will never be true, and intentions, both ranging over propositions and over actions. The idea underlying persistent goals is similar to that underlying our notion of goals. In the framework of Cohen

& Levesque agents intend to bring about a proposition if they intend to do some action that brings about the proposition. An agent intends to do an action if it has the persistent goal to have done the action. In our approach we do not use such a reduction technique.

Another important formalisation of motivational attitudes is proposed by Rao & Georgeff [32] in their BDI-architecture. Treating desires and intentions as primitive, Rao & Georgeff focus on the process of intention revision rather than the 'commitment acquisition' which is essential in our formalisation. Both desires and intentions in their framework suffer from the problems associated with logical omniscience, which we have avoided in our formalisation of goals. (Not in our treatment of wishes, though, but these only play a subsidiary role to get to goals.)

Perhaps the closest in spirit to our approach is that of Singh ([34]). Singh also aims at an integrated framework for describing intelligent agents. His framework rests mainly on branching-time temporal logic extended with 'dynamic-logic-like' operators that nevertheless have an explicit temporal interpretation, which makes the elaboration of the theory quite distinct from ours, and somewhat more involved to our taste. His focus is slightly different, as well: less on the belief-revisional attitudes and more on the intricacies of intentional notions, as related to nondeterminism and know-how. Furthermore it considers communications as based on speech act theory, which is beyond the scope of our model (But cf. [5] for a first integrating attempt in this direction). In future work we will try to obtain a more compact theory and give a treatment of communication along the lines of what has been done in the realm of distibuted systems rather than linguistics, although we certainly see the relevance of this for the enterprise of describing intelligent agents. But we also believe that somewhere the theory should be constrained in order to keep it useful. If one is not careful one ends up with a Theory of Everything, which will be very hard to use indeed for the specification of a practical agent system! As has been mentioned before, we consider it one of our main contributions to have proposed a single framework in which all of these aspects can be dealt with, to some considerable extent. In our view modal logic has proven to be an excellent tool for this purpose.

Finally we like to say some words on the restriction to deterministic actions in this paper. It has been shown in [14] that also nondeterminism can be treated within our framework by extending the set of action constructors by choice operators. In fact, in [14] we have considered two such operators: the internal non-deterministic operator $\oplus$ models those choices that are to be made by the agent, and is therefore supposed to behave *angelic*, i.e. as helpful as possible for the agent. Then, $\neg[do_i(\alpha \oplus \beta)]\varphi$ is equivalent to $\neg[do_i(\alpha)]\varphi \vee \neg[do_i(\alpha)]\varphi$: if agent i wants to avoid φ as a result of doing $(\alpha \oplus \beta)$, it is sufficient to either avoid φ as a result of α or as a result of β. The external non-deterministic choice operator $+$, on the other hand, has a *demonic* behaviour from the agent's perspective: since it must be prepared for the worst, it has to guarantee both $[do_i(\alpha)]\varphi$ and $[do_i(\beta)]\varphi$ in order to ensure that $[do_i(\alpha + \beta)]\varphi$ holds.

Wiebe van der Hoek (Universiteit Utrecht, The Netherlands)
Bernd van Linder (ABN-AMRO Bank, The Netherlands)
John-Jules Ch. Meyer (Universiteit Utrecht, The Netherlands)

REFERENCES

[1] C.E. Alchourrón, P. Gärdenfors, and D. Makinson. On the logic of theory change: partial meet contraction and revision functions. *Journal of Symbolic Logic*, 50:510–530, 1985.

[2] M.A. Brown. On the logic of ability. *Journal of Philosophical Logic*, 17:1–26, 1988.

[3] C. Castelfranchi. Personal communication.

[4] P.R. Cohen and H.J. Levesque. Intention is choice with commitment. *Artificial Intelligence*, 42:213–261, 1990.

[5] F. Dignum and B. van Linder. Modelling social agents: Communication as action. In J. Mueller, M. Wooldridge, and N. Jennings, editors, *Intelligent Agents III –Proceedings of the 3rd International Workshop on Agent Theories, Architectures, and Languages (ATAL-96)*, pages 83–93, 1996.

[6] D. Elgesem. *Action Theory and Modal Logic*. PhD thesis, Institute for Philosophy, University of Oslo, Oslo, Norway, 1993.

[7] R. Fagin, J.Y. Halpern, Y. Moses, and M.Y. Vardi. *Reasoning about Knowledge*. MIT Press, Cambridge MA, 1994. To appear.

[8] P. Gärdenfors. *Knowledge in Flux: Modeling the Dynamics of Epistemic States*. The MIT Press, Cambridge, Massachusetts and London, England, 1988.

[9] R. Goldblatt. *Axiomatising the Logic of Computer Programming*, volume 130 of *LNCS*. Springer-Verlag, 1982.

[10] R. Goldblatt. *Logics of Time and Computation*, volume 7 of *CSLI Lecture Notes*. CSLI, Stanford, 1992. Second edition.

[11] J. Hintikka. *Knowledge and Belief*. Cornell University Press, Ithaca, NY, 1962.

[12] W. van der Hoek. Systems for knowledge and beliefs. *Journal of Logic and Computation*, 3(2):173–195, 1993.

[13] W. van der Hoek, B. van Linder, and J.-J. Ch. Meyer. A logic of capabilities. In A. Nerode and Yu. V. Matiyasevich, editors, *Proceedings of the Third International Symposium on the Logical Foundations of Computer Science (LFCS'94)*, volume 813 of *Lecture Notes in Computer Science*, pages 366–378. Springer-Verlag, 1994.

[14] W. van der Hoek, B. van Linder, and J.-J. Ch. Meyer. Unravelling nondeterminism: On having the ability to choose (extended abstract). In P. Jorrand and V. Sgurev, editors, *Proceedings of the Sixth International Conference on Artificial Intelligence: Methodology, Systems, Applications (AIMSA'94)*, pages 163–172. World Scientific, 1994.

[15] W. van der Hoek and J.-J. Ch. Meyer. Possible logics for belief. *Logique & Analyse*, 127–128:177–194, 1989.

[16] W. van der Hoek and J.-J.Ch. Meyer. Graded modalities for epistemic logic. *Logique et Analyse*, 34(133-134):251–270, 1991.

[17] Z. Huang. Logics for belief dependence. In E. Börger, H. Kleine Büning, M.M. Richter, and W. Schönfeld, editors, *Computer Science Logic, 4th Workshop CSL'90*, volume 533 of *Lecture Notes in Computer Science*, pages 274–288. Springer-Verlag, 1991.

[18] T.W.C. Huibers and B. van Linder. Formalising intelligent information retrieval agents. In F. Johnson, editor, *Proceedings of the 18th BCS IRSG Annual Colloquium on Information Retrieval Research*, pages 125–143, 1996.

[19] H. Katsuno and A.O. Mendelzon. On the difference between updating a knowledge base and revising it. In P. Gärdenfors, editor, *Belief revision*, pages 183–203. Cambridge University Press, 1992.

[20] A. Kenny. *Will, Freedom and Power*. Basil Blackwell, Oxford, 1975.

[21] I. Levi. Direct inference. *The Journal of Philosophy*, 74:5–29, 1977.

[22] B. van Linder. *Modal Logics for Rational Agents*. PhD thesis, Utrecht University, 1996.

[23] B. van Linder, W. van der Hoek, and J.-J. Ch. Meyer. Communicating rational agents. In B. Nebel and L. Dreschler-Fischer, editors, *KI-94: Advances in Artificial Intelligence*, volume 861 of *Lecture Notes in Computer Science (subseries LNAI)*, pages 202–213. Springer-Verlag, 1994.

[24] B. van Linder, W. van der Hoek, and J.-J. Ch. Meyer. Tests as epistemic updates. In A.G. Cohn, editor, *Proceedings of the 11th European Conference on Artificial Intelligence (ECAI'94)*, pages 331–335. John Wiley & Sons, 1994.

[25] B. van Linder, W. van der Hoek, and J.-J. Ch. Meyer. Actions that make you change your mind. In A. Laux and H. Wansing, editors, *Knowledge and Belief in Philosophy and Artificial Intelligence*, pages 103–146. Akademie Verlag, 1995.

[26] B. van Linder, J.-J. Ch. Meyer, and W. van der Hoek. Formalising motivational attitudes of agents using the karo framework. Technical Report UU-CS-1997-03, Utrecht University, 1997.

[27] J.-J. Ch. Meyer and W. van der Hoek. *Epistemic Logic for AI and Computer Science.* Cambridge University Press, 1995.

[28] R.C. Moore. Reasoning about knowledge and action. Technical Report 191, SRI International, 1980.

[29] R.C. Moore. A formal theory of knowledge and action. Technical Report 320, SRI International, 1984.

[30] G. Plotkin. A structural approach to operational semantics. Technical Report DAIME FN-19, Aarhus University, 1981.

[31] D. Poole. A logical framework for default reasoning. *Artificial Intelligence*, 36:27–47, 1988.

[32] A.S. Rao and M.P. Georgeff. Modeling rational agents within a BDI-architecture. In J. Allen, R. Fikes, and E. Sandewall, editors, *Proceedings of the Second International Conference on Principles of Knowledge Representation and Reasoning (KR'91)*, pages 473–484. Morgan Kaufmann, 1991.

[33] R. Reiter. A logic for default reasoning. *Artificial Intelligence*, 13:81–132, 1980.

[34] M.P. Singh. *Multiagent Systems: A Theoretical Framework for Intentions, Know-How and Communications*, volume 799 of *Lecture Notes in Computer Science (subseries LNAI)*. Springer-Verlag, 1994.

[35] E. Thijsse. On total awareness logics (with special attention to monotonicity constraints and flexibility). In M. de Rijke, editor, *Diamonds and Defaults*, volume 229 of *Synthese Library*, pages 309–347. Kluwer Academic Publishers, 1993.

D. R. TRAUM

SPEECH ACTS FOR DIALOGUE AGENTS

1 INTRODUCTION

A *dialogue agent* is one that can interact and communicate with other agents, in a coherent manner, not just with one-shot messages, but with a sequence of related messages all on the same topic or in service of an overall goal. Following the basic insights of *speech act theory*, these communications are seen not just as transmitting information but as actions which change the state of the world. Most of these changes will be to the mental states of the agents involved in the conversation, as well as the state or context of the dialogue. As such, speech act theory allows an agent theorist or designer to place agent communication within the same general framework as agent action. In general, though, communicative action requires a more expressive logic of action than is required for something like the single-agent blocks world, familiar in classical AI planning. For one thing, there are multiple agents, and there is also a possibility of simultaneous and fallible action.

In studying speech acts, the focus is on pragmatics rather than semantics — that is, how language is used by agents, not what the messages themselves mean in terms of truth-conditions in a model (see [66] for a good introduction to issues in natural language pragmatics). As with other aspects of pragmatics such as implicature and presupposition, an important concern is what can be inferred (perhaps only provisionally or with a certain likelihood) as a result of the performance of a speech act. While much of speech act work has been analyzing interaction in natural language, speech acts are also a convenient level of analysis for artificial communication languages. While the rules for interpreting whether a particular act has been performed will be different for an artificial language (presumably simpler, with less concern about vague and ambiguous utterances), the range of actions that agents will wish to perform will still be roughly the same (although perhaps not as many fine gradations will be necessary for simpler agents).

In this chapter, we consider the role of speech acts in allowing an agent to participate in a dialogue with another agent. The bulk of the paper will be a review of some of the most important work on speech acts and its implication for dialogue agents. In Section 2, we review the foundational work on speech act theory, stemming from work in the philosophy of language. In Section 3, we consider work in Artificial Intelligence aimed at providing computational accounts of speech acts, primarily accounts that could be used by agents to engage in communication —

M. Wooldridge and A. Rao (eds.), Foundations of Rational Agency, 169–201.

both performing and interpreting such acts. In most of that work, the emphasis is on the performance or recognition of single utterances. We therefore turn, in Section 4, to work on expanding speech act theory to also include other aspects of dialogue. In Section 5, we consider some work in which the agent communication language itself is formulated in terms of speech act concepts. Such languages make identification of speech acts somewhat easier, but still contain many of the same issues as languages in which speech acts have a less transparent relation to surface form. In Section 6, we reconsider some of the issues confronting speech act theory, in light of the presented work. Finally, in Section 7, we begin to sketch a particular theory of speech acts (that presented in [95], pointing out some of the extra elements required beyond a BDI-theory of agency.

2 FOUNDATIONAL PHILOSOPHICAL SPEECH ACT WORK

Work on speech acts begins with philosophers of language interested in issues in natural language pragmatics. In particular, difficulties with the primary view of the meanings of sentences as merely truth-functions. Pragmatics (as opposed to semantics) concentrates on *utterances* (the performance) rather than sentences (the objects produced), and also is generally concerned with other aspects of the context of production and interpretation.

2.1 Austin

Austin observed that utterances are not just descriptions of states of affairs, but are used to *do* things [16]. Under felicitous circumstances, utterances can change the mental and interactional state of the participants. Speaking is *acting*, and in speaking the speakers are performing *speech acts*. There are multiple types of action performed in speaking, and Austin distinguished several. *Locutionary acts* are the act of saying something, including a *phonetic act* – producing certain noises, a *phatic act* – producing words belonging to a vocabulary in a constructions conforming to a grammar, and a *rhetic act* – using the product of the phatic act with a particular sense and reference.

Illocutionary acts are those acts performed *in* saying something, for example, asking or answering a question, giving some information, etc. Most of the subsequent work on speech acts has been on illocutionary acts. Illocutionary acts are taken to be composed of an *illocutionary force*, which specifies the type of action (e.g., requesting, suggesting, warning, apologizing, informing), and a *propositional content* which specifies the details of the action (e.g., what it is that the hearer is being requested to do). Illocutionary acts are not always directly deducible from the locutionary acts which generate them. *Indirect speech acts* are those in which the act performed is other than what would be expected from a compositional account of the content. Austin gives a bridge example in which an utterance of "I bid three clubs" is used to inform a partner that the speaker has no diamonds. As this example illustrates, utterances can be used to achieve multiple functions. In this case the direct (conventional) *bid* act is performed, as well as the indirect *inform*.

Perlocutionary acts are those which are performed *by* saying something – actions which achieve effects which are special to the particular situation of utterance rather than the conventional nature of the communication. Examples include persuasion, surprise, and deterrence. Although a speaker might intend these, it generally takes more than just the utterance to achieve them.

Austin spent much of his effort on *performatives*, in which the utterance itself creates a result. *Explicit performatives* are those utterances in which the result is described in the utterance. In English, the word "hereby" often accompanies performative utterances. Not all speech act verbs can be used performatively, however, e.g., we don't have the conventional form in English, "I insult you." As others (e.g., [32]) have noticed, some such declarations would be self-defeating, such as "lie". Austin considers the relation of performative utterances and other pragmatic phenomena such as *implicature* and *presupposition*, in which other facts can be derived, but not through the monotonic entailments of ordinary logic.

Austin classifies the illocutionary acts into several categories, based primarily on illocutionary force: *verdictives, exercitives, commissives, expositives,* and *behavitives*. *Verdictives*, which are basically giving a (perhaps preliminary or tentative) verdict about something. *Exercitives* exercise a right, such as appointing, or nominating. *Commissives* commit the speaker to do something. *Behavatives* all have something to do with attitudes and social behavior, such as apologizing and congratulating. *Expositives* have to do with how an utterance fits into a conversation or argument, such as clarifying or postulating.

2.2 Searle

Searle extends and refines Austin's work on illocutionary acts. He observes that Austin's decomposition of speech acts into *illocutionary force* and *propositional content* shows up in the different kinds of negation that can be performed in a sentence [87]. For example, the sentence "I promise to come" has two negations: a *propositional negation*, "I promise not to come", in which the illocutionary act is the same (promise), but in which the content is negated, and "I do not promise to come", in which the propositional content is the same, but the illocutionary act is no longer a promise, but a *refusal* to make a promise. Searle further decomposes the *propositional content* of an act as a combination of *predicating* and *reference* acts. Neither of these stand alone but are performed only in concert with illocutionary acts.

Probably the most important contribution was an attempt to provide necessary and sufficient conditions for the performance of illocutionary acts. He presented these as *constitutive rules* (like the rules which define the games of football or chess) of various sorts. *Normal input-output conditions* concern the conditions of intelligible speaking and understanding, including knowing the conventions of languages, paying attention, etc. *Propositional content conditions* describe restrictions on the content, e.g., for a promise the content must be a future action. *Preparatory conditions* involve the constraints on the world that make the speech act useful. *Sincerity conditions* involve alignment of the speaker's actual attitudes (belief, desire, etc.) with the attitudes expressed by the act. *Essential conditions* involve the speaker's intentions in performing the act – what she was *trying* to do. Searle also adds a condition based on Grice's notion of non-natural meaning [45], that the effect of the act is in part produced by the hearer's recognition that the utterance is intended to produce this effect by means of the recognition of the intention.

Searle also improves Austin's classification of types of illocutionary acts mentioned above [88]. Austin's classification is fairly haphazard, based more on similarity of illocutionary verbs than the acts themselves. Searle points out that there is no necessary correspondence between illocutionary acts, and the illocutionary verbs that a particular language chooses to describe these acts. Searle proposes an alternate taxonomy based on the purposes of the acts. To do this, Searle presents an analysis of several dimensions along which speech acts can vary, as outlined below:

1. **The point (purpose) of the act.** These correspond to the *essential conditions* described above, such as attempting to get the hearer to do something (request), represent the world (statement), or commit (promise). The point is the main, although not only component of illocutionary force. The force also includes other notions such as strength of the point, and the other dimensions.

2. **Direction of fit between words and the world.** Statements try to fit word to the way the world is, whereas commitments and requests try to fit the (future state of the) world to the way the words are. Searle uses an example (credited to [14]) of a list of groceries, and differences in how it might be constructed and used. If it is a shopping list given by a spouse, it is used by the shopper to fit the world (the groceries purchased) to the words (the list). If, on the other hand, it is made by an observer of the shopper, indicating what was actually bought, it would be a fit of word to the world. Although identical lists might be produced, the difference is critical for repairing mistakes. In the case of word to world fit, if the shopper really bought pork chops rather than bacon, the observer can simply cross off "bacon" and write "pork chops" on his list. The shopper, however, must go back to the store and exchange his purchases, to repair an error.

3. **Expressed psychological state** Different acts will express different mental attitudes on the part of the speaker. An assertion expresses a belief, while a request expresses a desire for the addressee to act. This dimension relates most closely to Searle's *sincerity condition* – a speaker might not actually

have these attitudes, but is nonetheless expressing them in performing the speech acts.

4. **Strength of point** This is related to the above – attitudes can generally be held with varying degrees of strength. Thus one can have a tentative or firm belief, and could suggest, request, or command.

5. **Status of the participants** The social relationship of the speakers may also play a role in the strength of the force. Thus a general can order a private, but a private may only suggest or request. These differences and the next also relate to Searle's *preparatory conditions*.

6. **Interest of the participants** This would distinguish boasts from laments, or promises from threats. While Searle separates this from condition (3), we can see this as relation to (expressed or assumed) desires of the speaker and addressee.

7. **Relations to the rest of the discourse** This is how discourse relations such as *answer* or *deduce* can be distinguished from other statements.

8. **Propositional content** Aspects of the propositional content that play a role in illocutionary force, e.g., that a promise or request can only involve future action, while a statement could be about the past. These correspond to Searle's *propositional content conditions*, though they are also closely related to dimension (2).

9. **Speech Acts vs. general acts** Some things must be performed as speech acts, whereas others might be performed by speech acts or other kinds of action. Examples include "classify", "estimate" and "diagnose".

10. **Extra-linguistic institutions** This includes features of the social and institutional context, going beyond language. Thus, only individuals with a certain position within an institution can marry people or bid three no-trump. Searle says that (perhaps contra-Austin) not all acts need such an institutional position, claiming that one need only obey the rules of language to make a statement or promise. However, if one considers more general agents, one could say that even for these actions, an institutional position is required: though merely as an agent within society. It is difficult to know what exactly Searle means by "the rules of language", but clearly syntactic and semantic well-formedness is not sufficient to perform a promise – there must be some notion of the language producer having undertaken a commitment.

11. **performativeness** Whether there is an illocutionary verb that has a performative use. "boast" and "threaten", for example, do not.

12. **style of performance** This distinguishes announcing from confiding.

Searle uses these features, mainly (1), (2), and (3) to produce a more consistent speech act taxonomy. Searle's categories include: *representatives*, which commit the speaker to the truth of an expressed proposition, *directives*, which involve getting the hearer to do something, *commissives*, which involve committing the speaker to some course of action, *expressives*, which convey a psychological state of the speaker, and *declarations*, which bring about the correspondence of the world to the word.

Searle's speech act theory was later formalized within a possible worlds semantics, and unified with a Tarskian tradition, along with Vanderveken [90, 103, 104].

3 AI MODELS OF SPEECH ACTS

A difficulty with the early philosophical work on speech acts was that there did not yet exist very formal accounts of action and mental states that could be used to design more precise definitions for speech acts. Moreover, such definitions can be rather difficult to test in the abstract, without observing what is really at stake in actual communication episodes. With the rapid progress in the 1970s on reasoning about actions and plans, it became possible to give a more operational account of speech acts, in terms of AI planning operators.

Bruce was the first to try to account for Speech Act theory in terms of AI work on actions and plans [20]. He defined natural language generation as *social action*, where a *social action* is one which is defined in terms of beliefs, wants, and intentions. He also presented *Social Action Paradigms* which showed how speech acts could be combined to form larger discourse goals. He showed how acts such as **Inform** or **Request** could be used in achieving intentions to change states of belief.

3.1 Allen, Cohen, and Perrault

Probably the most influential AI work on speech acts to date has been the work at University of Toronto by Allen, Cohen, and Perrault. Cohen and Perrault [38] tackled the problem of language production, defining speech acts as plan operators which affect the beliefs of the speaker and hearer.

In setting up the program they set forth some important goals for work on speech acts. They write that any account of speech acts should answer the following questions:

- Under what circumstances can an observer believe that a speaker has sincerely and successfully performed a particular speech act in producing an utterance for a hearer?

- What changes does the successful performance of a speech act make to the speaker's model of the hearer, and to the hearer's model of the speaker?

- How is the meaning (sense/reference) of an utterance x related to the acts that can be performed in uttering x?

They also suggest that a theory of speech acts based on plans should specify the following:

- A planning system: a formal language for describing states of the world, a language for describing operators, a set of plan construction inferences, a specification of legal plan structures. Semantics for the formal languages should also be given.

- Definitions of speech acts as operators in the planning system. What are their effects? When are they applicable? How can they be realized in words?

Cohen and Perrault's models of mental states consist of two types of structures: *beliefs* and *wants*. *Beliefs* are modal operators which take two arguments: an agent who is the believer, and a proposition which is believed. They also follow Hintikka [57], augmenting the belief structure to include quantified propositions. Thus an agent can believe that something has a value without knowing what that value is, or an agent can believe another agent knows whether a proposition is true, without the first agent knowing if it's true or not. *Wants* are different modal operators which can nest with beliefs. *Wants* model the goals of agents.

Perrault and Cohen then proceeded to make a first stab at satisfying these issues. The planning system they use is a modified version of STRIPS [43]. They maintain STRIPS's method of dealing with the frame problem: assuming that nothing can change the world except the explicit changes mentioned by the effects of an operator. They describe two different types of preconditions, both of which must hold for the action to succeed. *cando* preconditions indicate propositions which must be true for the operator to be applicable. *Want* preconditions are meant to cover sincerity conditions. In order to successfully perform an action, the agent (speaker) must want to do that action. They model the speech acts REQUEST and INFORM, within their planning system, including various forms of questions as requests for information (or to inform), and consider the relations of requests to informs of desires.

Figure 1 shows a slightly simplified example of the kinds of plan structures that an agent might build in deciding to issue a request, given the desire for some action to be performed. Here, one agent, S, wants anther agent, John, to perform some action α. S realizes that if John has a desire to perform α, then John may perform it. Now, S must find and perform some action that will have as an effect that John wants to perform α. Here, an assumption of cooperativity is required, such that if John realizes that S wants him to do α, John himself will want to do α, as well. And it is assumed that the direct effect of the request is a belief that S wants John to perform α. The upshot is that S can perform this request and then rely on John's rational processes of perception, reasoning, and deliberation to intend and perform α.

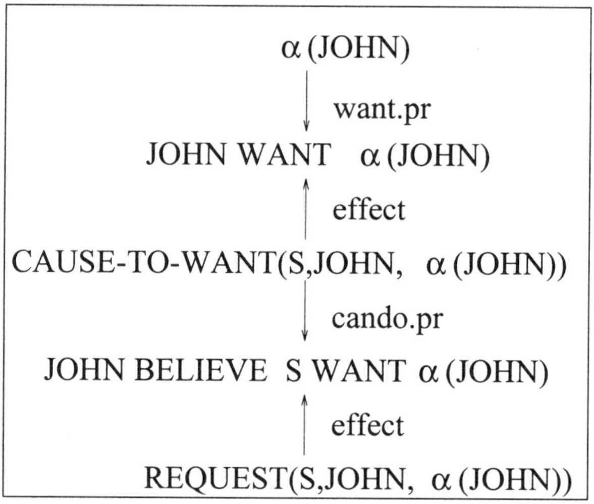

Figure 1. Cohen & Perrault (79) Plan for Request

This model has a wide range of applicability in cooperative situations, and has also been used to explain why an answer follows a question (i.e., if α is the performance of some speech act).

Allen and Perrault [8] use essentially the same formalism as Cohen and Perrault, but for a slightly different purpose. They investigate the role of plan inference and recognition in a cooperative setting. They show how the techniques of recognizing another agent's plans is important in interpreting utterances. For one thing, it can allow one to recognize an indirect speech act in a coherent and relevant manner. For example, Uttering (1) as a request for the information about when the train leaves.

(1) Do you know when the Windsor train leaves?

To do this, they bring in a level of *surface speech acts*, roughly comparable to Austin's locutionary act, which corresponds to the surface form, whereas the illocutionary act corresponds to the speaker's actual intention. Illocutionary acts are realized by surface speech acts, with an additional requirement that the hearer must recognize the speaker's intent. The planning system is again, basically a STRIPS system. There are preconditions and effects, and a body, which is a specification of the operator at a more detailed level, for instance decomposing an illocutionary act into a surface speech act.

The same logic was also used to provide an account of cooperative responses, where a hearer can deduce the speaker's plan and provide help even when it was not requested. For instance, providing the next step in a deduced plan, or detecting

REQUEST(speaker,hearer,act)
body: MB(hearer, speaker,
speaker WANT hearer DO act)
effect: hearer WANT hearer DO act

SURFACE-REQUEST(speaker,hearer,act)
effect: MB(hearer, speaker,
speaker WANT hearer DO act)

Figure 2. Allen (83) Speech Act Operators

obstacles in the speaker's plan that may be unknown to the speaker, and providing help.

This work was later expanded and improved in [3], where the speech act operators in Figure 2 were proposed.

While the Allen and Perrault model nicely handled the role of plan and intention recognition in recognizing indirect speech acts, it was not particularly helpful in relating the role of various aspects of the surface form. In fact, some linguists have proposed that indirect speech acts like (1) are really conventional. Hinkelman elegantly combined both linguistic and plan-based information [56, 55] to yield the best hypothesis for a current utterance. This was done by using linguistic cues to produce partial speech act templates, while plan inference was used to rule out certain possibilities and choose among those that were remaining.

3.2 Non-monotonic Theories of Speech Acts

Perrault [77] takes as a starting point the problem that the utterance itself is insufficient to determine the effects of a speech act. All effects of utterance actions are based in part on the prior context, including the mental states of the agents, as well as what was actually uttered. However, formalizing the precise conditions which must hold is a tricky endeavor, because of the many possible contingencies. Thus an axiom stating the effects of an utterance in declarative mood must take account of the possibilities of lies, failed lies, and irony as well as standard information-giving acts. Perrault's approach is to state the effects in terms of Default Logic [80], so that the simple, most common effects can be derived directly, unless there is some defeater. He has a simple axiomatization of belief, intention and action, along with some normal default rules, including a *Belief Transfer* rule which says that if one agent believes that another agent believes something the first agent will come to believe it too, and a *Declarative* rule, which states that if an agent said a declarative utterance, then it believes the propositional content of that utterance. This simple schema allows Perrault to derive expected consequences for the performance of a declarative utterance in different contexts.

Although the formalization is simple and elegant, it still contains a number of serious difficulties. Foremost is the lack of a serious treatment of belief revision. Although intuitively, speech acts are used to *change* beliefs, Perrault's framework can only handle the case of new beliefs being added. As well as not allowing the kind of discourses in which one agent would try to change the beliefs of another, the logic also has the strange property that one agent can convince itself of anything it has no prior beliefs about merely by making an utterance to that effect in the presence of another agent! The logic also does not lend itself to a computational implementation, since one would need a complete, inductive proof scheme to make all of the necessary deductions.

Appelt and Konolige [15] reformulate Perrault's theory in terms of Hierarchic Autoepistemic Logic [59]. This reformulation has the advantages of implementability and the ability to order the defaults to overcome the problems that Perrault had with normal default logic, but it also loses the simplicity of Perrault's framework. It is hard to see whether Appelt and Konolige are trying to describe something from the point of view of an ideal observer or from a participant in the conversation. In formulating their theory, they also resort to some unintuitive devices such as the beliefs of an utterance.

McRoy and Hirst use a uniform *abductive* approach for the production and recognition of speech acts, given aspects of the context as well as an (assumed) identification of surface form [75]. This approach lets them reason about mistakes in the interpretation process in a manner that is also amenable for reasoning about cancellable presuppositions and implicatures.

3.3 Dynamic Logic Approaches

One of the problems with work based on STRIPS is that it does not provide a fine enough characterization of time and change. Some researchers have thus moved to other logics of action which provide more powerful representational abilities, as well as a clearer semantics. The most explicit of these so far are based on variants of dynamic logic [53].

Cohen and Levesque

Cohen and Levesque have been attempting to solve a number of problems relating to formal characterizations of speech acts, through the use of a detailed logic of action and mental attitudes. [33] lays out the framework of the basic theory of rational action. It is based on a dynamic modal logic with a possible worlds semantics. They give axiomatizations for modal operators of beliefs and goals, and then derive intentions as persistent goals, those to which an agent is committed to either bring about or realize are unachievable.

Cohen and Levesque [34] use this logic to show how the effects of illocutionary acts can be derived from general principles of rational cooperative interaction. They claim, contrary to [90], that communicative acts are not primitive. They define what it means for an agent to be sincere and helpful, and give characterizations of imperatives and requests. They claim that recognizing the illocutionary force of an utterance is not necessary, that all that is important is that the hearer do what the speaker want, not that he recognize which act the speaker performed as a part of this process. They thus appear to be claiming that illocutionary acts should be seen as descriptive models of action, not as resources for agents. They conclude with a description of how Searle and Vanderveken's conditions on acts can be derived from their rational agent logic.

[32] extends the framework to handle Performatives. They define all illocutionary acts as attempts. Performatives like "I request you to . . . " are acts which have a request component and an assertion component, and the assertion component is made true merely by the *attempt*, not the success of the action. Thus *request* is a performative verb, while *frighten* is not (because it requires a successful attempt and the success is beyond the control of the speaker), and *lie* is paradoxical when used performatively, because the explicit mention defeats the aim.

Sadek

Sadek uses a similar logic of rational action [82] as the basis for semantics of speech acts [81, 83]. The basic logic differs from that of Cohen and Levesque in that possible worlds are seen as time points, not whole sequences of time points. Also, Sadek aims not only at a logic that can act as a meta-theory of the rational action of agents, but one that could be used by agents themselves in reasoning to accomplish their desires and intentions. The logic also allows planning acts to occur as a result of simple reasoning about the agent's own mental state, without any extra-logical procedures. This rational agent architecture and application to communicative acts has been implemented as the basis for the dialogue management components of the AGS spoken dialogue system [84, 19].

Sadek's communicative actions are formulated in terms of their feasibility preconditions, and intended perlocutionary effects. Each of these can lead to further planning (i.e., reasoning). The intended perlocutionary effects of an act are not (always) the actual effects of the performed action, but more the goal for which the action is planned. Some of these goals will also require other actions before being satisfied. In some ways, thus, these acts are *dialogue acts* rather than just simple speech acts. An example of Sadek's formalization of agent `i` requesting agent `j` to perform action `a` is shown in Figure 3.[1]

Discussion

While the approaches of Cohen and Levesque, and Sadek have the great benefit of clarity and semantic precision, they still contain some features which make them difficult to use as a general basis for modeling the use of speech acts. For

[1] `FP(a)[i\j]` means that all the feasibility conditions on the requested action a are included, where `i` and `j` are switched and only `i`'s mental state is kept. `PG` means persistent goal. See [81] for precise details.

$\langle i, REQUEST(j,a)\rangle$

FP: FP(a)[i\j] ∧ K(i,Agent(j,a)) ∧¬K(i,PG(j,Done(a)))
PE: Done(a)

Figure 3. Sadek (91) Speech Act Operator

one thing, the model of belief and other attitudes is based on a model logic that incorporates omniscience of logical consequence. While this might be acceptable in a metalogic, it is clearly not sufficient for an agent of limited means. Bretier and Sadek [19] get around this by using an incomplete forward chaining reasoner which is able to handle the important inferences relating to dialogue act planning. While this is adequate for the simple kinds of interaction used in the AGS system – querying a knowledge base about potential information servers, it remains to be seen if this method is powerful enough for more general interactions, such as more general collaborative problem solving.

Another difficulty is that the notion of time and sequence of events is not rich enough to model dialogue situations. The model of action inherited from dynamic logic allows only sequences of events. It also relies on predicates like **done(a)**, which means that action **a** was the last thing that happened. These together preclude adding temporally overlapping actions, since there would no longer be a clean action to serve as a transition from one world (or state) to the next.

A final difficulty is that the logic, being based on a notion of belief, goals, choice, and derived intention, is not adequate to model the whole range of speech acts. One thing that is missing is a notion of obligation or commitment to another, that is the main ingredient of commissives, for instance. Cohen and Levesque's *commitment* operator is more like a goal or intention than social commitment.

4 EXTENDING SPEECH ACTS TO DIALOGUE

The most interesting use of speech acts and pragmatics generally, is not the analysis of particular utterances, but the way that utterances are combined in a coherent interaction to further some other purposes. Here language is often intertwined with other kinds of action. While engaging in dialogue was always the aim of most of the speech act work, much of the early emphasis, particularly in the computational work was on the interpretation of isolated sentences (in context). Here we describe some work that extends the analysis towards dealing with dialogue phenomena.

4.1 Dialogue Function as Action

Litman and Allen extend Allen and Perrault's work to include dialogues rather than just single utterances, and to have a hierarchy of plans rather than just a single plan [68, 69]. They describe two different types of plans: domain plans and discourse plans. Domain plans are those used to perform a cooperative task, while discourse plans, such as *clarification* and *correction*, are task-independent plans which are concerned with using the discourse to further the goals of plans higher up in the intentional structure. They also use a notion of *meta-plan* to describe plans (including discourse plans) which have other plans as parameters. Using these notions, Litman and Allen are able to account for a larger range of utterances than previous plan-based approaches, including sub-dialogues to clarify or correct deficiencies in a plan under discussion.

Carberry and Lambert [25, 64, 63], make a further distinction, adding *problem solving* plans to the discourse and domain plans of Litman and Allen. Problem solving plans are ways of building up actual domain plans, and are usually achieved by way of discourse plans.

Cohen and Levesque extend the work described in Section 3.3, above, into a theory of *joint intention* and multi-agent action. [35] presents an analysis of why confirmations appear in task-oriented dialogue. Using their theory of joint intentions developed in [65], they state that the participants in one of these task-oriented dialogues have a joint intention that the task be completed. It is part of the definition of joint intention that if one party believes the object of intention to be already achieved or to be unachievable, that party must strive to make the belief mutual. It is this goal of mutual belief which drives the agent to communicate a confirmation. Although this is perhaps the first attempt in the computational literature which is explicitly concerned with a plan-based account of the generation of confirmations, it is noticeably lacking in several respects. It has no mention of how the intention to make something mutually believed turns into an intention to perform a confirmation. There is also some distance still from the logic to actual utterances. It is also not explained just what would count as a confirmation, and how one might recognize one.

4.2 Multiple Levels of Interactions

Several researchers have attempted to model the different kinds of phenomena in dialogue with different strata of action. Realizing that there is a range of action phenomena in dialogue, including but not limited to the sentence-level actions that have been concentrated on in speech act work, they extend the model to include multiple strata, even within the discourse act set. We consider some of those here.

One of the influential early classifications comes from the analysis of classroom dialogues [93], later modified in [39] and [94]. This was a system of 4 *ranks*[2], from

[2]according to the terminology of [51], *ranks* are hierarchical, such that an item in one rank can be built up of constituents at a lower level, like phrase and sentence in linguistics. *Levels*, on the other

Discourse Level	Act Type	Sample Acts
Sub UU	**Turn-taking**	take-turn, keep-turn, release-turn, assign-turn
UU	**Grounding**	Initiate, Continue, Ack, Repair, ReqRepair, ReqAck, Cancel
DU	**Core Speech Acts**	Inform, YNQ, Check, Eval Suggest, Request, Accept, Reject,
Multiple DUs	**Argumentation**	Elaborate, Summarize, Clarify Q&A Convince Find-Plan

Table 1. Traum & Hinkelman (92) Conversation Act Types

smallest to largest: act, move, exchange, transaction. Moves are speech acts used toward particular purposes in dialogue. Exchanges are a set of acts, including some by each speaker, such as a question, answer, and feedback unit. Transactions are whole subdialogues including multiple exchanges aimed toward achieving a particular task.

Several researchers have followed up on this line of work, calling the exchange structure instead a *dialogue game*, e.g., [71, 60, 26, 1]. This has a better parallelism with the idea of moves, and generally the distinction between moves and acts is dropped. Agents generally plan at the level of games, and then execute moves to further those games along the path they desire.

Novick, realizing that the traditional speech acts are insufficient for modeling dialogue control, introduces several levels of what he calls *meta-locutionary acts*, to contrast with Austin's locutionary, illocutionary, and perlocutionary acts. The meta-locutionary acts include levels for turn-taking, repair of mutual models, reference/information, and attention. Agents used plans at each of these levels to collaborate on a task.

Traum and Hinkelman introduced a scheme which incorporates some features of these other approaches. Rather than being based on ranks, this included four *levels* of acts, including turn-taking acts to coordinate who is speaking, *grounding acts* to coordinate the flow of mutual understanding, the traditional *core speech acts*, and *argumentation acts*, to handle higher-level coherence. These levels are summarized in Table 1. One difference from the other dialogue act taxonomies is that here, following [30], even the traditional speech acts are seen as multi-agent actions, since they require participation by both agents (by performance of grounding acts), before they have their full effects, such as a mutual belief or an established obligation.

Bunt [23, 24] describes several kinds of *dialogue acts*. Dialogue acts are seen as functions which update the dialogue context. Context is also divided into several

hand, cover different types of phenomena.

types, including *linguistic* – the surrounding utterances in the dialogue, *semantic* – the underlying task and domain, *physical* – the location of the participants and the interaction in space and time, *social* – the type of situation and roles the participants play with respect to each other, including rights and obligations, and *cognitive* – the mental states of the participants. Each of these types has both *global* and *local* aspects – the former are roughly constant throughout the dialogue, while the latter can change, through the performance of dialogue acts, among other things. Bunt then divides dialogue acts into *task-oriented acts* and *dialogue control acts*. All acts affect the linguistic and cognitive context, but the former also changes the semantic context, while dialogue control acts affect the social or physical context. Bunt only considers those task-oriented acts involved in information-seeking dialogues, namely information seeking, and information providing acts [22].

On the other hand, Bunt considers a wide range of classes of dialogue control acts. Following [13], there are several types of feedback acts. Bunt considers *perception* and *interpretation* acts, which correspond to the grounding acts of [101], but also evaluation acts, including acceptance and agreement, and *dispatch* acts, such as the fulfillment of a request, which appear at the argumentation level. Bunt notes, (see also [29]) that giving positive feedback at one level implies problems at a higher level, while giving negative feedback at one level implies success at lower levels. Bunt also has classes of *social obligations management* acts, which place or relieve pressure on one party to do something, and a range of *interaction management acts*, including acts for turn-management, own communication management, time management, contact management, and discourse structuring.

While all of these functions certainly occur in dialogue, it is sometimes difficult to see what the principles are for clustering them. In fact, Bunt's own classification of some of these types changes, e.g., from [23] to [24]. For instance, Bunt's contact management function could easily be seen as the lowest level of feedback, as in [13, 28], while *own communication management* (the production of speech repairs) is not necessarily closely related to interaction, while feedback clearly *is*. Likewise, a variety of acts introduce and dispose of obligations [99, 10].

Allwood has also written extensively about *Communicative Acts*, generalizing speech acts to cover other, non-verbal actions that work in a similar manner [9, 11]. Allwood rejects many of the distinctions in Austin's speech act theory, preferring, like [34], to reason directly from utterance form and context to function and effect.

Allwood distinguishes two kinds of effects of speech acts: *expressive*, which serve to indicate things about the speaker or previous state of the world, and *evocative*, which tend to (or are aimed at) bring about changes in the hearer or the world. Most utterances will have both of these kinds of effects. Table 2 shows some of the acts with the main expressive and evocative functions.

In addition, each utterance will be associated with a number of obligations, both for the sender and the receiver [10]. The sender related obligations are generally like the sincerity conditions of Searle's speech act theory [87]. For example, if an agent sends a request for information, that agent is obliged to actually wonder what the answer is, need the information, and believe that the receiver can provide

Communicative Act type	Expressive function	Evocative function
Statement	belief	belief judgment
Question	desire for information	the desired information
Request	desire for X	X

Table 2. Allwood (95) Functions of Some Communicative Acts

it. The receiver related obligations are similar to those described in [95] and in Section 7.2, below – mainly to evaluate the utterance and report on the evaluation. Allwood considers several types of feedback, based on four basic hierarchical communicative functions [13]: *contact*: whether the interlocutor is willing and able to continue the interaction, *perception*: whether the interlocutor is willing and able to perceive the message, *understanding*: whether the interlocutor is willing and able to understand the message, and *attitudinal reaction* - covering responses to the message such as acceptance or rejection.

The communicative function is not restricted to the content (termed main message), but also covers other aspects such as speech management (or own communication management) [12], and interactive communication management.

5 SPEECH ACT BASED AGENT COMMUNICATION LANGUAGES

Speech acts, as the abstract level of how utterances change the interactional state, are thus a good level of description for describing any agent communication, not just communication in natural language. As such, some researchers have also decided that a language based on speech acts would, by itself, be a good agent communication language. Thus the sometimes difficult process of speech act interpretation is bypassed, as the agents directly identify the speech act they are performing as a part of the message itself. Some researchers studying pragmatic aspects of natural language processing (e.g., [26, 76, 106]) have used this method as a way of testing their systems without having to do natural language interpretation. Several schemes based on speech acts have also been proposed, however, as agent communication languages in their own right, e.g. [92].

5.1 KQML

A recent proposal for a universal agent communication language is the *Knowledge Query and Manipulation Language* (KQML) [50, 62]. This was devised over several years by a committee of researchers from different groups interested in having heterogeneous agents communicate in a flexible manner. Some of the requirements for an agent communication language and how KQML attempts to meet these needs are described in [72]. The underlying agency model is one of knowledge-based agents that manipulate knowledge bases, but can send queries and answers to other agents, or broker services to other providers. The main idea of this language is based on Austin's *performatives*. Each message sent or received is a performative, with an identifier stating which kind of action it is, as well as other parameters identifying aspects of the content, such as what language and ontology is being used. Agents can then pass messages around without actually having to understand any details of the content language. KQML is also purposely underspecified in terms of the range of performatives allowed: in order to be KQML compliant, an agent must only be able to send and receive messages of the required syntax, and if it does use any of the reserved performatives, they have to have the same meaning. Designers are free to add extra performatives or not implement some of the reserved set. Some of the performatives are very familiar, such as *ask-if* and *tell*, whereas others are concerned more with the way messages are sent (all at once or one by one in a stream), or the way other agents can act as intermediaries for messages.

While KQML has been used in a number of groups as an agent communication language [44], there are still some problems with the specification. First it is lacking a precise semantics, which may cause divergences in the way performatives are used by different agents/groups. A first step towards this is provided by [61], but this only covers a few performatives. Cohen and Levesque [37] point out some further problems, in particular, vagueness of some performatives, mis-identified performatives (*achieve and broker* instead of *request*), and missing some performatives, including especially commissives. While KQML is flexible enough so that agents could avoid using the mis-identified performatives while adding the missing ones, it is disappointing that some of the basic building blocks of communication are left out.

5.2 Speech Acts as an Interface Language for Computer-mediated Collaboration

Some researchers have also advocated a limited interface based on speech acts even for human-human communication. There are generally two reasons for this: first, the computer mediator is doing more than just passing messages; it is also trying to do some analysis or updating of state, and thus the designers want to restrict the flexibility of communication to what the interface can handle (e.g., [17]). Another reason is to actually force explicitness, under the theory that this will make cooperation more productive [108].

5.3 Discussion

A number of issues arise in the attempt to bypass language interpretation and determine the speech act type directly from surface features of the message, by having agents explicitly signal the speech act type. As Sadock notes [85], the enterprise of speech act theory can be described as explicating the connections in (2).

(2) FORM $\overset{\longleftrightarrow}{\text{grammar}}$ CONTENT $\overset{\longleftrightarrow}{\text{pragmatics}}$ EFFECT

Thus, researchers are trying to trivialize the left relation by having agents communicate with forms precisely correlated with contents. While such limited languages can certainly reduce the ambiguity and vagueness of interpretation, it is arguable whether they can cut out such things as indirect speech acts and perlocutionary effects. Thus, even though a certain form is given a precise semantics in terms of it's direct effect, given that agents know something about how other agents operate, they could use particular forms for ulterior purposes. Other agents could even recognize such behavior, and novel conventions of interpretation can spring up (e.g., according to the theories of [67]). Such is almost undoubtedly the case in the computer-mediated circumstances, where humans have a limited channel in which to express all of their communicative and social desires. The only thing that might stop it in the case of artificial agents is the limitations of the agents themselves.

6 SPEECH ACT THEORY RECONSIDERED

Given the above discussion, there are a number of issues that arise in evaluating how best to view language as action, and whether speech act theory is a useful way to view the communication of agents. The most basic question is whether speech acts should be afforded some sort of ontological status as an intermediary between performances (e.g., language utterances) and effects (e.g., changes in mental state). In spite of their initial attractiveness, some have questioned the utility of this intermediary, because of a number of factors, especially the fact that there is (still) no small commonly agreed upon set of acts that can circumscribe the range of communicative interaction. Another difficult issue is the *multifunctionality* of utterances. As described above, a single utterance can introduce a number of different kinds of change. If one were to introduce a different act for each bundle of changes, there would be a large number of acts indeed! One can get around this by adopting a multi-stratal approach of hierarchical and/or orthogonal acts, such as those described above, but still, even within a particular stratum, a single utterance may perform multiple functions. E.g., someone might be simultaneously informing about the current state while suggesting a future course of action.

Another issue is whether speech acts are to be viewed objectively – as things that actually happened, according to some specified occurrence constraints, or as subjective entities, corresponding to a particular agent's views of what has been performed. Related to this is the question of what conditions beyond just the intention to perform an act are required to have the act actually be achieved. In other words, what distinguishes an attempt from successful performance of the act.

While such a level as speech acts might not be strictly necessary, they are still useful for agents in planning communication. It may be easier for agents to plan and recognize concrete bundles of function than trying to move directly from utterance features to context change. This is true even if agents do not even have the same precise ontology of action, or if it is not always possible to discriminate which action has been performed.

7 SPEECH ACTS IN A MULTI-AGENT ACTION THEORY

One of the main stumbling blocks for adequately formalizing speech acts is that it requires a very complex expressive logic of action and interaction. Speech acts always have at least two participating agents, and thus require a multi-agent logic. Moreover, in the most general case, action can proceed synchronously with other action (e.g., speech by the other agent). In addition, providing conditions and effects on performance of actions requires sophisticated reasoning about the mental states of agents. We sketch aspects of such a theory, using the conversation acts of [101, 95].

7.1 Reasoning about Multi-agent Action

A speech act theory which can account for conversations must include at least the following extensions to classical planning (e.g. STRIPS):

- temporal reasoning, including reasoning about overlapping and simultaneous actions
- uncertainty: attempted actions may fail to achieve their desired results, unexpected results may follow
- multiple agents, each with individual mental states
- cooperation among agents
- real-time resource-bounded reasoning
- integration of planning and acting

There is a large amount of research dedicated to addressing these problems, much more than can be summarized here. [97] explores some of the complexities involved in reasoning and acting in a richer environment. The annual European workshops on Modeling Autonomous Agents in a Multi-Agent World (MAAMAW) (reprinted in e.g., [40, 41, 107]) contain a variety of approaches to these problems.

7.2 Social Attitudes

One of the key features of *speech acts* as opposed to physical actions is that their main effects are on the mental and interactional states of agents, rather than on the state of some external domain. The attitudes of belief, desire, and intention are familiar to agency theories, (see for example chapters [79]s in this volume) but for a comprehensive theory of speech acts we must also go beyond these and consider *social attitudes*, which require component parts on the behalf of multiple agents. A big issue for these attitudes is whether they are basic or can be composed from more primitive individual attitudes such as belief and intention. While the holding of individual attitudes can certainly be deduced from holding of social attitudes, we will treat Mutual Belief and Obligations as basic, while proposing a compositional definition of executing a joint plan.

Mutual Belief and Grounding

Mutual belief is one commonly assumed social attitude. Most of the theories of speech acts as plans described above have as some of the main effects of speech acts the addition of some new *mutual beliefs*. Mutual beliefs are also taken to be some of the prerequisites for felicitous utterance of speech acts. But just what are mutual beliefs? This section reviews some of the proposals for how to represent the properties of mutual beliefs in terms of simpler beliefs, and how one could acquire new mutual beliefs. As with *belief* and *knowledge*, there have been a variety of names for a cluster of related concepts, including "mutual knowledge", "common knowledge", "mutual belief", and "shared information". In the discussion below, we generally use the term used by the author under discussion, but treat these terms as synonymously meaning *mutual belief*, in which what is mutually believed by a group of agents is not necessarily actually true.

Formulations of Mutual Belief

While people agree for the most part about the intuitions underlying the phenomenon of mutual belief, there have been a variety of different ways proposed of modeling it. Barwise [18] compares model theories for three different formulations.

Schiffer uses what Barwise calls "the *iterate* approach" [18, p. 202]. He defines mutual knowledge between two agents A and S of a proposition p, $K^*_{SA}p$ as [86, p. 30]: $K_S p \wedge K_A p \wedge K_S K_A p \wedge K_A K_S p \wedge K_S K_A K_S p \wedge K_A K_S K_A p \wedge \cdots$ It is thus an infinite conjunction of nested beliefs. This approach has since been adopted by many others, including Allen [3] and Perrault, who provides an elegant default logic theory of how to obtain each of these beliefs given prior knowledge and a conversational setting [77].

Barwise credits Harman with the *fixed-point approach*. Harman formulates mutual knowledge as "knowledge of a self-referential fact: A group of people have mutual knowledge of p if each knows *p and we know this*, where *this* refers to the whole fact known" [54, p. 422]. As Barwise points out, the fixed-point approach is strictly stronger than the iterate approach, because it includes as well the information that the common knowledge is itself common knowledge. It also replaces an infinite conjunction with a self-referential one.

The final approach discussed by Barwise is the *shared-situation approach*. He credits it to Lewis. Lewis formulates rules for common knowledge as follows [67, p. 56]:

Let us say that it is *common knowledge* in a population **P** that **X** if and only if some state of affairs **A** holds such that:

1. Everyone in **P** has reason to believe that **A** holds.
2. **A** indicates to everyone in **P** that everyone in **P** has reason to believe that **A** holds.
3. **A** indicates to everyone in **P** that **X**.

This schema is also used by Clark and Marshall, and is apparently the one which Barwise himself endorses.

[31] uses a *belief spaces* approach to model belief. Each space contains a set of propositions believed by an agent. Nested belief is represented by nested spaces. There is a space for the system's beliefs (SB) which can contain a space for the system's beliefs about the user's beliefs (SBUB) which in turn can contain a space for the system's beliefs about the user's beliefs about the system's beliefs (SBUBSB). If Cohen were to adopt the iterated approach directly, it would require an infinity of belief spaces. Instead, he takes the space one deeper than the deepest which contains any non-mutual beliefs, and points it to its parent space, thus creating a loop, where each even nesting is the same as every other even nesting. Now each of the nested beliefs in the iterated approach can be generated or seen to be present in his belief spaces, by iterating through the loop. This approach shares some features with the fixed-point approach (the self-referentiality) and it allows quick determination of whether mutual belief exists (by searching for a loop), unlike the iterated approach, but it is in fact not as strong as the fixed-point approach because the higher-order implications of the fixed-point approach, such as mutual belief about the mutual belief, cannot be represented.

A slight modification is to add a separate kind of space, a *mutual belief* space to represent mutual beliefs. This is the approach taken by Bruce and Newman [21]. The Rhetorical knowledge representation system [7] also uses a mutual belief space, but disallows nested beliefs within a mutual belief space, giving essentially the power of Cohen's system. This also seems to be the approach used by Maida [70].

How can Mutual Belief be Achieved?

While it is still somewhat controversial how best to formally model mutual belief, an interesting question is how such mutual belief gets established among interacting agents. While it has long been known that is impossible to guarantee the establishment of mutual knowledge in an environment where message transmission could fail [52], most formulations of speech acts have gone to the other extreme, and assumed mutual belief after the performance of any utterance within a shared situation (e.g., as the effect of a single agent speech act). However, examining actual conversation reveals that there is a process of feedback that accompanies initial utterances, and, in task-oriented spoken language, it is generally only after some sort of acknowledgment that an assumption of mutual belief is made. Furthermore, lack of understanding can be signaled with some sort of repair or request for repair. In cases in which the original speaker does not receive any feedback, one can observe requests for acknowledgment or repetitions and refashionings of the original contribution in an attempt to elicit some kind of feedback.

While the assumption of mutual belief resulting directly from a single utterance can be seen simply as an idealization (like modeling belief with a modal logic such as S5, with the resulting properties of logical omniscience), it is one with unfortunate consequences if used as the basis for models of speech acts and inter-agent communication. The consequences are twofold: first, a reasoner will make incorrect predictions about the mental state at particular times, and more importantly, the agent will be unable to recognize the relevance of or necessity for performing the feedback acts which actually establish the mutual understanding.

Clark and Shaefer refer to the process of reaching mutual belief (or common ground) as *grounding* [30]. They present a descriptive model, in terms of presentation and acceptance phases that allow them to track the augmentation of common ground as the conversation proceeds. Their model is not well-suited for an on-line agent involved in dialogue, however, since it requires examination of subsequent spans of text in order to determine the boundaries of these phases.

In [98, 95], we developed a computational account of the grounding process. This account was based on speech act theory, using actions to introduce, acknowledge, and repair material. Traditional speech acts, such as *inform*, and *request* are now seen as multi-agent actions, which require participation by both parties to have their full effects (such as the mutual belief that one speaker wanted the other speaker to believe something).We introduced a level of dialogue structure called *discourse units* (DUs), at which these core speech acts are completed. These DUs are built up by single-utterance *grounding acts*. Recognizing the fact that multiple types of action occur in conversation, we extended speech act theory to the multi-level *conversation act* theory, described in [101]. As well as the grounding and core speech acts, there are also levels to model turn-taking behavior and higher order coherence of dialogue. A finite automaton was used to track the state of a DU, given a sequence of grounding acts in conversation. This model could also be used to predict possible subsequent acts as well as determine which act(s) must be performed in order to have a grounded DU (which would thus realize the effects of the constituent core speech acts).

Obligations

We claim that *Obligations* are necessary for modeling many social situations including natural language conversation. For example, they are necessary for capturing the effects of some speech acts, such as requests [99]. Obligations represent what an agent *should* do, according to some set of norms; its formal aspects are examined using Deontic Logic (e.g., [105, 73]). Generally, obligation is defined in

source of obligation	**obliged action**
S_1 Accept or Promise A	S_1 achieve A
S_1 Request A	S_2 address Request: accept **or** reject A
S_1 YNQ whether P	S_2 Answer-if P
S_1 WHQ P(x)	S_2 Inform-ref x
utterance not understood or incorrect	repair utterance
S_1 Initiate DU	S_2 acknowledge DU
Request Repair of P	Repair P
Request Acknowledgment of P	acknowledge P

Table 3. Traum & Allen (94) Sample Obligation Rules

terms of a modal operator often called *permissible*. An action is *obligatory* if it is not permissible not to do it. An action is *forbidden* if it is not permissible.

Just because an action is obligatory with respect to a set of rules R does not mean that the agent will perform the action. So we do not adopt the model suggested by [91] in which agents' behavior cannot violate the defined social laws. If an obligation is not satisfied, then this means that one of the rules must have been broken. We assume that agents generally plan their actions to violate as few rules as possible, and so obligated actions will usually occur. But when they directly conflict with the agent's personal goals, the agent may choose to violate them (and perhaps suffer the consequences of not meeting its obligations). Obligations are quite different from and cannot be reduced to intentions and goals. In particular, an agent may be obliged to do an action that is contrary to its goals (for example, consider a child who has to apologize for hitting her younger brother). [42] use obligations in a similar way, noting also that authority (such as a pre-existing hierarchical relationship) can be important in the ability to force obligations on others.

In [99] we argued that obligations play an important role in accounting for many of the interactions in dialog. For example, Table 3 shows the obligations resulting from the performance of speech acts. Obligations do not replace the plan-based model of speech acts (e.g., [38, 8]) but augment it. The resulting model more readily accounts for discourse behavior in adversarial situations and other situations where it is implausible that the agents adopt each others' goals. The obligations encode learned social norms, and guide each agent's behavior without the need for intention recognition or the use of shared plans at the discourse level. While such complex intention recognition may be required in some interactions, it is not needed to handle the typical interactions of everyday discourse.

The deliberation process in a social situation must take obligations into account, in addition to goals and intentions. In forming new intentions, sometimes an agent will choose to pursue its obligations rather than its goals. It is important for the agent to reason about both of these notions, so that this choice can be made explicit (either in the agent design or by the agent itself). There is also an illuminating relationship between this deliberation process and the notion of initiative in dialogue. Following the initiative of the other can be seen as an *obligation-driven* process, while leading the conversation will be *goal-driven*.

There is another related concept that also occurs as the result of many speech acts: a commitment to certain facts being true [10]. These relate to the sincerity conditions postulated by Searle. Thus when an agent makes a claim, the agent is committed to the truth of that claim. Unlike the obligations considered above, there is no direct connection between this and future actions. The agent is obliged to keep these commitments to others accurate, but, if the agent comes to believe these are not accurate, there is no special preference for making them true – an equally valid approach is to just inform the other agent that the reported facts no longer hold, assuming the question is still of interest. This is to be contrasted with the kind of commitment to (or *intending that*) future states that is related to intention, and requires some action when the agent believes the state will not hold.

Coordinated Activity, Shared Plans and Joint Intention

Another common social attitude is that of *Joint intention* [35] or *shared plan* [47]. These concepts are used to model the propensity of a collaborative team to act. The intuition here is that it is more than just a collection of individual intentions and beliefs that is responsible for the coordinated teamwork activity. Although most agree that some sort of collective intentional attitude is useful in formalizing an account of cooperative behavior, it is still fairly controversial what the properties of such an attitude should be and how this collective intention is related to the individual intentions of the participants. Important questions include: how do shared intentions guide individual action, and how can individual beliefs and intentions come together to form shared intentions? This section reviews some proposals for such attitudes.

Lewis [67] defined a *Convention* as a situation in which there is some regularity R in behavior in a population, and everyone conforms to R, everyone expects everyone else to conform to R, and everyone prefers to conform to R, given that everyone else will. A typical example is which side of the road to drive on. In England and Japan it is the left side, in America and Europe, the right. It doesn't really matter to the drivers which side to drive on, as long as everyone agrees. Coordinated activity is thus seen as individual intention in a state of mutual knowledge about norms. Knowledge of conventions serve to make it in the mutual self-interest of each of the members of the population to follow along.

Grosz and Sidner [47] take basically the same viewpoint. They extend Pollack's definition of a Simple Plan [78], to that of a *SharedPlan*, which is formalized as a set of mutual beliefs about the executability of actions and the intentions of particular agents to perform parts of that action. They also present some conversational default rules based on cooperativeness to use communication to add to the shared beliefs. The shared plan formalism is further developed in [48, 49].

Searle [89] starts with the intuition that collective intention is not just a summation of individual intentions. He wants to distinguish between just following a convention and actual cooperative activity. He postulates that *we-intentions* are a primitive form of intentionality, not reducible to individual intentions. There is still a problem of how *we-intentions* can produce the individual intentions necessary for an individual to act.

Cohen and Levesque present their own theory, not in terms of individual intentions (which also aren't primitive in their theory) but in terms of mutual belief and weak mutual goals [65, 36]. Their formulation says that the individuals each have the goals to perform the action until they believe that either it has been accomplished or becomes impossible. Also in the event of it becoming completed or impossible, the agents must strive to make this belief mutual. This framework is also used to explain certain types of communicative behavior such as confirmations as described above.

In [95], we developed a similar notion, that of agents executing a multi-agent plan. This is an extension of the notion of executing a plan described in [100]. A group of agents $\{A_i\}$ is executing a multi-agent plan **MP** iff:

1. Each A_i is executing a single-agent plan $\mathbf{MP}_i$, which has as its actions A_i's actions from **MP**, and as its constraints the constraints of **MP**, as well as the occurrence of all actions by other agents (thus A_i will be committed to the occurrence of the actions of others).

2. Each A_i is obliged to the other agents to perform her own actions as part of the multi-agent plan.

This formulation has several differences from the other works mentioned. First, no mutual belief is stipulated as a necessary component of the multi-agent plan execution. While mutual beliefs may sometimes be important for collaboration, and particularly for decisions about adopting plans and repairing plan executions, they are not strictly necessary for this kind of teamwork. In this framework, it is the personal commitment to the occurrence of the actions of others, and the obligations *to* those others (as well as the personal intention to perform the action) that forms the glue binding the collaborating team together. While any agent may break the team at any time by dropping these commitments and intentions, the obligations will remain until the agent is released, and it is this which motivates such actions as letting another know that an action has been performed or is deemed impossible. Even with notions of mutual belief of intentions (as in the SharedPlan formalism of [47, 48]), or commitment to inform an agent if an action is performed or impossible (as in the joint intentions of [35]), it is hard to see, in practice, what keeps an agent adhering to these commitments when its personal goals diverge.

7.3 *Defining Speech Acts*

Given the rich logic of mental state and action partially developed above, how can we use this to give precise definitions of speech acts? There are actually several important components for speech acts. One is the essential characteristics of what *is* and *isn't* a particular act. Another important criteria is how one can recognize whether such an act has been performed. Since speech acts have as part of their defining characteristics (both enabling conditions and effects) aspects of the mental states of communicating agents, a particular observer (even a participant) may not have access to all the necessary information, and so may not be able to tell (with certainty) whether a particular act has occurred, or whether a particular bit of dialogue is the performance of a particular kind of act. A third component is the use of speech acts in planning – how can agents plan to use speech acts to accomplish their (non-linguistic) goals, and how are these plans related to actual behavior and the realization of such acts? We consider each of these in turn for some of the speech act types of [101], shown in Table 1.

The turn-taking acts can be easily described using an interval temporal logic, such as that of [2, 4, 5]. Of interest are periods of each speaker's speech and silence, and the turn, itself. Events such as starting, continuing and stopping speaking, as well as other features such as prosody and expectations will be interpreted as turn-taking acts and serve to delimit individual turns. Even these simple acts would be difficult with a simple dynamic logic, since overlapping intervals of the two agents speaking must be considered.

Since, according to the theory of [101], all core speech acts must be grounded to have their full effect, it will be instructive to first look at the grounding acts. We use the formulation of executing a multi-agent plan described in Section 7.2 to formally model the grounding process described earlier in that section. Agents involved in a task oriented dialogue are assumed to be executing a specialization of the abstract plan recipe shown in Figure 4. This abstract recipe will be called CR (for Communication Recipe). Our claim is that successful execution of (a specialization of) this recipe will result in the (mutually assumed) mutual belief between the two agents that INITIATOR(CR) has communicated CONTENT(CR) to RESPONDER(CR). Agents engaged in conversation can be modeled as executing multi-agent plans that are specializations of this recipe. We will call any plan which has as its recipe a specialization of CR a *conversation plan*.

The acts of presenting and acknowledging the content are broken into some indeterminate number of conceptual sub-acts, about at the granularity of the propositions in [58]. The only constraints on performance of these is that for a particular piece of the content, the presentation must come before the acknowledgment, and both must occur to achieve mutual belief. Constraints on exactly what types of executions can present and acknowledge the above contents will be determined by conventions of the particular language used and the communicative contexts. For any given execution, some parts of the content will be expressed explicitly as part of the compositional conventional meaning of the utterance, and others will

$$\textbf{Actions}(\text{CR}) = \{\text{present}_i \mid \text{present}_i = \text{Present}(\text{Agent1}_i, \text{Content}_i, \text{Recipient1}_i, \text{t1}_i)\} \cup \{\text{ack}_i \mid \text{ack}_i = \text{Ack}(\text{Agent2}_i, \text{Content}_i, \text{Recipient2}_i, \text{t2}_i)\}$$

Constraints(CR) =	{
Temporal constraints	$\forall i(\text{Before}(\text{t1}_i, \text{t2}_i))$,
Agent constraints	$\forall i(\text{Agent1}_i = \text{Recipient2}_i = \text{INITIATOR}(\text{CR}),$
	$\text{Agent2}_i = \text{Recipient1}_i = \text{RESPONDER}(\text{CR}))$,
Object Constraints	$\text{CONTENT}(\text{CR}) = \cup_i \text{Content}_i\}$

Figure 4. Traum (94) Plan Recipe for Communication (Recipe CR)

act	effects
Promise(A, B, Act)	Obliged(A, B, Act)
Accept(A, B, Act)	Committed(A, B, Cando(A, Act))
Request(A, B, Act)	Obliged(B, A, Do(A, Accept(Act) $\vee$ Reject(Act)))
	Committed(A, B, Believe(A, Cando(B, Act)))
	Committed(A, B, Desire(A, Do(B, Act)))
YNQ(A, B, ϕ)	Committed(A, B, $\neg$Knowif(A, ϕ))
	Committed(A, B, Desire(A, Knowif(A, ϕ)))
Assert(A, B, ϕ)	Committed(A, B, ϕ)

Table 4. Sample Speech Act Effects

be presented implicitly by conventions of situated meaning and Gricean implicatures. In [95], the grounding acts from [98] were given formal specifications as executions related to conversation plans. As well as the components for executing a multi-agent plan, additional definitions are given in a situation logic for actions of starting, continuing, completing, and repairing the execution of a plan.

The content of the grounding actions described above will include the speech act type (illocutionary force) of the actions as well as the contents of core speech acts. Thus, the performance of each core speech act could also be decomposed as the performance of a conversation plan, with appropriate constraints on the contents. All core speech acts thus include the effects in (3). In addition, some of the other effects are shown in Table 4. Here, *Committed* is meant to be a social commitment to a state of affairs, rather than the individual commitment of [33].

(3) MB(A,B,occurs(act)) $\wedge$ MB(A,B,holds(effect(act))).

The actual effects of interest will generally come from reasoning about the reasoning and deliberation of other agents, such as B coming to believe ϕ, or even B coming to believe that A believes B, given that A commits herself to its truth.

The definitions of the argumentation acts will be sequences of core speech acts, with constraints on the timing and content. E.g., that the answer actually provides information asked for by the question.

7.4 Planning Speech Acts

Definitions of speech acts only capture part of the reasoning necessary for planning speech acts and utterances. No matter how exhaustive the list of effects, agents may still care not about the direct effects, but the extended perlocutionary effects springing from the performance of the act in context. Likewise, agents may choose to perform acts which have only a certain probability of producing the effect desired. Acts can also be planned at different levels – one may plan a whole argumentation act (or dialogue game), or merely a single speech act.

Because of the on-line, interactive nature of dialogue, it is generally not very useful to plan very far in advance. Often, an agent can not predict the future actions of others with very much accuracy. Thus it usually only pays to plan only a few utterances in advance, while reacting to the actions of others in a manner so as to move towards achievement of the agent's goals. Some kinds of dialogues, such as arguments or negotiations may require more involved planning, while often casual conversation can be completely unplanned except for the very next utterance.

7.5 Recognizing Speech Acts

An agent will combine features of an input utterance with aspects of the current context (including the mental states of agents, as well as the visual field and the set of previous utterances) to decide what acts have been performed.

One big question for recognizing speech acts is whether one should be trying just to recognize the acts themselves or the intentions (e.g., plans) that motivated the other agent to produce the act. Often, as in the case of indirect speech acts, one cannot easily separate these questions – one may come to decide that an agent has performed a request by deciding that the agent wanted one to do something. Still, intentions and performed actions can sometimes diverge, e.g., an agent can mistakenly or unintentionally perform an action, which may carry with it all of the responsibilities and other effects as if it were intended. Likewise, intentions which were not (fully) realizes will not carry the burden of imposing obligations and social commitments.

Sometimes, especially when agents are cooperative, deducing the intention is actually more important/useful than what was actually done. Moreover, deducing *why* an agent performed the act can lead to helpful behavior (or avoidance of traps in a competitive dialogue).

The question arises, however, as to how much of the plan it is necessary to infer. McRoy, for instance, challenges the utility of the deep intention recognition in [8], claiming instead that considering the possible actions and their immediate effects (including expectations set up for future utterances) is sufficient, especially when combined with a facility to repair erroneous conclusions [74, 75].

While recognizing the intentions and plans of a speaker can play a useful role in act interpretation, and may in fact be crucial to an agent's actual success in some domains, one must also remember the on-line nature of dialogue in this case as well. More prominence should be placed on interpreting the act itself, and performing this interpretation in real time. Since the speaker is easily accessible by the interpreting agent, if there are difficulties understanding either the act itself, or the motivations, the agent can query to find out what was meant or why it was said, through the grounding process.

8 CONCLUSIONS

In summary, we believe that speech acts are a good link between the mental states of agents and purposeful communication. While a comprehensive theory of speech acts will strain most contemporary theories of rational agency, requiring a very expressive theory of action and mental state, it also provides a good testbed for a theory of agency in a multi-agent world. While many of the specific concerns of natural language speech act work, such as the interpretation of ambiguous utterances, may be alleviated by using more restricted agent communication languages, many of the same issues will arise for cooperative systems that must infer intentions of others.

ACKNOWLEDGMENTS

The author's ideas on speech acts reported above have been greatly aided by conversations with many others, especially: James Allen, Jens Allwood, Harry Bunt, Peter Heeman, Elizabeth Hinkelman, Massimo Poesio, Teresa Sikorski, and Richmond Thomason. The author was partially supported while writing this paper by the U.S. Army Research Office under contract/grant number DAAH 04 95 10628 and the U.S. National Science Foundation under grant IRI-9311988. Some of the work described above was developed in collaboration with James Allen and supported by ONR/DARPA under grant number N00014-92-J-1512, by ONR under research grant number N00014-90-J-1811, and by NSF under grant number IRI-9003841.

David Traum (University of Maryland, USA)

REFERENCES

[1] Gabriella Airenti, Bruno G. Bara, and Marco Colombetti. Conversation and behavior games in the pragmatics of dialogue. *Cognitive Science*, 17:197–256, 1993.

[2] James F. Allen. Maintaining knowledge about temporal intervals. *Communications of the ACM*, 26(11):832–843, November 1983.

[3] James [F.] Allen. Recognizing intentions from natural language utterances. In Michael Brady and Robert C. Berwick, editors, *Computational Models of Discourse*. MIT Press, 1983.

[4] James F. Allen. Time and planning. In R. Pelavin J. Allen, H. Kautz and J. Tenenberg, editors, *Reasoning About Plans*. Morgan Kaufmann, 1991.

[5] James F. Allen and George Ferguson. Actions and events in interval temporal logic. *Journal of Logic and Computation*, 4(5), 1994.

[6] James [F.] Allen, James Hendler, and Austin Tate, editors. *Readings In Planning.* Morgan Kaufmann, 1990.
[7] James F. Allen and B. W. Miller. The RHET system: A sequence of self-guided tutorials. Technical Report 325, Department of Computer Science, University of Rochester, July 1991.
[8] James F. Allen and C. Raymond Perrault. Analyzing intention in utterances. *Artificial Intelligence*, 15(3):143–178, 1980.
[9] Jens Allwood. *Linguistic Communication as Action and Cooperation.* PhD thesis, Göteborg University, Department of Linguistics, 1976.
[10] Jens Allwood. Obligations and options in dialogue. *Think Quarterly*, 3:9–18, 1994.
[11] Jens Allwood. An activity based approach to pragmatics. Technical Report (GPTL) 75, Gothenburg Papers in Theoretical Linguistics, University of Göteborg, 1995.
[12] Jens Allwood, Joakim Nivre, and Elisabeth Ahlsen. Speech management: On the non-written life of speech. Technical Report (GPTL) 58, Gothenburg Papers in Theoretical Linguistics, University of Göteborg, 1989.
[13] Jens Allwood, Joakim Nivre, and Elisabeth Ahlsen. On the semantics and pragmatics of linguistic feedback. *Journal of Semantics*, 9, 1992.
[14] G. E. M. Anscombe. *Intention.* Basil Blackwell, 1957.
[15] Douglas Appelt and Kurt Konolige. A nonmonotonic logic for reasoning about speech acts and belief revision. In *Proceedings of Second International Workshop on Non-Monotonic Reasoning*, pages 164–175, 1988.
[16] J. A. Austin. *How to Do Things with Words.* Harvard University Press, 1962.
[17] M. Baker and C. Lund. Flexibly structuring interaction in a CSCL environment. In *Proceedings of the European Conference on AI in Education.*, pages 401–407, 1996.
[18] Jon Barwise. *The Situation in Logic*, chapter 9: On the Model Theory of Common Knowledge. CSLI Lecture Notes: Number 17. Center for The Study of Language and Information, 1989.
[19] P. Bretier and M. D. Sadek. A rational agent as the kernel of a cooperative spoken dialogue system: Implementing a logical theory of interaction. In J. P. Müller, M. J. Wooldridge, and N. R. Jennings, editors, *Intelligent Agents III — Proceedings of the Third International Workshop on Agent Theories, Architectures, and Languages (ATAL-96)*, Lecture Notes in Artificial Intelligence. Springer-Verlag, Heidelberg, 1996.
[20] Bertram C. Bruce. Generation as a social action. In *Theoretical Issues in Natural Language Processing-1*, pages 64–67, 1975. Also appears in [46], pp. 419-422.
[21] Bertram C. Bruce and Denis Newman. Interacting plans. *Cognitive Science*, 2:195–233, 1978.
[22] H. C. Bunt. Information dialogues as communicative action in relation to partner modelling and information processing. In M.M Taylor, F. Neel, and D. G. Bouwhuis, editors, *The Structure of Multimodal Dialogue*. Elsevier Science Publishers B.V., 1989.
[23] Harry Bunt. Context and dialogue control. *Think Quarterly*, 3:19–31, 1994.
[24] Harry Bunt. Interaction management functions and context representation requirements. In *Proceedings of the Twente Workshop on Langauge Technology: Dialogue Management in Natural Language Systems (TWLT 11)*, pages 187–198, 1994.
[25] S. Carberry. *Plan Recognition in Natural Language Dialogue.* The MIT Press, Cambridge, MA, 1990.
[26] Jean Carletta. *Risk-taking and Recovery in Task-Oriented Dialogue.* PhD thesis, University of Edinburgh, 1992.
[27] Herbert H. Clark. *Arenas of Language Use.* University of Chicago Press, 1992.
[28] Herbert H. Clark. Managing problems in speaking. *Speech Communication*, 15:243 – 250, 1994.
[29] Herbert H. Clark and Edward F. Schaefer. Collaborating on contributions to conversation. *Language and Cognitive Processes*, 2:1–23, 1987.
[30] Herbert H. Clark and Edward F. Schaefer. Contributing to discourse. *Cognitive Science*, 13:259–294, 1989. Also appears as Chapter 5 in [27].
[31] Phillip R. Cohen. *On Knowing What to Say: Planning Speech Acts.* PhD thesis, University of Toronto, 1978. Reproduced as TR 118 Department of Computer Science, University of Toronto.
[32] Phillip R. Cohen and Hector J. Levesque. Performatives in a rationally based speech act theory. In *Proceedings ACL-90*, pages 79–88, 1990.
[33] Phillip R. Cohen and Hector J. Levesque. Persistence, intention, and commitment. In P. R. Cohen, J. Morgan, and M. E. Pollack, editors, *Intentions in Communication.* MIT Press, 1990.
[34] Phillip R. Cohen and Hector J. Levesque. Rational interaction as the basis for communication. In P. R. Cohen, J. Morgan, and M. E. Pollack, editors, *Intentions in Communication.* MIT Press, 1990.

[35] Phillip R. Cohen and Hector J. Levesque. Confirmations and joint action. In *Proceedings IJCAI-91*, pages 951–957, 1991.

[36] Phillip R. Cohen and Hector J. Levesque. Teamwork. *Nous*, 35, 1991.

[37] Phillip R. Cohen and Hector J. Levesque. Communicative actions for artificial agents. In *Proceedings of the First International Conference on Multi-Agent Systems (ICMAS-95)*, pages 65–72, June 1995.

[38] Phillip R. Cohen and C. R. Perrault. Elements of a plan-based theory of speech acts. *Cognitive Science*, 3(3):177–212, 1979.

[39] M. Coulthard, M. Montgomery, and D. Brazil. Developing a description of spoken discourse. In M. Coulthard and M. Montgomery, editors, *Studies in Discourse Analysis*, pages 1–50. Routledge & Kegan Paul, 1981.

[40] Y. Demazeau and J. P. Muller, editors. *Decentralized A.I.* Elsevier Science Publishers B. V., 1990.

[41] Y. Demazeau and J. P. Muller, editors. *Decentralized A.I. 2*. Elsevier Science Publishers B. V., 1991.

[42] F. Dignum and B. van Linder. Modeling social agents: Communication as action. In J. P. Müller, M. J. Wooldridge, and N. R. Jennings, editors, *Intelligent Agents III — Proceedings of the Third International Workshop on Agent Theories, Architectures, and Languages (ATAL-96)*, Lecture Notes in Artificial Intelligence. Springer-Verlag, Heidelberg, 1996.

[43] Richard E. Fikes and Nils J. Nilsson. STRIPS: A new approach to the application of theorem proving to problem solving. *Artificial Intelligence*, 2:189–208, 1971. Also appears in [6].

[44] Tim Finin, Richard Fritzson, Don McKay, and Robin McEntire. Kqml as an agent communication language. In *Proceedings of the Third International Conference on Information and Knowledge Management (CIKM'94)*, 1994.

[45] H. P. Grice. Meaning. *Philosophical Review*, 66:377–88, 1957.

[46] Barbara J. Grosz, Karen Sparck Jones, and Bonnie Lynn Webber, editors. *Readings In Natural Language Processing*. Morgan Kaufmann, 1986.

[47] Barbara J. Grosz and Candace L. Sidner. Plans for discourse. In P. R. Cohen, J. Morgan, and M. E. Pollack, editors, *Intentions in Communication*. MIT Press, 1990.

[48] Barbara J. Grosz and Sarit Kraus. Collaborative plans for complex group action. *Artificial Intelligence*, 86(2):269–357, 1996.

[49] Barbara J. Grosz and Sarit Kraus. The Evolution of SharedPlans. In this volume.

[50] External Interfaces Working Group. Draft specification of the kqml agent-communication language. available through the WWW at: http://www.cs.umbc.edu/kqml/papers/, 1993.

[51] M. A. K. Halliday. Categories of the theory of grammar. *Word*, 17:241–92, 1961.

[52] J. Y. Halpern and Y. Moses. Knowledge and common knowledge in a distributed environment. *Journal of the ACM*, 37(3):549–587, 1990.

[53] D. Harel. *First Order Dynamic Logic*. Springer-Verlag, 1979.

[54] Gilbert Harman. Review of *linguistic behaviour* by jonathan bennett. *Language*, 53:417–424, 1977.

[55] Elizabeth A. Hinkelman. *Linguistic and Pragmatic Constraints on Utterance Interpretation*. PhD thesis, University of Rochester, 1990.

[56] Elizabeth A. Hinkelman and James F. Allen. Two constraints on speech act ambiguity. In *Proceedings ACL-89*, pages 212–219, 1989.

[57] Jaakko Hintikka. *Knowledge and belief; an introduction to the logic of the two notions*. Cornell University Press, 1962.

[58] Jerry Hobbs. Ontological promiscuity. In *Proceedings ACL-85*, pages 61–69, 1985.

[59] Kurt Konolige. Hierarchic autoepistemic theories for nonmonotonic reasoning: preliminary report. In *Proceedings of Second International Workshop on Non-monotonic Reasoning*, pages 42–59, 1988.

[60] Jacqueline C. Kowtko, S. Isard, and G. Doherty. Conversational games within dialogue. In *Proceedings of the ESPRIT Workshop on Discourse Coherence*, 1991.

[61] Yannis Labrou and Tim Finin. A semantics approach for kqml – a general purpose communication language for software agents. In *Proceedings of the Third International Conference on Information and Knowledge Management (CIKM'94)*, 1994.

[62] Yannis Labrou and Tim Finin. A proposal for a new kqml specification. Technical Report CS-97-03, Computer Science and Electrical Engineering Department, University of Maryland Baltimore County, 1997.

[63] Lynn Lambert. *Recognizing Complex Discourse Acts: A Tripartite Plan-Based Model of Dialogue*. PhD thesis, University of Delaware, 1993. Reproduced as TR 93-19 Department of Computer and Information Science, University of Delaware.
[64] Lynn Lambert and Sandra Carberry. A triparite plan-based model of discourse. In *Proceedings of the 29th Annual Meeting of the Association for Computational Linguistics*, pages 47–544, 1991.
[65] Hector J. Levesque, Phillip R. Cohen, and Jose H. T. Nunes. On acting together. In *Proceedings AAAI-90*, pages 94–99, 1990.
[66] Stephen C. Levinson. *Pragmatics*. Cambridge University Press, 1983.
[67] David K. Lewis. *Convention: A Philosophical Study*. Harvard University Press, 1969.
[68] Diane J. Litman. *Plan Recognition and Discourse Analysis: An Integrated Approach for Understanding Dialogues*. PhD thesis, University of Rochester, 1985. Reproduced as TR 170 Department of Computer Science, University of Rochester.
[69] Diane J. Litman and James F. Allen. Discourse processing and common sense plans. In P. R. Cohen, J. Morgan, and M. E. Pollack, editors, *Intentions in Communication*. MIT Press, 1990.
[70] Anthony S. Maida. Belief spaces: Foundations of a computational theory of belief. Technical Report Cs-84-22, Pennsylvania State University Department of Computer Science, December 1984.
[71] William C. Mann. Dialogue games: Conventions of human interaction. *Argumentation*, 2:511–532, 1988.
[72] J. Mayfield, Y. Labrou, and T. Fini. Desiderata for agent communication languages. In *Working Notes AAAI Spring Symposium on Information Gathering from Heterogeneous, Distributed Environment.*, March 1995.
[73] L. Thorne McCarty. Permissions and obligations: An informal introduction. Technical Report LRP-TR-19, Dept. of Computer Science, Rutgers University, 1986.
[74] Susan McRoy. *Abductive Interpretation and Reinterpretation of Natural Language Utterances*. PhD thesis, University of Toronto, 1993. Reproduced as TR CSRI-288 Department of Computer Science, University of Toronto.
[75] Susan W. McRoy and Graeme Hirst. The repair of speech act misunderstandings by abductive inference. *Computational Linguistics*, 21(4):5–478, 1995.
[76] David Novick. *Control of Mixed-Initiative Discourse Through Meta-Locutionary Acts: A Computational Model*. PhD thesis, University of Oregon, 1988. also available as U. Oregon Computer and Information Science Tech Report CIS-TR-88-18.
[77] C. Raymond Perrault. An application of default logic to speech act theory. In P. R. Cohen, J. Morgan, and M. E. Pollack, editors, *Intentions in Communication*. MIT Press, 1990.
[78] Martha E. Pollack. Plans as complex mental attitudes. In P. R. Cohen, J. Morgan, and M. E. Pollack, editors, *Intentions in Communication*. MIT Press, 1990.
[79] A. Rao and M. Wooldridge Foundations of Rational Agency. In this volume.
[80] R. Reiter. A logic for default reasoning. *Artifical Intelligence*, 13(1,2):81–132, April 1980.
[81] M. D. Sadek. Dialogue acts are rational plans. In *Proceedings of the ESCA/ETR workshop on multi-modal dialogue*, 1991.
[82] M. D. Sadek. A study in the logic of intention. In C. Rich, W. Swartout, and B. Nebel, editors, *Proceedings of Knowledge Representation and Reasoning (KR&R-92)*, pages 462–473, 1992.
[83] M. D. Sadek. Communication theory = rationality principles + communicative act models. In *Proceedings of the AAAI Workshop on Planning for Interagent communication*, 1994.
[84] M. D. Sadek, A. Ferrieux, and A. Cozannet. Towards an artificial agent as the kernel of a spoken dialogue system: A progress report. In *Proceedings of the AAAI Workshop on Integration of Natural Language and Speech Processing*, 1994.
[85] J. M. Sadock. Comments on Vanderveken and on Cohen and Levesque. In P. R. Cohen, J. Morgan, and M. E. Pollack, editors, *Intentions in Communication*, pages 257–270. MIT Press, 1990.
[86] Stephen R. Schiffer. *Meaning*. Oxford University Press, 1972.
[87] John R. Searle. *Speech Acts*. Cambridge University Press, New York, 1969.
[88] John R. Searle. A classification of illocutionary acts. *Language in Society*, 5:1–23, 1976.
[89] John R. Searle. Collective intentions and actions. In P. R. Cohen, J. Morgan, and M. E. Pollack, editors, *Intentions in Communication*. MIT Press, 1990.
[90] John R. Searle and Daniel Vanderveken. *Foundations of Illocutionary Logic*. Cambridge University Press, 1985.
[91] Yoav Shoham and Moshe Tennenholtz. On the synthesis of useful social laws for artificial agent societies. In *Proceedings AAAI-92*, pages 276–281, 1992.

[92] Candace L. Sidner. An artificial discourse language for collaborative negotiation. In *Proceedings of the forteenth National Conference of the American Association for Artificial Intelligence (AAAI-94)*, pages 814–819, 1994.
[93] J. M. Sinclair and R. M. Coulthard. *Towards an analysis of Discourse: The English used by teachers and pupils.* Oxford University Press, 1975.
[94] Anna-Brita Stenstrom. *Questions and Responses.* Lund Studies in English: Number 68. Lund : CWK Gleerup, 1984.
[95] David R. Traum. *A Computational Theory of Grounding in Natural Language Conversation.* PhD thesis, Department of Computer Science, University of Rochester, 1994. Also available as TR 545, Department of Computer Science, University of Rochester.
[96] David R. Traum. A reactive-deliberative model of dialogue agency. In J. P. Müller, M. J. Wooldridge, and N. R. Jennings, editors, *Intelligent Agents III — Proceedings of the Third International Workshop on Agent Theories, Architectures, and Languages (ATAL-96)*, Lecture Notes in Artificial Intelligence. Springer-Verlag, Heidelberg, 1996.
[97] David R. Traum and James F. Allen. Causative forces in multi-agent planning. In Y. Demazeau and J. P. Muller, editors, *Decentralized A.I. 2*, pages 89–105. Elsevier Science Publishers B. V., 1991.
[98] David R. Traum and James F. Allen. A speech acts approach to grounding in conversation. In *Proceedings 2nd International Conference on Spoken Language Processing (ICSLP-92)*, pages 137–40, October 1992.
[99] David R. Traum and James F. Allen. Discourse obligations in dialogue processing. In *Proceedings of the 32th Annual Meeting of the Association for Computational Linguistics*, pages 1–8, 1994.
[100] David R. Traum and James F. Allen. Towards a formal theory of repair in plan execution and plan recognition. In *Proceedings of the 13th Workshop of the UK Planning and Scheduling Special Interest Group*, September 1994.
[101] David R. Traum and Elizabeth A. Hinkelman. Conversation acts in task-oriented spoken dialogue. *Computational Intelligence*, 8(3):575–599, 1992. Special Issue on Non-literal language.
[102] W. Van Der Hoek and B. Van Linder and J-J.CH. Meyer. An integrated Modal Appraoch to Rational Agents. In this volume.
[103] Daniel Vanderveken. On the unification of speech act theory and formal semantics. In P. R. Cohen, J. Morgan, and M. E. Pollack, editors, *Intentions in Communication.* MIT Press, 1990.
[104] Daniel Vanderveken. *Meaning and Speech Acts.* Cambridge University Press, 1990-1991.
[105] G. H. von Wright. Deontic logic. *Mind*, 60:1–15, 1951.
[106] Marilyn A. Walker. *Informational Redundancy and Resource Bounds in Dialogue.* PhD thesis, University of Pennsylvania, 1993.
[107] E. Werner and Y. Demazeau, editors. *Decentralized A.I. 3.* Elsevier Science Publishers B. V., 1992.
[108] Terry Winograd and Fernando Flores. *Understanding Computers and Cognition.* Addison-Wesley Publishing Company, Inc., 1986.

A. HADDADI

COMMUNICATION AMONG RATIONAL AGENTS

ABSTRACT: In this article we identify and give the formal semantics of a set of *generic speech acts* that can be performed by agents in communicating about a potential cooperation, and demonstrate how agents may reason about when, with whom, and what to communicate in this process. The aim of this exercise is to highlight various facets of study on communication and the diversity of issues to be considered in designing agents that can intelligently, flexibly and most of all, effectively communicate with one another. This is particularly important for the design and development of standards and protocols for communication among heterogeneous agents that often share little or no commonality in their internal design. As is a tradition in AI, a thorough theory of communication in multi-agent systems takes its input from a diverse range of disciplines and this article attempts to integrate some of the more important, seemingly separate viewpoints, into a unified study of communication among rational agents.

1 A LITERATURE SURVEY

1.1 Intentional Systems

There are two main traditions in the study of rationality. In decision theory which has its roots in Economics, rationality refers to selecting a choice that results in an optimal "gain" or "profits". Generally, the studies in decision theory model what could happen if a given option is selected, by associating probabilities and utilities[1] to each possible outcome. A rational choice is one that maximises expected utility.

The other category stems from philosophy, where rationality generally refers to "obeying the rules of logic". In practice, this involves developing a system of logic by describing the premises of the system in a set of basic axioms, and then devising inference rules which determine if a particular proposition follows from the given premises. A rational choice then, is one which the arguments for its selection logically follows from the premises of the system. In contrast to the work in decision theory which models rationality in terms of how best to maximise some quantitative criteria, logical rationality is a qualitative account, and therefore the corresponding approaches try to understand and model the interplay between various qualitative criteria. Another way of looking at this is that decision-theoretic approaches enable rational *decision making*, while logical approaches enable rational *reasoning*.

The theoretical studies based on the logical tradition, model agents as *intentional systems*, that is, systems whose behaviour is modelled in terms of some

[1] Utility of a state or an outcome is a numerical value resulting from subtracting the cost of reaching that state from the value of being in that state.

M. Wooldridge and A. Rao (eds.), Foundations of Rational Agency, 203–226.

mentalistic notions. For instance, one could express an agent's problem-solving mechanism in terms of what it may *believe*, how it may go about achieving its *goals* and *wishes*, which *skills* and *capabilities* it possesses, and relate its behaviour to its individual characteristics and attributes like its *preferences* and its particular *attitudes* in a particular context. By far, the best account of intentional systems is given by Dennett [14], who identifies six conditions which respectively describe an agent with increasing levels of sophistication:

1. rationality
2. intentionality
3. stance
4. reciprocity
5. communication
6. consciousness

The first three are mutually interdependent and are used for defining a *basic intentional system*.

As was mentioned earlier, in this perspective, rationality generally refers to obeying the rules of logic. *Intentionality* then refers to the system's intentions as rationally following from its beliefs about the world and from its high level preferences. This requires the modelling of the entities in the world in an appropriate level which brings us to the third condition: *stance*. Dennett distinguishes three categories: *physical stance* which refers to modeling entities based on the information of their physical state together with knowledge of the laws of nature; *design stance* refers to modeling entities based on the information of their functionality; and *intentional stance* refers to modeling a system as an intentional system, i.e., as having beliefs, intentions and so forth (at least from the outset). Prediction from intentional stance assumes rationality by the system, but not necessarily perfect rationality. Here one can see the interdependency of *intentionality* condition and the intentional *stance*.

Whereas the first three conditions are mutually interdependent and they are not ordered, the remaining conditions are ordered. The fourth condition, *reciprocity*, is the capacity in intentional systems to exhibit higher-order intentions. In other words, this requires an agent to reciprocate the stance and have beliefs, desires and intentions about beliefs, desires and intentions of other agents.

The next condition is the ability for meaningful *communication*. Gmytrasiewicz and Durfee [21] note that communication is related to the notion of reciprocity since in order to effectively communicate, an agent has to consider how its utterances are used by the hearer to update its beliefs, including the beliefs about the speaker's beliefs and so on. Finally, *consciousness*, according to Dennett, refers to an agent's ability to modify its own high level preferences [15]. This could be related to agents with learning capabilities, which although very important, will not be considered in this article.

What does all these lead to? For agents to be able to engage in meaningful communication, they need to be able to reason about their own attitudes and those of others (i.e., reciprocate the stance), and for this we need to model intentional attitudes and understand their role in *practical reasoning*. Expressed in a much simplied form, practical reasoning means reasoning about actions under the consideration of what an agent believes to be possible in a given situation and what it desires to bring about. There are significant works on practical reasoning, the best account of which is given by Bratman in [3]. But why practical reasoning? Agents communicate by message passing, but associated with each message is a particular speech act which among other factors that will be discussed later, differentiate the nature of one message from another. For instance the act of *requesting* something is different from the act of expressing your *agreement*. An agent therefore must reason about in order to select a speech act that best expresses its intended effect on the hearer. An agent can do this just like it would choose among other actions, and therefore reasoning about communicative actions can be taken as an integral part of practical reasoning.

1.2 Intentional Systems

As with intentional attitudes and their role in practical reasoning, Shoham and Cousins [41] divide intentional attitudes into *informational, motivational, social* and other emotions. Informational attitudes include notions such as '*knowledge*', '*belief*' and *awareness*'. The meaning of these attitudes are by now well understood, and there is a whole body of work in AI that have attempted to model these attitudes (for a survey see [48]).

Motivational attitudes influence the way an agent selects one among the various possible actions available to it. Examples of these attitudes are '*plans*', '*intentions*', '*desires*', and '*preferences*'. The meaning of many of these attitudes are less clear than those for knowledge and belief. There are however a substantial amount of work that have tried to model the interplay of these attitudes and incorporate them in the design and development of agent architectures in practice. Most of these activities are based on Bratman's philosophical analysis of practical reasoning and the role of informational and motivational attitudes in this process. In fact, the use of the term 'agent' is reminiscent of the era when these activities were first initiated.

Bratman [3] analyses the role of *desires*, *preferences*, and *intentions* which in his opinion are *pro-attitudes*. He distinguishes pro-attitudes from knowledge and belief in that pro-attitudes in their general sense play a *motivational* role and it is in concert with beliefs that they can move us to act. He further distinguishes pro-attitudes such as '*desires*' and '*preferences*' as being *potential influencers of conduct*, from '*intentions*' which are *conduct-controllers*. In fact intentions play a central role in this process, tying together beliefs, desires, preferences and plans.

An intention represents a preferred desire (or a chosen end) which an agent believes to be possible to achieve in a given situation. These type of intentions are known as *future-directed* or *prior* intentions. Once an intention is adopted an agent must find the means of satisfying it by filling in the appropriate plans with subplans and means of fulfilling the subplans. Bratman calls this *means-end analysis*. The plans or actions selected as a result of means-end-analysis are also intentions but they are referred to as *present-directed* intentions or *intentions to act*.

The most important property of intentions is *commitment*, that is, an agent has a measure of commitment to the desire it endevours to bring about and the actions it intends to perform in order to accomplish those intended ends. Commitments therefore shape an agent's later conduct in that its future considerations must take into account and be consistent with its current intentions. Intentions also control an agent's intra- and inter-coordination of activities. For instance my intention of writing this report this week, leads to my collecting literature material, organising it in a coherent form and writing the report. Any other task such as going to lengthy meetings in this period, must be scheduled in a way that allows me to complete this report in time, or else I have to reconsider and reschedule or even give up my intention of writing the report. Bratman states that although intentions need to have certain stability and resist being reconsidered or abandoned, it would be also irrational to treat one's prior plans as irrevocable. This has motivated several lines of research investigating policies that should govern reconsideration of intentions, and the measure of commitment considered as rational for agents in practical applications (see for instance [49]).

Bratman's work has lead to various theories which attempt to capture properties of intentions and its relation to other attitudes, which in turn have lead to various practical agent architectures that in some form or another incorporate this reasoning mechanism in the design of rational agents. Some of these works have been reviewed in [26] which gives an overview and comparison of a range of agent architectures and in [24] which discusses the underlying theories together with a more detailed overview of Bratman's work on practical reasoning. The formal framework presented in this article is a variant of Rao and Georgeff's BDI formalism [36] which is based on Bratman's account of practical reasoning. In this formalism, an agent is modelled in terms of its *beliefs, desires* and *intentions*, and reasons in terms of the interplay of these attitudes. A detailed account of this formalism will be given in later sections.

The third category of mental attitudes is *social* attitudes which relates to the motivational attitudes, but gives social, moral and organisational reasons for behaving in a certain way. '*Obligations*' , *guilt*, moral, legal and social *responsibility* are examples of such attitudes. The study of these attitudes are important in a multi-agent setting. Agents have responsibilities towards others when committed to working together, or merely because they have to share or compete for resources if they share a common environment. Whether agents choose to cooperate or compete are influenced by social attitudes. These attitudes play a strong role in reciprocity, the fourth element in Dennett's description of "*agenthood*". Next subsection will give an overview of the role of some of these attitudes in practical reasoning in a multi-agent environment.

Although the role of other emotions such as '*joy*', '*fear*' and '*anger*', in practical reasoning is even less clear than motivational and social attitudes, emotional attitudes have recently raised attention in study of systems that must interact with users, in particular, in educational and entertainment applications where the system (often an animated character) must emulate human-like gestures, and give the impression of having human-like emotions. Since in this article we are concerned with communication among artificial agents in more critical applications, emotional attitudes fall out of the scope of material discussed.

1.3 Reciprocity

As was mentioned earlier, the major role of intentions is that they move an agent to act, and control the agent's future conduct. Similarly, Searle [40] identifies two components to an action (of for example raising an arm): a "*mental*" component and a "*physical*" component. The mental component both represents and causes the physical component. Agents in multi-agent systems can engage in collaborative activities either because no one agent can accomplish the task alone, or in order to obtain a better overall performance. According to the philosophical studies, collaborative activity of a group of agents is also caused by some mental component that ultimately leads the members to act, and controls their conduct in the collaborative activity. Conte [10] uses the term "*group minds*" and "*collective unconscious*" to refer to this mental component. In this view, group mind is not part of any individual member and is external to the individuals.

A number of researchers use the notions of *we-intentions* [46], *group intentions* [42], or *joint intentions* [9] [38] in order to refer to the mental component that represents and causes the actions of the group members in the collaborative activity. In essence, they all attempt to describe joint intentions in terms of individual intentions with mutual belief that they all intend to participate. This view does not however support Conte's theory that there is a group mind independent of individuals.

Searle [40] also objects to this view of collective intention but on a different ground. He argues that collective intention is not reducible to individual intentions of the group members. In his view, collective intention is a primitive which leads to individuals' intention-to-act, that is, intend the *means* of achieving the collective intention. For example consider two agents who wish to search the web for information on behalf of their user, by one agent collecting samples of user web-browsing habits, building user profile and getting feed back from the user, and the other agent actually searching the web and filtering out information of interest using the profile compiled by the first agent. In this case the two agents have the collective intention of collaboratively aiding the user to find relevant information which in turn leads to individual intentions of building user profile by one agent, and searching the net and filtering the information by the other agent. Searle also rejects Conte's argument by stressing that any theory of collective intentionality must be consistent with the fact that: "*since a society consists of only individuals, there cannot be a group mind or group consciousness. All consciousness is in individual minds, in individual brains.*"

It appears that in talking about collective activities, we are dealing with two phenomena: (i) an external phenomenon that exists outside individual members' mind, which controls the group's activity before and after collaboration commences, and prescribes the code of conduct within the group, and (ii) an internal phenomenon that exists inside individual members' mind which leads to individual's action as part of achieving the collective goal. In other words, there is a social/organisational/macro phenomenon and an individual/micro phenomenon.

Conte and Chastelfranchi [11] focus on the notion of *norms* as the key issue. Norms provide rules of conduct for a group of agents (e.g., how things are normally done in an organisation). They state that norms are a social as well as a mentalistic notion, linking individual agents to the collective minds and social structures. In other words they see norms as the missing micro-macro link.

Jennings talks of individual's *responsibility* [28] towards the group members and argues for certain general *conventions* that the team members should obey. A convention "*provides abstract guidance for the actions of the agents engaged in cooperative action. It describes the circumstance under which an agent should review its progress towards a joint goal, and indicates the course of action to follow if, for example, problems are detected, or the goal is achieved.*" [49].

It appears that norms and conventions are the type of external phenomenon that exist with respect to group activities. In other words, norms and conventions are separate phenomena, independent of the states of individual agents' minds (i.e. wishes, beliefs, desires, intentions and so forth). They bind the individuals' actions into a coherent group activity. Joint intentions on the other hand are internal phenomenon. They lead to actions of the individual members of the group. It is the *knowledge* of the existing norms and conventions that are internal to the agents.

Some authors treat joint intentions as a separate notion from the conventions and the norms involved in collaborative activities (see for example [38]), and some explicitly integrate norms and conventions as part of the definition of joint intentions (see for instance [29] and [49]). In this article we take the first approach and define joint commitment of agents to cooperate, and show how conventions can guide the code of conduct before and after cooperation commences.

1.4 Communication

While reasoning about cooperation relies on modelling other agents as intentional systems, due to the temporal nature of many intentional attitudes as well as the fact that these attitudes are local to the individual agents, the need for communication is at times inevitable. Therefore, after reciprocity Dennett associates the next level of sophistication to the ability for a "meaningful" communication, which ultimately relies on the preceding steps.

Research in communication in multi-agent systems has in many parts benefited from the insights in Speech Act Theory. Speech Act Theory is primarily concerned with natural language utterances. Austin [1], one of the pioneers in this field, noticed that the sentences uttered by human during communication do not always simply assert a fact, but actually perform some action and that utterances accomplish actions by having certain forces. There are three types of actions associated with an utterance:

1. *Locution*: the physical act of uttering the sentence;
2. *Illocution*: the action of conveying the speaker's intent to the hearer; and
3. *Perlocution*: any actions caused as a result of uttering the sentence. This includes (if not limited to) the actions taken by the hearer upon the occurrence of the utterance.

For example, "shut the door" is the actual locution, with the illocutionary force that the hearer shuts the door, and whatever the hearer does as a result is the perlocution. While often a speaker utters a sentence performing an illocution, hoping that it would have certain intended effect on the hearer, the actual effect on the hearer's mental states and the actions that the hearer may perform as a result, is beyond the control of the speaker.

In the multi-agent systems literature, a speech act has mainly been associated with its illocution, which is characterised as having two parts: an *illocutionary force* and a *proposition* [39]. The illocutionary force distinguishes for example a *command* from a *promise*, and the proposition describes the state of the world that is, respectively, commanded or promised [43]. Cohen and Levesque [7] argue that the explicit recognition of the illocutionary force by the hearer is unnecessary. Instead, many properties of illocutionary acts can be derived by the hearer from the speaker's and hearer's mental states, especially from their beliefs, intentions and the common knowledge of the communicating parties about certain conventions. Therefore they aim to employ insights from linguistics theory on human communication to artificial systems, by characterising the conditions under which illocutionary force of an utterance may be *inferred*. For instance the sentence "can you pass the salt" may be inferred as being an indirect request to give the salt to the speaker, under the conditions that both the speaker and the hearer are aware of this linguistic/social convention. This may be referred to as *indirect speech act* [43]. Unless communicating with a human, reasoning about ambiguities of messages exchanged is unnecessary in artificial systems as long as the semantics of messages are unique both to the sender and receiver. As a result, while locutionary acts may be designed directly and explicitly into the syntax of messages, their illocutionary force must be determined by giving speech acts precise formal semantics.

A more common impact of speech acts is seen in *message types*, initially motivated by Searle's [39] classification of illocutionary forces into *assertives, directives, commissives, permissives, prohibitives* and *declaratives*. Examples of each category are given in the following:

Searle's classification suggested that speech acts can be grouped together into classes in such a way that the speech acts in the same class share the same condition of satisfaction. But his classification was criticised basically because the basis

Assertive	"The TV is on"
Directive	"Switch the TV on!"
Commissive	"I will switch the TV on"
Permissives	"You may switch the TV on"
Prohibitive	"Nobody is allowed to switch this TV on"
Declarative	"I call this TV set the Mind Trap"

of the classification is semantically coarse [34], attempting to put together speech acts of varying strengths, and differing pragmatic effects [43]. Ballmer and Brennenstuhl [2] for instance attempted to classify all German speech act verbs into Searle's speech act classes. They faced enormous difficulties in this process, as with addition of more verbs, they had to re-analyse all previously classified verbs.

The problem lies on the basis upon which the conditions of satisfaction of each group may be characterised. There are a variety of pragmatic factors that could take part in this classification such as roles and social statures, norms and conventions of the system and the language, which are in any case difficult to be taken as isolated and general factors to base the classification upon. For example, one could distinguish between a *command* and a *request* in that a command is usually issued to a sub-ordinate, while a request can be issued to almost anyone, although both may be classified under directives. According to Singh [43], "*the constraint on when requests and commands are issued and satisfied may be stated that capture their non-semantic aspects properly*".

Instead of such a rigid classification, Ballmer and Brennenstuhl [2] suggested to group the verbs that are similar in meaning, into *semantic categories*, and group the semantic categories into *models* according to semantic similarity. For example a model of an 'enquiry', consists of a 'question' and an 'answer'. The models also include an ordering of the categories according to their *temporal relationship* and *degree of effectiveness*. For example an 'answer' always comes after a 'question', and 'threatening' is stronger than 'warning'. The categories and the models including the temporal ordering between them not only provide a more pragmatic categorisation of speech acts, but also suggest that semantics of communication must be studied in the form of dialogues rather than single messages in isolation. In multi-agent systems studies, this has been the motivation for designing and developing dialogue protocols.

A number of protocols have been developed to support cooperation for specific type of problem. The well known Contract Net Protocol [44], [12], the hierarchical protocol [16], and PERSUADER [45] are fine examples of such protocols. Each protocol is designed to incorporate a unique method of cooperation and specifies the possible negotiation steps within that method. For example, the Contract Net Protocol is designed for task-allocation, the hierarchical protocol of Durfee et al. is aimed at coordination, and PERSUADER is a protocol for negotiation in a specific context.

There have also be attempts on developing representation and languages to design and specify dialogue steps and protocols in general. To describe cooperation in multi-agent systems, deGreef et al. [13] have devised a general specification language that is transformable into an already implemented language. The purpose of this language is to abstract away from details which are irrelevant to the interacting agents, so that designers can easily define the interaction knowledge. Winograd and Flores [47] use a network of speech acts consisting of straightforward conversations for action. This network is a state transition dialogue specifying the possible future state of a dialogue from a current state. This effort is aimed at structuring dialogues and analysing conversations. Campbell and d'Inverno [6] proposed a graphical representation, where protocols not only represent the communication between two agents, but also the reasoning states. Bearing in mind that these approaches have been taken with other goals in mind, neither deals with the problem of giving semantics to message types. This is important especially if heterogeneous agents, that is agents developed by different authorities or those differing in their internal design, have to communicate with one another and engage in dialogues. Lack of proper semantics leads to ambiguities, and can result in unnecessary communication failures merely because the meaning of each speech act is open to interpretation.

In developing standards for communication among agents there are many pragmatic factors to be considered, for instance the syntax of messages exchanged, and representational and specification issues in developing dialogues and in developing protocols that ensure correct execution of the dialogue states by the communicating parties. Formal theories of communication in this context is there to support definitions of concepts, precision, and verification of the suggested models.

Knowledge Query Manipulation Language (KQML) [19] is a language intended as a standard for communication among agents. A message in this language consists of a message type which in KQML terminology is called a *performative*, and a parameter list. The parameter list consist of $\langle keyword, expression \rangle$ pairs, where the keyword is the identifier for the expression that follows it. There are a number of defined performatives, each specific performative is associated with a number of reserved parameters, some mandatory, and some optional. Taken from [31], Fig.1 shows two examples of KQML messages exchanged between two agents A and B.

Although currently the number of performatives defined are limited, missing for example the class of performatives required to communicate for system administrative purposes, the language is in an appropriate level of abstraction adequate for encoding messages to be communicated by agents. The components of the language (i.e., *reply-with* and *in-reply-to*) allow agents to precisely reference messages when responding to previously received messages. Furthermore, agents can formulate complex queries by nesting multiple expressions. For instance the main content of a message may itself be another KQML message, and the content of this second message my be the expression encoded in the application language. The major draw back of KQML performatives is the lack of formal semantics, which can lead to ambiguities regarding how a performative is intended to be used. As experience has shown us, in practice those that have adopted KQML as a standard for their agent communication language do not necessarily use the given performa-

```
(ask-all   :sender        A
           :receiver      B
           :in-reply-to   id0
           :reply-with    id1
           :language      Prolog
           :ontology      foo
           :content       "bar(X,Y)"
```

```
(tell      :sender        B
           :receiver      A
           :in-reply-to   id1
           :reply-with    id2
           :language      Prolog
           :ontology      foo
           :content       "[bar(a,b), bar(c,d)]"
```

Figure 1. Two Examples of KQML Messages

tives and their reserved parameters as they were originally intended by the KQML developers. Having realised this problem there have been attempts in giving the formal semantics, first of which has been reported in [32]. This work hasn't so far been satisfactory.

This can be attributed to the fact that the KQML effort started out with the aim of developing standards for querying databased and knowledge-based systems, but later found to be also useful as a language for communication among agents. For cooperative activities however, agents need to be able to reason about and communicate not only their beliefs or knowledge but also motivational attitudes such as plans, goals, intentions and commitments, which influence and control reasoning about actions (i.e., practical reasoning). This in turn requires us to assume a minimal but sufficiently rich set of attitudes, develop formal theories to capture their relationship, and precisely define their input to practical reasoning. Speech acts then can be treated like any other actions whose conditions of execution and effects are expressed in terms of mental states, allowing an agent to reason about and select among them.

In agent systems therefore, communication is more than simple syntactic utterances in dialogues. Singh [43] for instance has developed a theory to capture an objective semantics of the messages exchanged, by specifying the satisfaction conditions of speech acts in Searle's classification. For instance, his theory specifies that "*a directive is satisfied just if (1) its proposition becomes true at a moment in the future of its being said, and (2) all along the scenario from now to then the hearer has the know-how, as well as the intention to achieve it.*".

He states that his theory is an objective criteria with which to evaluate the correctness of the different scenarios that are the possible runs of executions of a multi-agent system. But he also includes that we require *prescriptive* specifications that tell agents what to do given their beliefs and intentions, so that only correct scenarios may emerge. This nicely captures one of the objectives of this book and will be discussed in chapter 4.

Therefore, to achieve meaningful dialogues, we require to specify under what conditions individual agents may choose to perform a particular speech act. In the work presented in this article, these conditions are specified in terms of beliefs, goals and intentions of individual agents and their belief about the intentional states of their communicating partners. Messages or rather communicative actions will not however be considered in isolation, but in the context within which a dialogue may take place, and according to the rules and conventions governing the interactions in that context. This means that reasoning about communication does not only involve reasoning about when and what speech act to perform, but also what is involved after a message is received, (i.e., the perlocutionary effects of the speech act).

2 INTRODUCTION

Cooperative problem solving is considered as an important feature of multi-agent systems. The majority of studies on this topic have identified *joint intentions* as the mental component that causes the cooperative activity of a group of agents. The existing theories have largely characterised such joint attitudes from an observer's perspective (*external* perspective) (e.g., [33], [42]). These approaches have aimed at developing theories *about* agents, and for this purpose they are very rich. To model and design practical reasoning by individual agents, such an approach must enable the theory to be used *by* agents [49]. For this purpose, we take an *internal* perspective, describing the formalisms from the perspective of individual agents on their own mental states, their environment and the other agents. Internal theories typically identify the key data structures and describe the relationship between them. Therefore, coupling of the theory to the resulting design concepts is more straightforward.

The formal apparatus and the underlying concepts are largely identical to the BDI model for single *situated agents* in [36], but extends it to agents in multi-agent systems. Following their concept of situated agents, we adopt the view that agent-oriented reasoning can be expressed by the triple "*chance* + *choice* + *commitment*" [37]. While chances refer to possible future states of the environment, choices correspond to the decisions an agent makes based on what it believes to be possible and desires to bring about. Commitment then expresses *when* an agent pursues a *chosen* course of action and *under what conditions* it should give up its persistent to those choices. This view forms the fundamental concept behind reasoning about choosing appropriate communication acts which we view as part of the *means-ends reasoning* process. This process involves reasoning about appropriate means (plans) of achieving some desired ends (goals). A complete treatment of our formal framework will not be provided in this article. The formal syntax and semantics of the language and its properties can be found in [22] and [24].

Next section gives an informal description of our agent specification language. This language is used to describe a number of key constructs that take part in the means-end reasoning process of an agent. This process is then specified in a set of rules which characterise under which conditions individual and agent to agent commitments may be formed and revised. As part of this process, a subset of these rules specify an agent's reasoning about communication.

3 THE LOGICAL FRAMEWORK

The language of formulations is a first-order, multi-modal, branching-time logic (a first-order variant of CTL* [17]). The world is modelled as a time tree with a single past and a branching future. Two types of formulae can be expressed in a world: *state formula* which is evaluated at a specific time point, and *path formula* which is evaluated over a specific path (sequence of time points). The formulae of the language are constructed using the usual operators of first-order logic: $\neg, \wedge, \vee, \Rightarrow, =, \exists, \forall$; the temporal operators: $\bigcirc$ (next), $\diamond$ (eventually), and $\square$ (always), which operate over state and path formulae; and the path operators E (optional) and A (inevitable) of branching time logic.

Each agent is capable of performing a number of actions. These actions are structured into *plans*, each aimed at achieving a certain state of affairs. Every agent is assumed to have a *library* of pre-compiled plans. The constructs operating on actions and plans are informally defined as follows: $(\mathsf{Does}\ z\ p)$ states agent z's attempt in executing plan p next. $(\mathsf{Achieved}\ z\ \varphi)$ and $(\mathsf{Achieves}\ z\ \varphi)$ denote past successful and future attempt of agent z in achieving φ. $(\mathsf{Has\_Plan}\ z\ p\ \varphi)$ states, z has a plan p in its plan library that if executed successfully, would achieve φ.

Beliefs, desires and intentions of an agent are respectively modelled by associating a set of *belief-, desire-* and *intention-accessible* worlds to each situation. These worlds refer to the worlds that an agent *believes* to be possible, *desires* to bring about, or *commits* to achieve. We distinguish *goals* from *desires* in two ways: (a) goals are *chosen* desires of an agent that are consistent; and (b) the agent should believe that the goal is achievable. A full account of modelling of these attitudes, together with their properties may be found in [36]. The two most important axioms related to our work are: (i) *belief-goal compatibility* axiom which states that if an agent adopts an option as a goal, it must believe that it is possible to achieve that option[2]:

[2] R&G [36] define the well-formed formulae that contain no positive occurrences of 'inevitable' (A) outside the scope of the modal operators BEL, GOAL, or INTEND *O-formulae*. In all the formulae in

Axiom 1 $(\mathsf{GOAL}\ z\ \varphi) \Rightarrow (\mathsf{BEL}\ z\ \varphi)$.

And (ii) *goal-intention compatibility* axiom which states that if an agent commits to an option, this option must be one of its goals.

Axiom 2 $(\mathsf{INTEND}\ z\ \varphi) \Rightarrow (\mathsf{GOAL}\ z\ \varphi)$.

The following abbreviation is introduced to express that an agent either believes that a formula φ holds, or that it believes that it does not hold[3], in other words, it has a belief about the value of φ:

Definition 1 $(\mathsf{Aware}\ z\ \varphi) \stackrel{def}{=} (\mathsf{BEL}\ z\ \varphi) \vee (\mathsf{BEL}\ z\ \neg\varphi)$.

Finally, the operator $\mathcal{B}$ is a variant of the *before* operator in temporal logic, meaning that if sometime in the future φ holds, ψ will hold before φ holds:

Definition 2 $\psi\ \mathcal{B}\ \varphi \stackrel{def}{=} \Diamond\varphi \Rightarrow \Diamond(\psi \wedge \Diamond\varphi)$.

4 MEANS-ENDS REASONING

Expressing the means-end reasoning process in terms of the interplay of *chance, choice* and *commitment*, means to an end may be decided to be carried out individually – as with *individual task achievement* in single-agent systems– or cooperatively. In the context of individual task achievement we are only concerned with commitment to the choice of plan of actions, i.e., individual intention to carry out a plan. In terms of cooperative activity, in this article we restrict our attention to *task delegation* and *task adoption*. In simplest form, task delegation refers to delegation of achievement of a task to another agent, and task adoption refers to adoption of achievement of another agent's task. Agents may engage in communication and possibly negotiation in order to come to some sort of agreement about each other's attitude towards a possible cooperation. But cooperation would only commence if the two agents have agreed on the terms of cooperation, believe that they both wish to cooperate and make a commitment. Hence here we are in addition concerned with agent to agent commitments.

From a pragmatic point of view, specification of commitments would have to explicitly address the following issues:

- The semantics of intentions including the conditions an agent is committed to maintain, (*maintenance conditions*),
- The conditions which lead to the formation of commitments, (*formation conditions*), and

this article, if applicable, formula φ is an O-formula.

[3]Note that this is not be confused with the the general logics of '*awareness*' in the literature, which involve logics of implicit and explicit beliefs (see [18]).

- The conditions under which a commitment must be re-evaluated, (*revision conditions*).

An extra requirement that may or may not be directly associates with a commitment is the specification of what an agent should do once it gives up a commitment [49], which will not be dealt with in this article. The approach in [8] and formalisms based on this approach (e.g., [33] and [49]) integrate these conditions in the semantics of intentions. From the practical point of view this is problematic, since one cannot say when an agent has an intention *per se*. One have to state in advance under what conditions they should be revised and in case of [49], for each specific commitment what the agent should do once it gives up that commitment. This would be satisfactory if these conditions could be generalised. With respect to individual intentions to act, the specification of these conditions is related to certain strategies that an agent must follow, and successful strategies can only be outlined with respect to the requirement of the application in question. With respect to commitments of agents to engage in a cooperative activity, specification of these conditions is also dependent on the macro aspects of the system, such as norms, conventions and laws of the "society" or the organisation in which the agents are operating. Therefore, for practical purposes we would require to disentangle these issues and identify what could be defined and specified in general and what is needed to be specified as part of the application in question.

Another problem is a technical one. Many approaches to defining joint intentions rely on the recursive notion of mutual beliefs. Under certain circumstances such a mutual state of affairs may not be attainable in practice (e.g., the coordinated attack problem [27]). To attain a state of mutual belief such that eventually the agents involved would indeed act on their committed terms, certain conventions must be established [42].

In this article we aim at a more pragmatic approach and start with introducing a convention and developing our theory within the framework of this convention.

Convention: Once a *potential for cooperation* is detected, the negotiation process commences. After a series of negotiation steps, eventually one of the negotiating parties (say agent x) finalises its terms of cooperation by making a *pre-commitment*, and communicates this to the other agent (say agent y). This finalisation marks the last steps of negotiation, that is, agent y now must either make a *commitment* or else opt out. We require that if y agrees on the final terms then it commits to x on the agreed terms and communicates this to x, after which pre-commitment of x becomes a full commitment.

Our theory models this process in terms of three reasoning constructs: *potential for cooperation, pre-commitment* and *commitment* of agents to one another. These constructs will be formally defined and their properties including their formation and revision conditions will be specified.

4.1 *Definitions*

The reasoning constructs are defined in terms of beliefs, goals and intentions of individual agents. While goals represent choices that an agent believes to be possible in the current situation, intentions represent commitment to some of these choices. Therefore, before forming an intention, an agent has to deliberate on the choices available to it. For individual intentions to act, these choices typically correspond to the choice of actions or plans to achieve certain goals. For agent to agent commitment, these choices also include a potential cooperating partner. The final decision on these choices are respectively represented by the following decision constructs.

- (Willing y φ) denotes that agent y *has chosen* to carry out a plan that if executed successfully would achieve φ.

- (Want x y φ) denotes that agent x has chosen agent y to be the actor of achieving φ.

These decision constructs will not be defined in our general theory, since their value should be evaluated by some deliberation procedures that need to be defined as part of the theory of the application in question. These procedures may be encoded in them some decision-theoretic approach that enables an agent to choose the best among the possible options available to it. THis is what was referred to as decision-theoretic rationality in section 1.1. However, there are some restrictions imposed on them as will be outlines later (see Axioms 4 and 5).

In the remainder of this article, we take the simple scenario of an agent x's attempt in delegating the achievement of one of its goals to another agent y, and an agent y's attempt in adopting the achievement of another agent's (here x's) goal. It is also important to note that hereafter in all the formulae involving agents x and y, $y \neq x$.

Since we take an internal perspective, the three reasoning constructs *potential for cooperation* (PfC), *pre-commitment* and *commitment* of two agents are defined once in the context of task delegation (agent x's perspective) and once in the context of task adoption (agent y's perspective). To distinguish the two, the corresponding operators have either the subscript 'd' (for delegation), or 'a' (for adoption). The formal definition of these states is given in Figure 2. In the context of task delegation, these constructs are defined informally as follows:

- Agent $\underline{x}$ sees a potential for cooperation with agent $\underline{y}$ with respect to its goal state φ, such that y achieves at least ψ, if (i) x has φ as a goal, (ii) x is not willing to achieve φ individually, (iii) x believes that one way of achieving φ is first to achieve ψ, and has ψ as a subgoal, (iv) x has it as a goal that y achieves at least ψ. and (v) up to now, x does not have any belief that y would disagree with these terms.

$$(\mathsf{PfC_d}\ x\ y\ \varphi\ \psi) \stackrel{def}{=} \begin{bmatrix} (\mathsf{GOAL}\ x\ \mathsf{E}\Diamond\varphi) \wedge \\ \neg(\mathsf{Willing}\ y\ \varphi) \wedge \\ (\mathsf{GOAL}\ x\ \mathsf{E}(\psi\ \mathcal{B}\ \varphi)) \wedge \\ (\mathsf{GOAL}\ x\ \mathsf{E}(\mathsf{Achieves}\ y\ \psi)) \end{bmatrix}.$$

$$(\mathsf{Pre\_commit_d}\ x\ y\ \varphi) \stackrel{def}{=} \begin{bmatrix} (\mathsf{PfC_d}\ x\ y\ \varphi) \wedge \\ (\mathsf{Want}\ x\ y\ \varphi) \end{bmatrix}.$$

$$(\mathsf{Commit_d}\ x\ y\ \varphi) \stackrel{def}{=} \begin{bmatrix} (\mathsf{INTEND}\ x\ \mathsf{E}(\mathsf{Achieves}\ y\ \varphi))) \wedge \\ (\mathsf{BEL}\ x\ \mathsf{E}\Diamond(\mathsf{INTEND}\ y\ \mathsf{E}(\mathsf{Achieves}\ y\ \varphi))) \end{bmatrix}.$$

$$(\mathsf{PfC_a}\ y\ x\ \varphi\ \psi) \stackrel{def}{=} \begin{bmatrix} (\mathsf{BEL}\ y\ (\mathsf{GOAL}\ x\ \mathsf{E}\Diamond\varphi)) \wedge \\ (\mathsf{GOAL}\ y\ \mathsf{E}(\psi\ \mathcal{B}\ \varphi)) \wedge \\ (\mathsf{GOAL}\ y\ \mathsf{E}(\mathsf{Achieves}\ y\ \psi)) \end{bmatrix}.$$

$$(\mathsf{Pre\_commit_a}\ y\ x\ \varphi) \stackrel{def}{=} \begin{bmatrix} (\mathsf{PfC_a}\ y\ x\ \varphi) \wedge \\ (\mathsf{Willing}\ y\ \varphi) \end{bmatrix}.$$

$$(\mathsf{Commit_a}\ y\ x\ \varphi) \stackrel{def}{=} \begin{bmatrix} (\mathsf{INTEND}\ y\ \mathsf{E}(\mathsf{Achieves}\ y\ \varphi)) \wedge \\ (\mathsf{BEL}\ y\ \mathsf{E}\Diamond(\mathsf{INTEND}\ x\ \mathsf{E}(\mathsf{Achieves}\ y\ \varphi))) \end{bmatrix}.$$

Figure 2. Formal definitions

- Agent $\underline{x}$ is *pre-committed* to agent $\underline{y}$ such that y achieves φ, if (i) x sees a potential for cooperation with y such that y achieves φ;[4], and (ii) x wants that y achieves φ.
- Agent $\underline{x}$ is *committed* to agent $\underline{y}$ achieving φ if, (i) x has committed to the choice of y achieving φ, and (ii) x believes that y will eventually commit to achieving φ.

In the context of task adoption, these constructs are defined analogously, as formally defined in Figure 2. The definition of agent to agent commitment provided a separate "semantics" including the *maintenance conditions* discussed earlier.

4.2 Axioms and Theorems

Due to the limited space, only some of the more important axioms and theorems of our theory will be outlined. A complete analysis may be found in [24].

The first theorem below (derivable from our formal framework) captures the fact that if an agent has a goal to achieve φ individually, then it must have a plan in its plan library which if executed successfully would achieve that goal, and has it as a goal to carry out that plan (similarly for intentions):

Theorem 1

$$(\mathsf{GOAL}\ x\ \mathsf{E}(\mathsf{Achieves}\ x\ \varphi)) \Rightarrow \exists p\ .\begin{bmatrix} (\mathsf{BEL}\ x\ (\mathsf{Has\_Plan}\ x\ p\ \varphi)) \wedge \\ (\mathsf{GOAL}\ x\ \mathsf{E}(\mathsf{Does}\ x\ p)) \end{bmatrix}.$$

[4]The representation $(\mathsf{PfC_d}\ x\ y\ \varphi)$ is used as a short form for $(\mathsf{PfC_d}\ x\ y\ \varphi\ \varphi)$. Similarly for $\mathsf{PfC_a}$.

An important property of intentions is that they should move an agent to act. Under the assumption that an agent can only perform one action at a time, the following axiom ensures that an agent executes an action next, if it intends *inevitably* to execute that action. Note that if this action is intended *optionally*, then at the time of execution the agent must deliberate and choose one action in order to act intentionally.

Axiom 3 $(\mathsf{INTEND}\ x\ \mathsf{A}\Diamond(\mathsf{Does}\ x\ \alpha)) \Rightarrow \mathsf{A}\Diamond(\mathsf{Does}\ x\ \alpha)$

Since the decision constructs Willing and Want express certain decisions that an agent makes on the choices available and possible to it, the following two axioms are introduced to ensure that these decisions are indeed made on the choices that an agent has as a goal.

Axiom 4 $(\mathsf{Willing}\ y\ \varphi) \Rightarrow (\mathsf{GOAL}\ y\ \mathsf{E}(\mathsf{Achieves}\ y\ \varphi))$.

Axiom 5 $(\mathsf{Want}\ x\ y\ \varphi) \Rightarrow (\mathsf{GOAL}\ x\ \mathsf{E}(\mathsf{Achieves}\ y\ \varphi))$.

Furthermore we require that if an agent chooses to achieve a goal individually, it has a plan and chooses to carry out that plan in order to achieve that goal:

Axiom 6 $(\mathsf{Want}\ y\ y\ \varphi) \Leftrightarrow (\mathsf{Willing}\ y\ \varphi)$.

An important axiom of our theory is Axiom7 which ensure that an agent does not make a pre-commitment on some terms of cooperation (here φ), if he already believes that the other agent is not going to make a pre-commitment under those terms. Note that here it is a delicate difference between '*not being aware of the other agent's attitude*' and actually '*holding a belief about the status of the other agent's attitude*'.

Axiom 7 $(\mathsf{BEL}\ x\ (\neg(\mathsf{Pre\_commit_a}\ y\ x\ \varphi)) \Rightarrow \neg(\mathsf{Pre\_commit_d}\ x\ y\ \varphi)$.

As was stated earlier, the common definition of joint intentions relies on the recursive definition of mutual beliefs, according to which agent x would only make a commitment to agent y, if it believes that y is also committed and believes that y believes that x is already committed and so on. Similarly for y. Therefore, it is questionable if the two agents ever end up making a commitment. Our convention makes use of a weaker condition for forming an agent to agent commitment. It require that both agents to be *at least* pre-committed *before* any of them makes a full commitment. As is expressed in Axiom 8, [5], it is sufficient for agent x to be pre-committed and believe that agent y is either pre-committed or is already committed, *before* he makes a full commitment.

Axiom 8

$$\left[\begin{array}{l}(\mathsf{Pre\_commit_d}\ x\ y\ \varphi) \wedge \\ \left(\begin{array}{l}(\mathsf{BEL}\ x\ (\mathsf{Pre\_commit_a}\ y\ x\ \varphi)) \veebar \\ (\mathsf{BEL}\ x\ (\mathsf{Commit_a}\ y\ x\ \varphi))\end{array}\right)\end{array}\right] \Leftrightarrow$$
$$\mathsf{A}\bigcirc(\mathsf{Commit_d}\ x\ y\ \varphi).$$

[5]Because of the space restrictions we will only outline the axioms in the context of task delegation.

This axiom specifies under what conditions an agent makes a commitment (next) (*formation condition*). Because this is an equivalence relation, it also states that whenever any of the conditions that originally led to the formation of the commitment no longer hold, the agent should drop its commitment. Therefore, this axiom also specifies some general *revision conditions*. This axiom 1 is consistent with the goal-intention compatibility axiom, i.e.,

$$(\mathsf{INTEND}\ x\ \mathsf{E}(\mathsf{Achieves}\ y\ \varphi)) \Rightarrow (\mathsf{GOAL}\ x\ \mathsf{E}(\mathsf{Achieves}\ y\ \varphi)).$$

One important theorem derivable from our model is Theorem 2, which states: as long as an agent is committed to the other agent and believes that the other agent is at least pre-committed, then it will believe that the other agent is either already committed, or will eventually make a commitment[6].

Theorem 2

$$\begin{bmatrix} (\mathsf{Commit_d}\ x\ y\ \varphi) \wedge \\ (\mathsf{BEL}\ x\ (\mathsf{Pre\_commit_a}\ y\ x\ \varphi)) \end{bmatrix} \Rightarrow (\mathsf{BEL}\ x\ \mathsf{A}\Diamond(\mathsf{Commit_a}\ y\ x\ \varphi)).$$

Finally, Theorem 3 captures the fact that if an agent has made a pre-commitment but comes to believe that the other agent will never make a commitment, then it will also not make a commitment, (i.e., drop its commitment if it is already committed).

Theorem 3

$$\begin{bmatrix} (\mathsf{BEL}\ x\ (\mathsf{BEL}\ y\ (\mathsf{Pre\_commit_d}\ x\ y\ \varphi))) \wedge \\ (\mathsf{BEL}\ x\ \neg\Diamond(\mathsf{Commit_a}\ y\ x\ \varphi) \end{bmatrix} \Rightarrow \mathsf{A}\bigcirc\neg(\mathsf{Commit_d}\ x\ y\ \varphi).$$

Therefore the definitions, the axioms and the resulting theorems specified the three issues associated with commitment, namely, *maintenance, formation* and *revision* conditions.

5 SPECIFICATION OF MEANS-END REASONING

The means-end reasoning process is modelled in terms of temporal state transitions. These transitions are specified in a set of rules. According to the conditions that hold in a given state, a rule specifies the next valid state to which the transition must proceed:

Definition 3 $\varphi \longleftrightarrow \psi \overset{def}{=} \Box(\varphi \Leftrightarrow \bigcirc\psi)$.

[6]This theorem is not immediately obvious, but because of space restrictions it cannot be appropriately dealt with in this article.

That is, whenever the process is in a state in which φ holds, it should proceed to a next state in which ψ holds. This is similar to the specification of *liveness properties* of concurrent programs [35], which is of the form $(\varphi \Rightarrow \Diamond\psi)$ where φ and ψ are past formulas. The difference is that since by these rules we aim to specify both the formation conditions and some of the general revision conditions, we use the equivalence relation instead of the 'implies' operator.

5.1 Formation of Commitments

The formation and some of the general revision conditions associated with agent to agent commitments were already specified in Axiom 7 (which is in any case in the form of our specification rules). But we still need to specify the formation and revision conditions for individual intentions to act:

$$\forall x\,.\,(x \neq y) \wedge \begin{bmatrix} \neg(\mathsf{Pre\_commit_a}\ y\ x\ \varphi) \wedge \\ (\mathsf{GOAL}\ y\ \mathsf{E}(\mathsf{Achieves}\ y\ \varphi)) \wedge \\ (\mathsf{Willing}\ y\ \varphi) \end{bmatrix} \longleftrightarrow$$

$$(\mathsf{INTEND}\ y\ \mathsf{E}(\mathsf{Achieves}\ y\ \varphi)).$$

According to this rule, an agent forms and maintains its commitment to achieve φ individually iff: (i) it has not pre-committed itself to another agent to adopt and achieve φ, (ii) has it as a goal to achieve φ individually, and (iii) is willing to achieve φ individually. The first condition is to distinguish the formation of intention to act in the context of individual task achievement, from that formed as a result of task adoption, that is, it ensures that if y has made a pre-commitment, it does not intend to act for it before both agents have made a commitment.

5.2 Formation of Goals To Communicate

By communication the agents can update their belief about whether, when and how they wish to cooperate. Having in mind that we take an internal perspective, an agent's goal to communicate is directed towards two objectives: (i) update of the other agent's belief, about its own attitude towards a possible cooperation; and (ii) update of its own belief about the other agent's attitude. These goals are of the form $(\mathsf{GOAL}\ z\ \delta)$, where $\delta = \rho\ \wedge\ (\mathsf{Does}\ z\ \gamma?)$, ρ denoting the first objective, $(\mathsf{Does}\ z\ \gamma?)$ the second objective[7] and z is either of agents x or y, depending on which agent's goal is being expressed.

In the context of task delegation the formation of some of these goals is specified in figure 3. For instance, based on our convention, in attempt to delegate the achievement of a goal φ (or part of it ψ) to another agent, four possible goals are identified: δ_1 is formed when x detects a potential for cooperation with y,

[7]Having the goal of $(\mathsf{Does}\ z\ \gamma_i?)$, expresses the fact that agent z has it as a goal to execute a plan in order to find out if γ_i (a fact about the other agent) holds.

$$\begin{bmatrix} (\mathsf{PfC_d}\ x\ y\ \varphi\ \psi) \wedge \\ \neg(\mathsf{Pre\_commit_d}\ x\ y\ (\varphi \vee \psi)) \wedge \\ \neg(\mathsf{Aware}\ x\ (\mathsf{PfC_a}\ y\ x\ \varphi\ \psi)) \wedge \\ (\mathsf{BEL}\ x\ \neg(\mathsf{Aware}\ y\ (\mathsf{PfC_d}\ x\ y\ \varphi\ \psi))) \end{bmatrix}$$
$$\rho_1 = (\mathsf{BEL}\ y\ (\mathsf{PfC_d}\ x\ y\ \varphi\ \psi)) \quad \gamma_1 = (\mathsf{PfC_a}\ y\ x\ \varphi\ \psi)$$

$$\begin{bmatrix} (\mathsf{Pre\_commit_d}\ x\ y\ \varphi) \wedge \\ \neg(\mathsf{Aware}\ x\ (\mathsf{Pre\_commit_a}\ y\ x\ \varphi\ \psi)) \wedge \\ (\mathsf{BEL}\ x\ \neg(\mathsf{Aware}\ y\ (\mathsf{Pre\_commit_d}\ x\ y\ \varphi))) \end{bmatrix}$$
$$\rho_2 = (\mathsf{BEL}\ y\ (\mathsf{Pre\_commit_d}\ x\ y\ \varphi))$$
$$\gamma_2 = (\mathsf{Pre\_commit_a}\ y\ x\ \varphi\ \psi)$$

$$\begin{bmatrix} (\mathsf{Commit_d}\ x\ y\ \varphi) \wedge \\ (\mathsf{BEL}\ x\ \neg(\mathsf{Aware}\ y\ (\mathsf{Commit_d}\ x\ y\ \varphi))) \end{bmatrix}$$
$$\rho_3 = (\mathsf{BEL}\ y\ (\mathsf{Commit_d}\ x\ y\ \varphi)) \quad \gamma_3 = \mathsf{true}$$

$$\begin{bmatrix} (\mathsf{BEL}\ x\ (\mathsf{PfC_a}\ y\ x\ \varphi\ \psi)) \wedge \\ \neg(\mathsf{Pre\_commit_d}\ x\ y\ \psi) \wedge \\ (\mathsf{BEL}\ x\ \neg(\mathsf{Aware}\ y\ \neg(\mathsf{Pre\_commit_d}\ x\ y\ \psi))) \end{bmatrix}$$
$$\rho_4 = (\mathsf{BEL}\ y\ \neg(\mathsf{Pre\_commit_d}\ x\ y\ \psi)) \quad \gamma_4 = \mathsf{true}$$

Figure 3. Communication Rules: Task Delegation

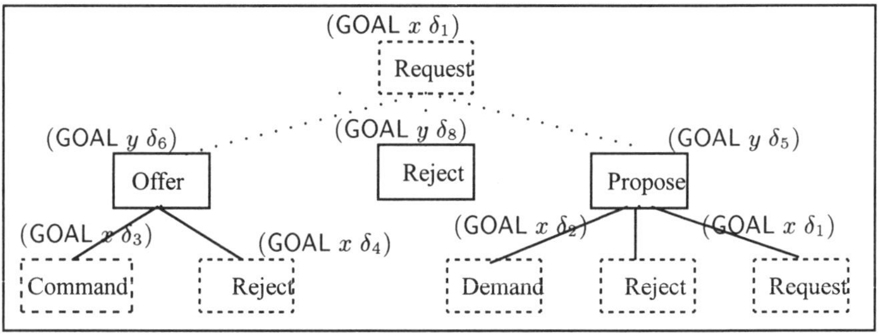

Figure 4. Demonstration

which initiates negotiation; δ_2 is formed when x has pre-committed to y achieving φ, which finalises the negotiation process, prompting y for a commitment or opting out; δ_3 is formed when x has made a commitment (as the response to y's pre-commitment), and according to our convention should communicate this to y; finally δ_4 is formed when x does not wish to cooperate with y. There are also analogous rules in the context of task adoption which have not been included here.

5.3 *Active Communication*

Upon adopting a goal to communicate, an agent has to enter another round of means-end reasoning in order to find an appropriate plan that if executed successfully would achieve that goal. Among the plans in the plan library of an agent, there are a number of so called *primitive communication* plans that either consist of a send action (i.e., $\{send(msg)\}$), or the sequence of actions 'send and wait until receive' (i.e., $\{send(msg1);\ true*;\ receive(msg2)\}$). Primitive plans are distinguished by the message type of the message to be sent.

Sending and receiving messages are taken to be primitive actions[8]. Roughly, a message consists of a *sender*, a *receiver*, a *message type* and a *content*. The content of a message is a well-formed formula of (another) communication language which contains the actual proposition to be communicated.

One goal of *sending* a message is to update the receiver's belief. In the context of this article, this refers to the update of the receiver's belief about the sender's attitude towards a possible cooperation. This corresponds to $(\mathsf{GOAL}\ z\ \rho)$.

Message types are used to express what the sender attempts to convey to the receiver and possibly prompt the receiver for a response. This corresponds to the $(\mathsf{GOAL}\ z\ (\mathsf{Does}\ z\ \gamma?))$.

The language admits the following set of message types: $\{$*'Request', 'Demand', 'Command', 'Propose', 'Offer', 'Report', 'Reject' and 'Accept'*$\}$, and consequently primitive communication plans. To simplify, with each type of goal (i.e., δ_1 to δ_9) we associate a unique message type, and consequently a unique communication plan as shown in the table below.

Request δ_1	Demand δ_2	Command δ_3	Reject δ_4	
Propose δ_5	Offer δ_6	Accept δ_7	Reject δ_8	Report δ_9

However, it is possible to associate more than one communication plan to a goal, but in that case an agent has to deliberate and choose and intend one of them.

There are a number of assumptions that we make about communication: (1) If $\gamma = \mathsf{true}$, the corresponding plan expression is only a *send* action meaning that the agent does not expect a response. (2) After receiving a message, the receiving agent believes the 'content' of the message. (3) After sending a message, the sending agent believes that the receiving agent has received the message and that the sending agent believes that the receiving agent believes the 'content' of the message.

An example scenario for the goals formed and the execution of the corresponding plans is demonstrated in Figure 4. In this example, to achieve its goal δ_1, agent x initiates the communication (and possibly negotiation process) by executing a *request* plan, i.e., sending a request to agent y, expressing its belief on a potential for cooperation, and waiting for a response from y. Upon receipt of this request, if y wishes to negotiate on the requested terms, it will form the goal δ_5, and *propose* its desired terms. If however, y agrees with the requested terms (or fixes some terms of its own), it will pre-commit to x, and communicate this by sending an *offer* and wait for x's response to commit or withdraw its request. Else, if y does not

[8] The formal syntax of messages can be found in [22].

favour any type of cooperation in the context of the request made, it will simply *reject* the request.

6 SUMMARY AND FUTURE DIRECTIONS

With the focus on agent to agent interactions to participate in a cooperative activity, we gave a brief overview of our internal theory. This theory allows a pragmatic approach to the specification of the reasoning processes behind communication. For this specification we identified three important reasoning states of *potential for cooperation, pre-commitment*, and *commitment* based on the beliefs, goals and intentions of individual agents. The process of reasoning was characterised as the interplay of *chance, choice* and *commitment*, and specified with a set of rules analogous to the *temporal implication* of concurrent programs. Finally, we demonstrated how our theory leads to the execution of appropriate communication plans.

Although our work extends Rao and Georgeff's single-agent BDI model to multi-agent systems, it is different from their follow up works [38] [30] which embody teams of agents and their joint attitudes. Since the objective of their work has been somewhat different, many important issues concerning reasoning about communication and cooperation are simplified by encoding and compiling most of the relevant information, at design time, into the structure of social plans. Our approach however, does not rely on existence of such social plans, and allows individual agents to reason about with whom and how they wish to cooperate, and actively choose appropriate communication plans as they proceed. It is perceived that since the meta-level plans in PRS [20] do the meta-level reasoning and decision making on the object-level plans, the model of reasoning specified in this article may be encoded in a set of *meta-level plans*.

This work was motivated by our earlier studies in [25] and forms the theoretical background and specification of the cooperation protocols described in [5]. These ideas were incorporated into the design of the general reasoning mechanism in our COSY agent architecture. The concepts behind this design directly maps onto the general concepts developed in the theory briefly reviewed in this article, (see [24] for details). The resulting protocols have also been implemented as plans of dialogue in dMARS, a BDI agent development and execution environment, and subsequently applied in some of our practical applications (see [4]. A detailed account of the practical work on protocols and the results will be reported in our future publications.

7 ACKNOWLEDGEMENT

The formal part of this work includes corrections and improvements on the work appeared in [23].

Afsaneh Haddadi (Daimler-Benz AG, Germany)

REFERENCES

[1] J. L. Austin. *How to do things with words*. Oxford University Press, 1962.

[2] T. Ballmer and W. Brennenstuhl. *Speech Act Classification: A Study in the Lexical Analysis of English Speech Activity Verbs*. Springer, 1981.

[3] M. E. Bratman. What is Intention? In P. R. Cohen, J. Morgan, and M. E. Pollack, editors, *Intentions in Communication*. MIT Press, 1990.

[4] B. Burmeister, A. Haddadi, and G. Matylis. Application of multi-agent systems in traffic and transportation. *IEE proceedings on Software Engineering*, 144(1):51–60, 1997.

[5] B. Burmeister, A. Haddadi, and K. Sundermeyer. Configurable Cooperation Protocols for Multi-Agent Systems. In *Proc. MAAMAW-93*, Neuchâtel, Switzerland, 1993.

[6] J. A. Campbell and M. d'Inverno. Knowledge Interchange Protocols. In Y. Demazeau and J. P. Müller, editor, *Decentralized A.I.* Elsevier/North-Holland, 1990.

[7] P. R. Cohen and H. J. Levesque. Rational Interaction as the Basis for Communication. Technical Report 433, SRI International, 1988.

[8] P. R. Cohen and H. J. Levesque. Intention Is Choice with Commitment. *Artificial Intelligence*, 42:213–261, 1990.

[9] P. R. Cohen and H. J. Levesque. TEAMWORK. Technical report, SRI International, Technote 504, 1991.

[10] R. Conte. Institutions and Intelligent Systems. In M. Jackson, P. Keys, and s. Cooper, editors, *Operational Research and the Social Sciences*, 1989.

[11] R. Conte and C. Castelfranchi. Norms as mental objects. From normative beliefs to normative goals. In *Reasoning about Mental States: Formal Theories and Applications, spring symposium*. AAAI Press, 1993.

[12] R. Davis and R. G. Smith. Negotiation as a Metaphor for Distributed Problem Solving. *Artificial Intelligence*, 20:63–109, 1983.

[13] P. de Greef, K. Clark, and F. McCabe. Toward a Specification Language for Cooperation Methods. In *GWAI-92 Workshop Supporting Collaborative Work Between Human Experts and Intelligent Cooperative Information Systems*, 1992.

[14] D. C. Dennett. *Brainstorms*. Harvester Press, 1981.

[15] D. C. Dennett. *Consciousness Explained*. Little, Brown & Co., 1991.

[16] E. H. Durfee and T. A. Montgomery. A Hierarchical Protocol for Coordinating Multiagent Behavior. In *Proc. AAAI-90*, 1990.

[17] E. A. Emerson. Temporal and Modal Logic. In J. van Leeuwen, editor, *Handbook of Theoretical Computer Science*, volume B. Elsvier Science Publishers B. V., 1990.

[18] R. Fagin and J. Y. Halpern. Belief, Awareness, and Limited Reasoning. In *Proc. of IJCAI-85*. Morgen Kaufmann, 1985.

[19] T. Finin, R. Fitzson, D. McKay, and R. McEntire. KQML as an Agent Communication Language. In *The proc. of the 3rd International Conf. on Information and Knowledge Management (CIKM'94)*. ACM Press, Novemebr 1994.

[20] M. P. Georgeff and F. F. Ingrand. Research on procedural reasoning systems. Technical report, AI Center, SRI International, Menlo park, California, 1988. Final Report, phase 1.

[21] P. J. Gmytrasiewicz and E. H. Durfee. Reasoning about Other Agents: Philosophy, Theory, and Implementation. In *Proc. of the 12th International Workshop on Distributed AI*, 1993.

[22] A. Haddadi. *Reasoning about Interactions in Agent Systems: A Pragmatic Theory*. PhD thesis, University of Manchester Institute of Science and Technology (UMIST), U.K., 1995.

[23] A. Haddadi. Towards a Pragmatic Theory of Interactions. In *Proc. of 1st. International Conf. on Multi-Agent Systems (ICMAS'95)*, 1995.

[24] A. Haddadi. *Communication and Cooperation in Agent Systems: A Pragmatic Theory*. Springer, LNAI series, No. 1056, 1996.

[25] A. Haddadi and K. Sundermeyer. Knowledge about Other Agents in Heterogeneous Dynamic Domains. In *ICICIS-93*, 1993.

[26] A. Haddadi and K. Sundermeyer. Belief, Desire, Intention Agent Architectures. In N. Jennings and G. O'Hare, editors, *Foundations of Distributed Artificial Intelligence*. Wiley Inter-Science, 1996.

[27] J. Y. Halpern and Y. Moses. Knowledge and Common Knowledge in a Distributed Environment. In *Proc. of the Third ACM Symposium on the Principles of Distributed Computing*, pages 50–61, 1985.

[28] N. R. Jennings. *Joint Intentions as a Model of Multi-Agent Cooperation.* PhD thesis, Queen Mary and Wesfield Colledge, Department of Electrical Engineering, University of London, technical report 92/18, 1992.
[29] N. R. Jennings. Specification and Implementation of Belief-Desire-joint-Intention Architecture for Collaborative Problem Solving. In *International Journal of Intelligent and Cooperative Information Systems*, 1993.
[30] D. Kinny, M. Ljungberg, A. Rao, E. Sonenberg, G. Tidhar, and E. Werner. Planned Team Activity. In C. Castelfranchi and E. Werner, editors, *Artificial Social Systems, Lecture Notes in Artificial Intelligence, 830.* Springer-Verlag, 1994.
[31] Y. Labrou. Proposed KQMl Specification Document. An excerpt from doctoral dissertation called Semantics for an Agent Coomunication Language, 1994.
[32] Y. Labrou and T. Finin. A semantic approach for KQML–a general purpose communication language for softwae agents. In *Proc. of the 3rd International Conf. on Information and Knowledge Management (CIKM'94).* ACM Press, 1994.
[33] H. J. Levesque, P. R. Cohen, and J. H. T. Nunes. On Acting Together. In *Proc. AAAI-90*, pages 94–99, 1990.
[34] S. C. Levinson. The Essential Inadequacies of Speech Act Models of Dialogue. In H. Parrett, editor, *Possibilities and Limitations of Pragmatics*, 1981.
[35] Z. Manna and A. Pnueli. Synthesis of communicating processes from temporal logic specifications. *ACM Transactions on Programming Languages and Systems*, 6(1):68–93, 1984.
[36] A. S. Rao and M. P. Georgeff. Modeling Rational Agents within a BDI Architecture. In *Knowledge Representation and Reasoning, (KR-91)*, pages 473–484, 1991.
[37] A. S. Rao and M. P. Georgeff. A Model-Theoretic Approach to the Verification of Situated Reasoning Systems. In *Proc. IJCAI-93*, pages 318–324, 1993.
[38] A. S. Rao, M. P. Georgeff, and E. A. Sonnenberg. Social Plans: A Preliminary Report. In E. Werner and Y. Demazeau, editors, *Decentralized A. I. 3.* Elsvier Science Publishers, 1992.
[39] J. R. Searle. *Speech Acts.* Cambridge University Press, 1969.
[40] J. R. Searle. Collective Intentions and Actions. In P. R. Cohen, J. Morgan, and M. E. Pollack, editors, *Intentions in Communication*, pages 401–415. MIT Press, 1990.
[41] Y. Shoham and S. B. Cousins. Logics of Mental Attitudes in AI. In G. Lakemeyer and B. Nebel, editors, *Foundations of Knowledge Representation and Reasoning, Lecture Notes in AI.* Springer Verlag, 1994.
[42] M. P. Singh. Group Intentions. In *Proceedings of the 10th International Workshop on Distributed Artificial Intelligence.* MCC Technical Report Number ACT-AI-355-90, 1991.
[43] M. P. Singh. *Multiagent Systems: A Theoretical Framework for Intentions, Know-How, and Communications.* Springer Verlag, Lecture Notes in Computer Science, volume 799, 1995.
[44] R. G. Smith. The Contract Net Protocol: High-Level Communication and Control in a Distributed Problem Solver. *IEEE Transactions on Computers C-29*, 1980.
[45] K. Sycara. Resolving Goal-Conflicts via Negotiation. In *Proc. AAAI-88*, 1988.
[46] R. Tuomela and K. Miller. We-intentions. In *Philosophical Studies 53*, pages 367–389, 1988.
[47] T. Winograd and F. Flores. *Understanding Computers and Cognition.* Addison-Wesley, 1986.
[48] M. J. Wooldridge. *The logical Modelling of Computational Multi-agent Systems.* PhD thesis, Department of Computation, UMIST, Manchester, 1992.
[49] M. J. Wooldridge and N. R. Jennings. Towards a Theory of Cooperative Problem Solving. In *Pre-proceedings of MAAMAW-94*, 1994.

B. J. GROSZ AND S. KRAUS

THE EVOLUTION OF SHAREDPLANS

1 INTRODUCTION

Rational agents often need to work together. There are jobs that cannot be done by one agent—for example, singing a duet or operating a computer network—and jobs that are more efficiently done by more than one agent—for example, hanging a door or searching the Internet. Collaborative behavior—coordinated activity in which the participants work jointly with each other to satisfy a shared goal—is more than the sum of individual acts [24, 8] and may be distinguished from both interaction and simple coordination in terms of the commitments agents make to each other [4, 10, 9]. A theory of collaboration must therefore treat not only the intentions, abilities, and knowledge about action of individual agents, but also their coordination in group planning and acting. It also must account for the ways in which plans are incrementally formed and executed by the participants.

Our long term goal is to provide a formalization of collaboration that supports the design and construction of collaborative computer systems, including systems that are able to collaborate with one another [13], systems that support groups of people working together, and collaborative systems for human-computer communication [16, 22]. The SharedPlan formalization of collaboration [10, 7] provides specifications that are normative, but it has been informed by an analysis of human collaborations [1]. The formalization is based on a mental-state view of plans [2, 21]: agents are said to have plans when they have a particular set of intentions and beliefs.

Agents may have different beliefs about the ways to do an action or to achieve some desired state. Pollack [21] argued that, as a result, a move away from a view of plans as primarily data-structure-like (i.e.,a plan as a set of [partially] ordered actions that when done in the appropriate conditions lead to a specified new state of the world) to a view of plans as mental state was necessary for plan inference.[1] Bratman [2, 3] also argues for a mental-state view of plans, emphasizing the importance of intentions to plans. Having a plan is not merely knowing how to do an action, but also having the intention to do the actions entailed. Bratman argues

[1] Pollack showed in particular that these differences among agent beliefs must be detected in plan recognition because they may affect the ways in which a question-answering system should respond to questions that are possibly incomplete or inaccurate.

M. Wooldridge and A. Rao (eds.), Foundations of Rational Agency, 227–262.

that intentions to do an action play three roles in rational action: having an intention to do an action constrains the other intentions an agent may adopt, focuses means-ends reasoning, and guides replanning.

To adequately model collaboration it is necessary both to accommodate differences among the beliefs of individual participants and to distinguish between knowledge about action performance and intention to act. Agents may differ not only in their beliefs about the ways to perform an action and the state of the world, but also in their assessments of the ability and willingness of an individual to perform an action. The three roles Bratman describes for intentions may be even more important for collaborative activity than for individual plans. Hence, the adoption of a mental-state view of plans is essential to the formalization.

Following Pollack, we will use the terms "recipe" and "plan" to distinguish between knowing how to do an action and having a plan to do the action. When agents have a SharedPlan to do a group action, they have certain individual and mutual beliefs about how the action and its constituent subactions are to be done. We use the term *recipe* [21, 1, 17] to refer to a specification of a set of actions, which we will denote as β_i $(1 \leq i \leq n)$, the doing of which under appropriate *recipe-constraints*, denoted as ρ_j $(1 \leq j \leq m)$, constitutes performance of α.[2] We use the meta-language symbol R_α to denote a particular recipe. The β_i may be either single-agent or multi-agent actions; they may also be complex, i.e., may comprise constituent subactions. In some cases we may need to refer to a partial (possibly empty) recipe for α; we will use the meta-language symbol R^p_α to denote such a partial recipe.

The SharedPlan formalization provides mental-state specifications of both *SharedPlans* and *individual plans*. SharedPlans are constructed by groups of collaborating agents and include subsidiary SharedPlans [16] formed by subgroups as well as subsidiary individual plans formed by individual participants in the group activity. The formalization distinguishes between complete plans—those in which all the requisite beliefs and intentions have been established—and partial plans. In addition to the propositional attitude of *intending to* do an action, it introduces the attitude of *intending that* a proposition hold.

In an earlier paper [10] we provide definitions for individual plans and SharedPlans, both complete and partial, and briefly describe the planning actions that play a role in expanding partial plans to complete ones. The focus of that paper was on specifying the capabilities to act and mental attitudes (beliefs and intentions) required for the participants in a group activity to have a collaborative plan. This paper complements that one by examining in more detail the dynamics of SharedPlans and the role of intentions-that in their evolution.

The remainder of this paper is organized as follows. The next section provides an overview of the formalization of partial SharedPlans and presents an example used throughout the paper to illustrate various features of the formalization and

[2]The indices i and j are distinct; for simplicity of exposition, we omit the range specifications in the remainder of this document.

its use in the design of automated collaborative agents. Section 3 details extensions to the formalization of intentions-that. Section 4 presents an architecture for the design of collaborating agents based on the formalization. Section 5 briefly concludes with mention of some important open problems.

2 BACKGROUND

2.1 *Basic Tenets of the SharedPlan Approach*

Because the world is dynamic and uncertain and an agent's knowledge is typically incomplete, recent efforts in the areas of planning and rational action have focused on ways to coordinate planning and acting and on methods for dealing with resource limitations [14, 5, 6, 11, 14, 26, inter alia]. Change, uncertainty, and incomplete knowledge also affect agents when they work together and thus must be accommodated by any theory of collaboration. Plans—whether of an individual or of a group—evolve over time. Although completed plans are the ultimate aim of planning, partial rather than complete SharedPlans best characterize the mental state of participants in a group activity. As a result, the evolution of SharedPlans is, in reality, the evolution of partial SharedPlans, and this paper focuses largely on partial SharedPlans.

Partial individual plans get expanded to more complete plans through means-ends reasoning about intended actions. Collaborative plans have a correlate of this reasoning, or plan-elaboration, process, but their expansion entails communication and negotiation as well as means-ends reasoning about the way in which to do the group action. In our formalization, means-ends reasoning is represented by the complex planning action Elaborate_Individual and the group planning activity is represented by the complex action Elaborate_Group.[3]

A basic claim of the SharedPlans formalization is that collaborative activity is rooted in the individual mental actions and domain actions of individual agents. This constraint holds of complex group planning actions as well as of domain actions. Thus, Elaborate_Group comprises individual planning actions of the group members. We use the the complex action I_Elaborate_Group to represent these individual actions.

In multi-agent activities participants not only do means-ends reasoning about their own actions, they also reason about how to coordinate with and support the actions of others in the group. These activities require plan-based reasoning that arises from the participants' attitudes of intentions-that toward the actions of others and of the group as a whole. To handle this aspect of the dynamics of SharedPlans we introduce a new complex planning action, Cultivate.

Two different complex planning actions — Elaborate_Group and Cultivate — therefore combine to provide the planning processes for group activities analo-

[3]In the formalization [10], Elaborate_Individual appears in the definition of the modal operator for intentions-to act, Int.To, and Elaborate_Group appears in the definition of SharedPlan.

gous to those that Elaborate_Individual provides for individual actions. Cultivate is the general process that is active whenever an agent has an intention-that, in the same way that Elaborate_Individual is active whenever an agent has an intention-to. Elaborate_Group is activated whenever the intention-that leads to a SharedPlan; it serves to oversee the completion of the plan.

This paper extends the original treatment of the modal operator for intentions-that, Int.Th, by providing additional axioms, specifying the minimal conditions for an agent having an intention-that, and describing various ways in which intentions-that can lead to actions in the support of a collaborative plan. It also presents an architecture for an automated collaborative agent based on the specification provided by the SharedPlan definitions. This architecture includes I_ELABORATE_GROUP and CULTIVATE processes which realize the correspondingly named complex actions of the formalization. The paper uses a slightly revised definition of partial SharedPlans. This new definition covers a wider range of transitional states in partial plans and provides for more generality in the planning processes.

2.2 A Collaborative Planning Scenario

To help explain the definition of partial SharedPlan (PSP), we will examine a scenario based on the Truckworld simulated environment [12] which was the test environment for an initial implementation [20]. This scenario is as follows (see Figure 1). MT_1 and MT_2 are mobile truck agents; ST_q is a stationary truck agent with crane-arms located at Q, and ST_b is a stationary truck agent with crane-arms located at B. G_q and G_b are gas tanks. In the scenarios we will consider, the four agents MT_1, MT_2, ST_q, and ST_b, have formed a collaborative team, TG, to move certain building material from Q to B. We assume initially that the agents are part-way in forming their plan for performing this group action, which we will refer to as MOVE39. They have identified a recipe for doing the action—i.e., they have figured out at a relatively high level of description how to do MOVE39—but they have not yet worked out the lower-level details. For instance, they may not have decided how to do some subactions in the recipe or who will do them. In summary:

Two types of trucks agents:
ST: stationary truck with crane arm for [un]loading only.
MT: mobile truck with arms that can [un]load

Two sites:
Q: quarry with raw materials [e.g., big rocks], a stationary truck, ST_q, and a gas pump, G_q.
B: building construction site, a stationary truck, ST_b, and a gas pump G_b.

Routes: The layout of the roads between Q and B is shown above. Route distances are given in the rightmost column.

R1:	Q → J1 → B	5 units
R2:	Q → J2 → B	8 units
R3:	Q → J1 → J2 → B	10 units

The road between J1 and B is sometimes closed. If so, the shortest route from Q to B (R1) is unavailable.

Actions:

Multi-agent actions: Load truck, Unload truck
Load truck requires one ST and one or more MTs. Unload truck requires two trucks (either two MTs or one MT and one ST). Loading is slower than unloading.

Single-agent actions: All Truckworld basic-level actions (e.g., drive a route, send-info by radio, receive-info by radio; also [un]loading related actions: pick-up, put-down, move).

The basic scenario for the TruckWorld is then as follows: Mobile truck MT_1 moves the first shipment of rocks from Q to B along R1 if possible; ST_q helps load, ST_b helps unload; MT_2 does the transport for the second shipment. In addition to this basic scenario, we will also refer to the following variations in what follows:

Variation 1: Helpful behavior. MT_1 helps ST_q unloading another truck to lessen the wait for ST_q to begin participating in MOVE39.

Variation 2: Cooperative replanning. MT_2 breaks down doing Shipm2; MT_1 takes over so that Shipm2 and hence MOVE39 can be done successfully.

Variation 3: Intention-conflict avoidance. MT_1 moves from ST_b's unloading area while getting gas so that MT_2 can unload.

Variation 4: CBA-enable. MT_2 runs out of gas; MT_1 can help by bringing gas.

Variation 5: CBA-helpful behavior. Truck MT_1 senses road closure at J1 and radios this back to subgroup TG2 (or directly to truck MT_2) which is at Q so that MT_2 will take R2 and not waste time.

Variation 6: Monitoring. ST_q monitors road conditions and forwards any important information to MT_1 or MT_2 while they are enroute and unable to obtain this information independently.

Collaborative activity requires a commitment on the part of the participants to the joint endeavor [4, 10, 15]. Thus, each of the team members in TG must be committed to the group's moving material from Q to B. In the basic scenario, the group has chosen a recipe which comprises two shipment actions, each of which will move some of the material. The first shipment action, Shipm1, will be performed by the subgroup TG1, comprising MT_1, ST_q and ST_b. TG1 must form a

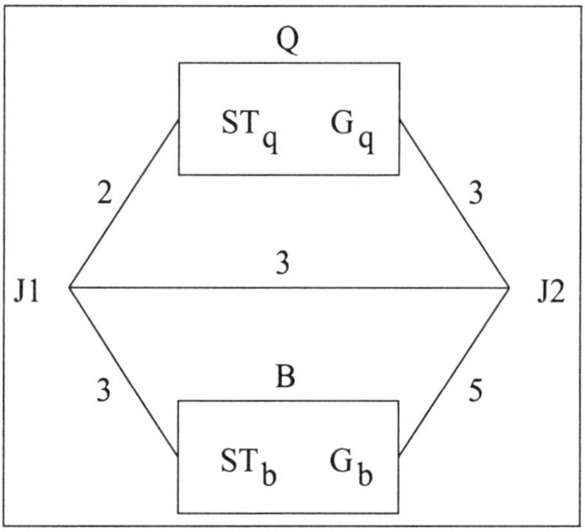

Figure 1. The TruckWorld Domain

SharedPlan for doing Shipm1. This SharedPlan is subsidiary to the SharedPlan for MOVE39; it may be either partial or complete. For illustrative purposes, we assume TG1 has decided to use a recipe with the following three subsidiary actions: a subgroup of TG1 loads a truck (this truck is one of the members of TG1), the truck drives to B, and a subgroup of TG1 unloads the truck.

In collaborative activity, plans for subactions also impose some constraints on the group doing the overall activity. In particular, the full group must agree on who will do the subsidiary action, must have confidence that the subgroup can and will do the action, and must be committed to the subgroup's success. However, it would be too burdensome (engendering unnecessary communication and computation) to require that agents outside the subgroup know all the details of how the subgroup will act. Thus, the formalization requires that the full group, TG, have mutual belief that the subgroup TG1 has formed a SharedPlan for Shipm1 and that TG1 can perform this action. It does not require though that MT_2, who is not in TG1, know the recipe TG1 has chosen; rather MT_2 must believe there is some recipe that the group TG1 can use. In addition MT_2 must be committed to TG1's being able to do the Shipm1 action.

For TG1 to complete their SharedPlan for Shipm1, they must decide who will do each subsidiary action in the recipe. For instance, they might decide, as in the basic scenario, that ST_q and MT_1 will load MT_1, MT_1 will drive to B, and then MT_1 and ST_b will unload MT_1. If so, MT_1 and ST_q must form a complete subsidiary SharedPlan to do the loading action; MT_1 must have an intention-to drive from Q to B and a plan for doing so; and MT_1 and ST_b must form a complete subsidiary SharedPlan to do the unloading action.

SP(P_α, GR, α, T_p, T_α, Θ_α, IC_α)

1. The group has a full shared plan:
(1) $[(\exists R_\alpha)$FSP(P_α, GR, α, T_p, T_α, R_α, Θ_α, IC_α)] $\otimes$

2. The group has a partial shared plan, and a full shared plan to complete it:
(2) $[(\exists P_{elab}, T_{elab}, R_{elab})$
(2a) PSP(P_α, GR, α, T_p, T_α, Θ_α, IC_α)$\wedge$
(2b) FSP(P_{elab}, GR, Elaborate_Group(P_α, GR, α, T_p, T_α), T_p, T_{elab}, R_{elab}, Θ_α, P_α)]

Figure 2. Definition of SP (SharedPlan).

We assume initially that TG has not yet determined who will do the second shipment action, Shipm2. We refer to recipe subactions in this status as "unresolved actions."[4] Collaborative activity imposes two constraints on participants with respect to unresolved actions. All of the group members must believe that some subgroup can do the unresolved action, and the group must have a way to choose a subgroup to do it. Thus, to resolve Shipm2 (making it a "resolved action"), TG must agree on a subgroup, TG2, to do the action and form beliefs and intentions analogous to those described above for Shipm1. In choosing who will do Shipm2, the members of TG must take into account what they know about the plans (especially the commitments made by members of TG1) for Shipm1.

The scenario variations described above will be used later in the paper to illustrate various features of the formalization and architecture. For some of these examples, we will assume that the recipe for Shipm2 is identical to the one for Shipm1, but that MT_2 will drive the load of rocks from Q to B for Shipm2.

2.3 *Partial Plans: Definition and Illustration Using Scenario*

The formalization uses two meta-predicates to which later sections of this paper refer. CBA ("can bring about") and CBAG ("can bring about group") are used to represent the beliefs agents have about their own and their collaborators' abilities to perform actions. The plan definitions and axioms use these meta-predicates within an embedding belief context so that omniscience is not required; a more extensive discussion of the reasons for the particular choice of meta-predicates may be found in our earlier paper [10]. CBA(G, α, R_α, T_α, Θ_α) represents that the agent G is able to do the action α using the recipe R_α at time T_α under constraints Θ_α. CBAG is defined analogously for groups. Formal definitions of these meta-predicates are in the earlier paper.

Figure 2 gives the definition of SharedPlans that shows the role of the action Elaborate_Group. When the agents' plan is only partial [the second clause], they must also have a second plan, which is complete (an FSP) for extending the parent

[4] In our earlier work, we referred to these actions as "unreconciled." We have changed the terminology to avoid conflict with the use of "reconcile" for intentions.

PSP(P_α, GR, α, T_p, T_α, Θ_α, IC_α)

0. The group GR has mutual belief that all members of the group are committed to the group doing α, i.e., that they each **intend-that** the group do α.

0a. Parameter identification: The group GR has mutual belief that all members of the group are committed (have an **intend-that**) to the identification of parameters necessary to α.

0b. Constraint satisfaction: The group GR has mutual belief that all members of the group are committed (have an **intend-that**) to making sure that the constraints for doing α (Θ_α) will hold.

1. The group GR mutually believe that there is a **recipe** for α, but their recipe for doing α may be partial; e.g., they may only have identified *some* of the subactions that need to be performed. If their recipe is only partial, they have a FSP to complete their partial recipe (using Select_Rec_GR).

For each subaction β_i in the partial recipe, one of (2) through (4) holds:

2. Core case (resolved actions β_r):

2a. Single-agent subaction: A member, G_k, of the group intends to do the subaction β_r; G_k may have either a complete or partial plan for doing it.

2b. Multi-agent subaction: A subgroup GR_k has a shared plan (SP) to do the subaction, but this plan may be only partial.

2b1. The subgroup GR_k has a SharedPlan for the subaction.

2b2. The group GR mutually believe that the subgroup GR_k has a SharedPlan for the subaction.

2b3. Every member of GR **intends-that** the group GR mutually believe that the subgroup GR_k is able to do the subaction.

2b4. The group GR mutually believe that all of its members are committed to the subgroup's success, i.e., **intend-that** the subgroup GR_k can do the subaction.

3. Contracting case: (omitted here; see [10])

4. Unresolved case (actions β_k): GR has not fully deliberated about the subaction β_k; no decision has been made about which agent(s) will do it.

4a1. Single-agent subaction:

4a1a The group GR mutually believe that some (perhaps as yet unidentified) group member G_k can do the subaction.

4a1b. The group GR has a full SharedPlan to select an agent to do the subaction (Select_Agent).

4a2. Multi-agent subaction:

4a2a. The group GR mutually believe that some (perhaps as yet unidentified) subgroup, GR_k, can do the subaction.

4a2b. The group GR has a full SharedPlan to select a subgroup to do the subaction (Select_Subgroup).

Figure 3. English description of the PSP (partial SharedPlan) definition.

partial plan [Clause (2b)]; that is, they must have agreed on the procedures they will follow to establish the additional beliefs and intentions needed to complete the parent plan. To explain this elaboration process, we must look at the definition of partial SharedPlan in some detail.

General form:
Clause 0 $\wedge$ Clause 1 $\wedge$
[Clause 2 $\otimes$ Clause 3] $\wedge$
Clause 4

Detailed form:
Clause 0 $\wedge$
$(\exists R_\alpha, \mathrm{B}, \mathrm{B}_{res}, \mathrm{B}_{unres}, \mathrm{C})[$
Clause 1 $\wedge$
$(\forall \beta_r \in \mathrm{B}_{res})$[Clause 2 $\otimes$ Clause 3] $\wedge$
$(\forall \beta_k \in \mathrm{B}_{unres})$Clause 4
]

Figure 4. The form of the PSP definition

Notation	Meaning	Comments
α	action	Also: β_i,β_r, β_k, δ, γ
IC_α	intentional context of α	
P_α	plan for α	
R_α	recipe for α	Also: R^f_α
R^p_α	partial (possibly empty) recipe for α	
T_i	time of i	i: action, plan, proposition
G	agent	Also G_i, G'
GR	group	Also GR_i
Θ_α	constraints of α	Also, ρ_j
$prop$	proposition	
A_{prop}	atemporal propositional content	

Table 1. Summary of notation used for special variables and constants.

Figure 3 gives an informal definition of partial SharedPlan that will be the basis for our discussion. Because the full definition is complex, Figure 4 gives the general form of the definition with the relevant quantifier scopings for each of the major clauses. The formal definition, provided in Appendix A, uses a first-order logic augmented with several modal operators and meta-predicates. To enable readers to connect the intention-that discussion and new axioms to the formal definition, clauses in the two definitions correspond exactly. Table 1 provides a summary of the notation used in the formalizations in this paper. Tables 3 and 4 in Appendix A lists the various operators, predicates and functions used in the formal definition, provides an informal description of their meaning, and indicates a reference for further information (either a section of this paper or the paper in which the original formalization is provided).

SharedPlans include as constituents individual plans formed by participants to do actions needed by the group activity. To distinguish among the different types of plans, the formalization defines five meta-predicates: FIP for *full individual plans*; PIP for *partial individual plans*; FSP for *full SharedPlans*; PSP for *partial SharedPlans*; and SP for SharedPlans of indefinite completeness. The plan meta-predicates make claims about the mental states of agents, but are not new modal operators. Each is defined in terms of intentions and beliefs of the agents who have plans of the given type [10].

As shown in the informal definition given in Figure 3, PSP's contain four main elements. Clauses 0 and 1 specify, respectively, the commitments required of the agents in the group to the joint action and the need for them to establish mutual belief in the existence of a (partial) recipe for the action. In the early stages of working together, many of the subactions of the partial recipe may be unresolved. The beliefs and intentions required for unresolved actions are represented in Clause 4. Clauses 2 and 3 represent the two possibilities for actions that have been resolved: either some group member (or subgroup) will do the action or the group will contract out the action.

In the Truckworld scenario, each of the members of TG must have an intention-that the group move material from Q to B (Clause 0); here "move" refers to the full complex of loading the material, transporting it, and unloading it at the destination. They have chosen a recipe (Clause 1) which comprises two shipment actions, each of which will move some of the material. The group has decided that the first shipment action, Shipm1 will be performed by the subgroup TG1, comprising MT_1, ST_q and ST_b. Thus, Shipm1 is a "resolved" action and falls under the "core case" of the definition (Clause 2). To meet the PSP specification, TG1 must have a SharedPlan for doing Shipm1 (Clause 2b1). This subsidiary SharedPlan may be either partial or complete; for illustrative purposes, we assume TG1 has decided to use the recipe described earlier. (A specific instantiation of this recipe is given in the basic scenario.) In addition, the full group, TG, must mutually believe TG1 has a SharedPlan for Shipm1 (Clause 2b2) and that they can (successfully) complete it (Clause 2b3); MT_2, who is not in TG1, may not know the recipe TG1 has chosen, but must believe there is some recipe that group can use.[5] In addition MT_2 must have an intention-that TG1 will be able to do the Shipm1 action (Clause 2b4).

The locus of activity in the evolution of SharedPlans is in the transition from the "unresolved case" (Clause 4 of the definition), to the "core case" (Clause 2) or the "contracting" case (Clause 3).[6] The action resolution process for the group requires that they identify the agent(s) who will do the constituent action and obtain adequate commitments from group members. The member (or subgroup) selected to do an action must commit to doing it and the other agents must commit to the member's (subgroup's) being able to do it. In each of these cases, the individual agents must reconcile the intentions-to and intentions-that entailed by these new

[5]The definition requires only that each agent intend-that this mutual belief holds at the time of the shipment. We discuss the reasons for this weaker constraint later.

[6]In this paper we do not discuss the contracting case further.

commitments with the other commitments they hold.

Clause 4 is also the principal way in which the PSP definition given here differs from the original definition. When designing the architecture for an automated collaborating agent, we realized that the original definition could be generalized. The Select_Agent and Select_Subgroup complex actions in the new definition allow for a wider range of strategies than the original definition, both for agent selection and for commitment formation.

In the Truckworld example, Shipm2 is unresolved and so falls under Clause 4. The formalization requires that each member of TG believe that some subgroup of TG can do Shipm2 (Clause 4a2a). In addition, TG must have a full SharedPlan to choose a subgroup to do this action (Clause 4a2b); that is, TG must have determined the procedure they will use to decide who will do Shipm2. To move Shipm2 from the unresolved to the core case (the job of the Select_Subgroup action), TG must agree on a subgroup, TG2, to do the action and form beliefs and intentions analogous to those described above for Shipm1. The members of TG2 do not need to know the recipe they will use, but they must at least agree on how they will figure this out (Clause 1).

The SharedPlan definitions also associate an intentional context parameter with each plan meta-predicate and intention operator.[7] The intentional context constrains replanning in the case of execution problems and guides recipe choice or completion. From a more theoretical standpoint, it provides the reason for an agent undertaking a plan and thus can be used to distinguish between two different uses of the same action.[8]

Table 2 lists the intentional context for the meta-predicates and operators relevant to this paper. To simplify the presentation, in this table and in the subsequent text, we elide those arguments of operators that are not germane to the discussion. The table shows, for example, that the SharedPlan of a group to do α and the partial recipe they have constitute the intentional context for the intention-to do subsidiary acts (first line) and the recipe selection plan (fourth line). The last entry in the table gives the intentional context for an action γ that indirectly brings about some intended proposition, *prop*. It covers situations like contracting in which *prop* results from an action, δ, done by another agent G_j (or, alternatively, by a group GR_j) and γ functions to get G_j (or GR_j) to do δ. In this case, both the intention-that for *prop* and the intention to do δ (or the SharedPlan for δ if it is a group activity) are in the intentional context for γ.

The plan meta-predicates also have a constraints parameter that encompasses three types of constraints: execution preconditions, recipe-constraints, and constraints imposed by intention-conflict resolution. Typically, recipe-constraints con-

[7] In the original definitions, the context parameter also encoded constraints on the action. In the current definitions (see Appendix A) we have two separate parameters, IC_α and Θ_α, for intentional context and constraints respectively. The motivation for each of these parameters was given previously [10].

[8] For example, the intentional context would distinguish between Searle's two types of business school graduates [24, pp. 404-405].

Intention/SP	IC
Int.To($G_k, \beta_i, \ldots$) Int.Th(G_k, Do(GR$_k, \beta_i, \ldots$), ...) PSP(P_{β_i}, GR$_k, \beta_i, \ldots$) FSP($P_{select.rec.g}$, GR, Select_Rec_GR(GR, α, ...)...) FSP(P_{elab}, GR, Elaborate_Group(P, GR, α, ...)...) FSP($P_{select.subgroup}$, GR, Select_Subgroup(GR, β_k, ...)...) FSP($P_{select.agent}$, GR, Select_Agent(GR, β_k, ...)...)	SP(GR, α, ...), R^p_α SP(GR, α, ...), R^p_α SP(GR, α, ...), R^p_α SP(GR, α, ...), R^p_α SP(GR, α, ...), R^p_α SP(GR, α, ...), R^p_α SP(GR, α, ...), R^p_α
Int.Th(G_j, id.params(GR, α, T_α), ...) Int.Th($G_i, \Theta_\alpha, \ldots$)	SP(GR, α, ...) SP(GR, α, ...)
Int.To($G_i, \gamma, \ldots$) where $prop$ is a direct effect of γ	Int.Th($G_i, prop, \ldots$)
Int.To($G_i, \gamma, \ldots$) where $prop$ is an indirect effect of γ via δ	Int.Th($G_i, prop, \ldots$), Int.To($G_j, \delta, \ldots$)/ SP(P_δ, GR$_j, \delta, \ldots$)

Table 2. Intentional Context for elements of SP(P_α, GR, $\alpha, T_p, T_\alpha, \Theta_\alpha$, IC$_\alpha$) with β_i in R^p_α.

cern time, location or other resources. For example, there is a time constraint between making dough and baking bread: the dough-making must precede baking. When reconciling intentions, an agent may discover it needs to further restrict the time, location or choice of other parameters of an action (e.g., only one of the family cars may be available for a trip to the movies). Constraints affect the selection of recipes for an action and may limit the resources available for doing subactions. Even though in general an agent may be able to perform an action, constraints may preclude its being able to do so in a particular situation.

Constraints of each type may derive from properties of a subaction β_i itself or from the embedding action α of which β_i is a constituent. However, not all the constraints on α will be constraints on β_i. Furthermore, there may be constraints on α that are not inherited by any of the actions it comprises. For instance, if an open-door action has the precondition *unlocked(door)* and two constituent subactions, *turn-handle* and *push*, then assuming doors have bolt locks that are separated from the door-handle mechanism, the precondition *unlocked* is not a precondition of either *turn-handle* or *push*. The formalization uses the function *inherit* to refer to the constraints that are inherited, the function *prec* to refer to an action's preconditions, and the function *con* to refer to the constraints associated with an action.

3 INTENDING-THAT

The intentional attitude, intending-that, was introduced into the SharedPlan formalization to account for the commitment participants in a group activity make to one another's actions and to their joint activity. Intentions-that, like intentions-to, serve both to constrain the intentions an agent adopts and to affect its plan-based reasoning. In this section, we first review the basic properties of intending-that from the initial formalization. We then extend the treatment by specifying the minimal conditions for an agent to have an intention-that and providing additional axioms for the intention-that modal operator, Int.Th. Finally, we enumerate the ways in which an intention-that may lead an agent to adopt an intention-to do some action in service of that intention-that. The extensions presented here were motivated by the ways in which intentions-that lead agents to undertake actions that are required by their group activity, restrict the actions rational agents consider performing, and cause agents to adopt beliefs needed for collaboration. These extensions were constrained by our use of the formalization in the design of an architecture for automated collaborative agents.

3.1 An Overview of Intentions-That in SharedPlans

Intending-that is an attitude that holds between an agent and a proposition. We write

$$\text{Int.Th}(G, prop, T_i, T_{prop}, \text{IC}_{prop})$$

to represent an agent G's intention at time T_i that a certain proposition *prop* hold in the intentional context IC_{prop}.[9] In our formalization, propositions have a time [interval] associated with them. We denote the time of proposition *prop* by T_{prop} and the atemporal propositional content by A_{prop}. For example, if it snows in Boston on Tuesday then *prop* is $snow(Boston, Tues)$, T_{prop} is Tuesday, and A_{prop} is the atemporal fact of snowing in Boston. In most of the cases we are concerned with, T_{prop} is a time that is in, or extends into, the future. One class of propositions that play central roles within the context of SharedPlans are propositions that represent agents' abilities to perform actions (using CBA and CBAG). For the CBA and CBAG propositions, T_{prop} (i.e., the time at which the beliefs about ability must hold) is the time of the action that the ability proposition concerns.

Intentions-that play a central role in a collaborative activity from its very inception. A collaboration is initiated when an agent, G, moves from simply desiring that some action occur to intending that some group of agents, GR, carry out that action, i.e., to $\text{Int.Th}(G, \text{Do}(\text{GR}, \alpha, T_\alpha, \Theta_\alpha), T_p, T_\alpha, \text{IC}_\alpha)$. The intention-that leads the agent G to take action, as described below, to get the group GR to form a SharedPlan for α. If the other members $G_i \in$ GR agree, then they too will have an intention-that of the form $\text{Int.Th}(G_i, \text{Do}(\text{GR}, \alpha ...)...)$. Because each G_i may have different reasons for agreeing to participate in the group activity and thus in forming the SharedPlan, the intentional contexts for the different agents' intentions-that may be different.

[9] As discussed in our earlier paper [10], the significant distinction between intentions-to and intentions-that is "not in the types of objects each relates, but in their connection to means-ends reasoning and in their different presumptions about an agent's ability to act in service of the intention." [p. 281]. The reader is referred to that paper for a more extensive discussion of the differences.

Group activity commitment (AC-IT): (Clause (0))
$$\text{Int.Th}(G_j, \text{Do}(\text{GR}, \alpha, T_\alpha, \Theta_\alpha), T_p, T_\alpha, \text{IC}_\alpha)$$

Parameter identification (PI-IT): (Clause (0a))
$$\text{Int.Th}(G_j, \text{id.params}(\text{GR}, \alpha, T_\alpha), T_p, T_\alpha, P_\alpha)$$

Constraints satisfaction (CS-IT): (Clause (0b))
$$\text{Int.Th}(G_j, \Theta_\alpha, T_p, T_{\Theta_\alpha}, P_\alpha)$$

Core Case (CC-IT): (Clauses (2a3) and (2b4))
$$\text{Int.Th}(G_j, (\exists R_{\beta_r})\text{CBA}(G_k, \beta_r, R_{\beta_r}, T_{\beta_r}, \text{inherit}(\Theta_\alpha, \beta_r) \wedge \bigwedge \rho_j \wedge \text{prec}(\beta_r)), T_p, T_{\beta_r}, P_\alpha)$$
$$\text{Int.Th}(G_j, (\exists R_{\beta_r})\text{CBAG}(\text{GR}_k, \beta_r, R_{\beta_r}, T_{\beta_r}, \text{inherit}(\Theta_\alpha, \beta_r) \wedge \bigwedge \rho_j \wedge \text{prec}(\beta_r)), T_p, T_{\beta_r}, P_\alpha)$$

MB **satisfaction (MB-IT):** (Clauses (2a2) and (2b3))
$$\text{Int.Th}(G_i, \text{MB}(\text{GR}, (\exists R_{\beta_r})\text{CBA}(G_k, \beta_r, R_{\beta_r}, T_{\beta_r}, \text{inherit}(\Theta_\alpha, \beta_r) \wedge \bigwedge \rho_j \wedge \text{prec}(\beta_r)), T_{\beta_r}), T_p, T_{\beta_r}, P_\alpha)$$
$$\text{Int.Th}(G_i, \text{MB}(\text{GR}, (\exists R_{\beta_r})\text{CBAG}(\text{GR}_k, \beta_r, R_{\beta_r}, T_{\beta_r}, \text{inherit}(\Theta_\alpha, \beta_r) \wedge \bigwedge \rho_j \wedge \text{prec}(\beta_r)), T_{\beta_r}), T_p, T_{\beta_r}, P_\alpha)$$

Figure 5. Intentions-that in SharedPlans

In the Truckworld scenario, MT_1 might have been scheduled (e.g., by the plant manager) to do the job of moving material from Q to B. MT_1 might then have recognized that it would need help to do this job. MT_1 might have subsequently formed, on its own, an intention-that the group TG, comprising the set of agents MT_1, MT_2, ST_q, and ST_b, perform the action MOVE39. MT_1 would then contact the other three agents and convince them to form a collaboration. As a result, each of MT_1, MT_2, ST_q, and ST_b would then have an individual intention-that TG moves the materials.

The SharedPlan formalization employs Int.Th in five settings, listed in Figure 5. The intentions-that have the following three major functions:

Decision-making influence: all intentions-that influence which other intentions an agent will adopt.

Planning instigation: some intentions-that cause agents to complete their plans by changing either their mental state or the state of the world; as a result, commitment to the collaborative activity may lead to planning.

Collaborative acting assistance: some intentions-that lead agents to assist others in the group in performing actions relevant to the group activity or to alert them when things go wrong.

To demonstrate these functions, we will examine each of the five settings given in Figure 5 in turn.

Group activity commitment: Group activity commitment intentions-that (AC-IT's) exert decision-making influence by constraining the adoption of new intentions to ones that are compatible with the joint activity. The conflict-avoidance axiom in our previous paper [10] represents one way in which this occurs. For instance, once truck MT_2 joins the MOVE39 collaboration, it will ordinarily not take on a job that would hinder completion of MOVE39.[10]

AC-IT's may instigate a range of planning activities, because one way in which to satisfy an intention of the form Int.Th(G_j, Do(GR, α, ...)) is to form a SharedPlan for doing the action α.[11] To form this SharedPlan, the agent and other members of the collaborating team must do an Elaborate_Group (see the SP definition in Figure 2). Each agent thus will undertake planning activities as part of its associated I_ELABORATE_GROUP. For example, ST_q's AC-IT for the MOVE39 action may lead it to various activities in pursuit of determining, with other group members, a recipe for doing MOVE39.

AC-IT's may lead agents to assist one another in their collaborative activity in two ways, by prompting helpful behavior and through cooperative replanning. Helpful behavior in the Truckworld domain is exemplified in Variation 1: MT_1 recognizes that ST_q cannot work on MOVE39 for several hours because it is busy unloading another truck. MT_1 is not occupied; by doing some unloading it can lessen the wait for ST_q to begin participating in MOVE39. Because this will lower the overall cost of MOVE39 without otherwise affecting MT_1's performance, MT_1 helps out.

Cooperative replanning is exemplified in Variation 2: MT_1 successfully delivers the first shipment of rocks and heads back to home base; MT_2 breaks down and cannot complete delivery of the second shipment; Shipm2 thus reverts to being unresolved; MT_1 takes over and completes the delivery.[12] MT_2's communication to MT_1 is motivated by its AC-IT: because it cannot complete the drive subaction of Shipm2, it knows that Shipm2 and thus MOVE39 cannot be done according to the original recipe. Thus, it no longer believes that the Do-proposition embedded within its AC-IT holds. However, MT_2 can do something that will re-establish its belief in this Do-proposition, namely recruit someone else to complete the drive action. It does so, based on reasoning that follows the specification of Axiom 2 presented later in this paper. Because MT_2 cannot do the action, MT_1 is not assisting MT_2 but replacing it as the agent of the drive subaction of Shipm2. Thus, MT_1's assistance is motivated by its AC-IT commitment to MOVE39, not by a core-case intention that.

[10] That is, it will not do so unless it decides to opt-out of the collaboration [10].

[11] An agent might decide to engage in some other form of group acting.

[12] MT_1 and MT_2 may be able to do the replanning required without consulting ST_q and ST_b. However, ST_q and ST_b must be informed of the agent change.

Parameter identification: Parameter-identification intentions-that (PI-IT's) lead agents to perform actions that establish knowledge preconditions of the group activity [18].[13] The formalization specifies that all agents must intend-that the parameters of the group action be identified [Int.Th(id.params)], not that each agent must intend to do some action that leads to the parameter being identified. Typically only some of the participants in a group activity will be involved in identifying the parameters; however all of them must be committed to the group knowing what they are. For example, ST_q might determine the time or place of the loading for Shipm1 and announce this to MT_1; ST_q thus does an action to determine the identity of this parameter, but MT_1 must also intend-that a time and place be determined and ST_q and MT_1 must, by communicating and negotiating as necessary, agree to the choices. Thus, all three of the functions of intentions-that may be invoked by PI-IT's.

Constraint satisfaction: Constraint-satisfaction intentions-that (CS-IT's) may engender planning actions to establish any of the three types of constraints — preconditions, recipe-constraints, or constraints imposed by intention conflict resolution — or to maintain them during the construction and execution of a SharedPlan. The planning instigated by precondition constraints for SharedPlans is similar to that done in single-agent planning. However, the collaborative setting introduces the possibility that different agents or subgroups may handle different preconditions. It concomitantly requires that the agents coordinate and communicate about the assignment of responsibility for establishing preconditions. CS-IT's also exert decision-making influence: the need to have certain conditions hold for one action (e.g., the constraint that MT_1 be at the quarry at 10 AM for Shipm1) may preclude other actions (e.g., MT_1 working on a job in a different location at that time).

Core Case: Core-case intentions-that (CC-IT's) exert decision-making influence by restricting collaborating agents' choices of the ways to do constituent actions and the resources that are used so that plans for these mesh with each other and with the SharedPlan overall [4, 10]. TG2 will not plan to use ST_q to load MT_2 for Shipm2 until after ST_q has finished loading MT_1 for Shipm1. A second example is provided by Variation 3: MT_1 has unloaded and is ready to refuel. It can stay where it is to refuel, but knows that if it does MT_2 will not be able to load until the refueling is done. Alternatively, it may move a short distance, thereby enabling MT_2 to begin unloading immediately.

CC-IT's are a major locus of motivation for agents to assist one another. If, as in Variation 4, MT_1 realizes that MT_2 is stuck enroute from Q to B and cannot complete its delivery without getting more fuel, and MT_1 believes it can provide that fuel on its way back to Q, a CC-IT will cause it to consider

[13] Lochbaum also argues for another knowledge precondition, *has.recipe*, but we treat *has.recipe* directly in the plan definitions.

doing so.[14] Another form of collaborative action assistance is communicating to avoid conflicts in resource use (and hence conflicting intentions). For instance, if MT_1 believes MT_2 may need to refuel and there is only a limited amount of gas in the pump, then MT_1 may ask MT_2 about its needs, and, to the extent possible, adjust the amount of gas it uses to leave sufficient gas for MT_2.

Mutual belief satisfaction: MB satisfaction intentions-that (MB-IT's) serve a different type of planning function. They cause agents to communicate about their abilities to do actions and their beliefs about other agent's capabilities so that the mutual beliefs required for complete SharedPlans can be established. By requiring only an intention-that these mutual beliefs be established rather than the existence of the mutual beliefs themselves, the partial SharedPlan definition [Clauses (2a2 and 2b3)] allows for a more smooth evolution of partial SharedPlans. In particular, so long as no one in the group believes it is impossible for G_k to perform the action, and everyone believes G_k is committed to doing it, the definition classifies the action as resolved. If there ceases to be mutual belief of an individual's or subgroup's ability to perform an action, it is more effective to attempt to re-establish this mutual belief before seeking a new agent, because the agent selection process is more complex than the process of establishing mutual belief of ability—in particular in its need for negotiation and intention-conflict resolution.

The three functions of intentions-that—decision-making influence, planning instigation, and collaborative acting assistance—are central to efficient collaborative activity. The decision-making influence of intentions-that is to prevent agents from making commitments that would preclude their being able to do the group action. The other two functions of intentions-that cause agents to take action in service of their SharedPlans. These functions raise a formal question: what beliefs and intentions must an agent have when it has an intention-that so that only appropriate and useful intentions-to are generated? The next two sections address this question.

3.2 Minimal Conditions for Intending-That

Axioms in the original SharedPlan formalization [10] constrain an agent's intentions by requiring that the agent not hold conflicting intentions. The formalization imposes additional minimal constraints on the attitude of intending-to: for an agent to intend to do a complex action it must have some knowledge about determining a way to do the action, it must believe there is some recipe for the action that it can find and carry out, and it must be committed to expanding its partial plan for the action to a complete plan. These requirements follow Bratman [2] in connecting intentions to means-ends reasoning; they also ensure that intending-to is closely related to acting.

[14]Whether MT_1 will do so or not depends on its other obligations and the cost, as discussed elsewhere [10].

This section provides an analogous constraint for intending-that: it specifies the minimal conditions on an agent's beliefs and intentions needed for it to hold an intention-that. To explain these constraints, we need to distinguish among three possible belief relations between an agent and a proposition *prop*: (a) at time T_{bel} the agent believes that *prop* holds: Bel$(G, prop, T_{bel})$; (b) at time T_{bel} the agent believes *prop* does not hold: Bel$(G, \neg prop, T_{bel})$; (c) at time T_{bel} the agent believes neither that *prop* holds nor that it does not: $\neg$Bel$(G, \neg prop, T_{bel}) \wedge \neg$Bel$(G, prop, T_{bel})$.

Because propositions include time, case (b) represents the situation in which an agent holds the strong belief that it is impossible for the atemporal fact A_{prop} to hold at time T_{prop}. Cases (a) and (c) thus represent the only situations in which, according to the agent's beliefs, *prop* might hold: in case (a) the agent believes it does hold and in case (c) there are some possible futures in which it might. The distinction of importance to SharedPlans is that between case (b) on the one hand and cases (a) and (c) on the other. We will say that an agent believes *prop* is *impossible* in case (b) and *possible* otherwise.[15]

We assume that beliefs persist by default, but do not formalize this here (cf., [25]). However, we note that an agent's beliefs about *prop* are not fixed. External events, in particular actions by another agent, may cause an agent to change its beliefs. Furthermore, an agent who does not have definite beliefs about *prop* (case (c)), may seek information or take actions that will change its belief about *prop*.

We impose two constraints on an agent's holding an intention-that toward a proposition. First, the agent can only have an intention-that toward propositions it believes are possible. Our axiomatization does not allow an agent to intend-that an impossible proposition hold. Second, if the agent believes that the intended proposition *prop* is possible but does not currently believe it holds and knows some actions α_j it can take to cause proposition *prop* to hold, then the agent must consider doing at least one of these actions.

Axiom 1: Restriction to Possible Propositions

The formalization constrains an agent's intentions-to do an action by requiring that the agent must be knowledgeable enough to do means-ends reasoning about the intended action. The first constraint on intending-that is analogous: an agent must believe it is possible for the intended proposition to hold.

Axiom 1, given in Figure 6, represents this constraint; it stipulates that an agent cannot intend-that *prop* if the agent believes *prop* is impossible. There are two main reasons for this constraint. First, if an agent intends-that *prop*, it will consider this intention before adopting any additional intentions. If the agent believes that *prop* is impossible, there is no benefit to avoiding actions or commitments

[15] The incorporation of time in propositions and tying the possible/impossible classification to time is important for the formalization. For instance, what matters is that MT_1 will be able to carry rocks to the building site at a particular time. If the rocks must be transported today and MT_1 is only available tomorrow, then the action that is needed cannot be done by MT_1.

that might make *prop* impossible. Second, because having an intention-that toward *prop* may, following Axiom 2, lead an agent to take actions that can directly or indirectly help to bring about *prop*'s holding, it is detrimental to have such an intention when *prop* is impossible; there is no point in wasting energy on lost causes. AC-IT and CC-IT intentions-that provide examples of cases in which the restriction that Axiom 1 imposes functions to limit unproductive collaborative planning actions.

AC-IT's represent an agent's commitment toward the joint activity; it is detrimental to commit to impossible plans. More specifically, the commitment to the joint activity will lead an agent to undertake planning actions in service of this joint activity (from Axiom 2), including attempting to establish the minimal conditions for a SharedPlan (i.e., having collaborators and a plan for finding a recipe); there is no point in an agent's planning for an action if it believes the performance of that action is impossible (i.e., if it believes the Do proposition in the AC-IT is impossible). For example, if time constraints preclude MT_2's doing the driving for Shipm1 in the Truckworld scenario, then there is no point in MT_2 attempting to be part of a group that forms a SharedPlan to do Shipm1. Axiom 1 would prevent MT_2 from doing so because MT_2 does not believe that $\text{Do}(\{MT_2, ST_q, ST_b\}, \text{Shipm1}, ...)$ is possible.

CC-IT's represent agents' commitments to each other's actions. If an agent G_j believes of one of its collaborators, G_k, that $\neg\text{CBA}(G_k, \beta_i, T_p, T_{\beta_i}, \Theta_{\beta_i})$, then G_j believes it is impossible for G_k to perform β_i. G_j's belief may be based on G_k's general abilities to do an action of the type of β_i, or may be a result of the constraints on this performance of β_i, i.e., the Θ_{β_i}. As a result, G_j believes that the group activity based on a SharedPlan with the current recipe and agent choices cannot succeed. It only makes sense to continue to collaborate in doing α, if G_j and G_k and perhaps other collaborators revise their plan. Because a SharedPlan in which G_k performs β_i cannot succeed, there can be no such plan. In this situation, G_j should drop its intention-that G_k be able to do the action (the CC-IT). Axiom 1 stipulates that an agent in such a situation will not have an intention-that. For instance, given the time constraints above, there is no point in MT_1 forming an intention-that of the form $\text{Int.Th}(MT_1, \text{CBA}(MT_2, \textit{drive}...)...)$ if MT_1, like MT_2, believes that $\text{CBA}(MT_2, \textit{drive}...)$ is impossible. However, the intention-that the group do the higher level action of which β_i is a constituent is still valid, and in fact should lead G_j to inform its partners of the problem.

Axiom 2: Actions Prompted by Intentions-that

It is common to distinguish between desiring an action and intending to do an action [3]. Axiom 2 (Figure 6) enforces a similar distinction between desiring that a proposition hold and intending that the proposition hold. It ties an agent's intending-that *prop* to actions that contribute to *prop*'s holding in a way similar to tying an agent's means-ends reasoning and intentions to do subactions to an agent's intending to do an action. In particular, if the agent intends-that *prop*, it

Axiom 1 :
$(\forall G, T_i)$ Int.Th$(G, prop, T_i, T_{prop}, \mathrm{IC}_{prop}) \Rightarrow \neg$Bel$(G, \neg prop, T_i)$

Axiom 2 :
If an agent intends-that *prop* and does not believe *prop*
$(\forall G, T_i)$
(1) [Int.Th$(G, prop, T_i, T_{prop}, \mathrm{IC}_{prop}) \wedge \neg$Bel$(G, prop, T_i)] \Rightarrow [$
and there is some action the agent can take that would lead to *prop* holding then
(2) $[(\exists \alpha, T_\alpha)$LEAD$(G, \alpha, T_\alpha, prop, \mathrm{con}(\alpha), T_i)] \Rightarrow [$
either there is some such action that the agent (potentially) intends to do
(3*a*) $[(\exists \alpha, T_\alpha)$LEAD$(G, \alpha, T_\alpha, prop, \mathrm{con}(\alpha), T_i) \wedge$
(3*b*) [Int.To$(G, \alpha, T_\alpha, T_i, \mathrm{con}(\alpha), prop) \vee$ Pot.Int.To$(G, \alpha, T_\alpha, T_i, \mathrm{con}(\alpha), prop)]] \bigvee$
or for every such action there is some conflict with the agent's other intentions
(4) $[(\forall \alpha, T_\alpha)$LEAD$(G, \alpha, T_\alpha, prop, \mathrm{con}(\alpha), T_i) \Rightarrow$
$[[(\exists \beta, T_\beta, \mathrm{IC}_\beta)$Int.To$(G, \beta, T_i, T_\beta, \mathrm{con}(\beta), \mathrm{IC}_\beta) \wedge$
Bel$(G$, CONF$(\alpha, \beta, T_\alpha, T_\beta, \mathrm{con}(\alpha), \mathrm{con}(\beta)), T_i)] \vee$
$[(\exists prop, T_{prop}, \mathrm{IC}_{prop})$Int.Th$(G, prop, T_i, T_{prop}, \mathrm{IC}_{prop}) \wedge$
Bel$(G$, CONF$(\beta, prop, T_\beta, T_{prop}, \mathrm{con}(\beta)), T_i)]]]]]$

LEAD$(G, \alpha, T_\alpha, prop, \Theta, T_i)$
(1) Bel$(G, (\exists R_\alpha, T_\alpha)$ CBA$(G, \alpha, R_\alpha, T_\alpha, \Theta), T_i) \bigwedge$
(2) [[Bel$(G$, Do$(G, \alpha, T_\alpha, \Theta) \Rightarrow prop, T_i)] \bigvee$
(3) [Bel$(G_1$, Do$(G_1, \alpha, T_\alpha, \Theta) \Rightarrow$
$[[(\exists \beta, G_2, R_\beta, T_\beta, T_i')$
$[T_\alpha \leq T_\beta \wedge$ single.agent$(\beta) \wedge$ CBA$(G_2, \beta, R_\beta, T_\beta, \mathrm{con}(\beta)) \wedge$
Pot.Int.To$(G_2, \beta, T_\beta, T_i', \mathrm{con}(\beta), \mathrm{IC}_\beta) \wedge T_i' < T_\beta \wedge$
[Do$(G_2, \beta, T_\beta, \mathrm{con}(\beta)) \Rightarrow prop]]] \vee$
$[(\exists \beta, \mathrm{GR}_2, R_\beta, T_\beta, T_i')$
$[T_\alpha \leq T_\beta \wedge$ multi.agent$(\beta) \wedge$ CBAG$(\mathrm{GR}_2, \beta, R_\beta, T_\beta, \Theta) \wedge$
$(\forall G_j \in \mathrm{GR}_2)$Pot.Int.Th$(G_j$, Do$(\mathrm{GR}_2, \beta, T_\beta, \mathrm{con}(\beta)), T_i', T_\beta, \mathrm{IC}_\beta) \wedge T_i' < T_\beta \wedge$
[Do$(\mathrm{GR}_2, \beta, T_\beta, \mathrm{con}(\beta))) \Rightarrow prop]]], T_i)]]$

Figure 6. Axiom schema encoding minimal conditions for intending-that

must be willing to consider undertaking actions that will help bring *prop* about.

Axiom 2 applies in the situation in which an agent is uncertain about whether proposition *prop* holds, that is, case (c) described above[16],

$$\neg\mathrm{Bel}(G, prop, T_{bel}) \wedge \neg\mathrm{Bel}(G, \neg prop, T_{bel})$$

either from a lack of knowledge alone (the agent just does not know whether *prop*) or, in the more interesting case, from circumstances in which the future is indeterminate with respect to *prop*. The agent may be in this state either initially when

[16] The axiom does not include $\neg$Bel$(G, \neg prop, T_{bel})$ in the antecedent because the intention-that and Axiom 1 guarantee this.

adopting the intention-that or later because some monitoring action indicates that *prop* may no longer hold. If the future is indeterminate, then the agent believes that should certain events, $e_1, ..., e_k$, occur then *prop* will hold, but other events may lead to $\neg prop$.

Axiom 2 stipulates that if an agent is uncertain about whether proposition *prop* holds and believes there are some actions $\alpha_1, ..., \alpha_m$ any of which might lead to *prop* holding (i.e., the α's each could cause one of $e_1, ..., e_k$ to occur), then it must either [Clause 3] intend to do, or be actively considering doing, one of these α's[17] or [Clause (4)] it must have considered all of the α's and determined that it could not do any of them.

Axiom 2 is of the form Clause 1 $\rightarrow$ [Clause 2 $\rightarrow$ (Clause 3 $\vee$ Clause 4)]. It uses the meta-predicate CONF which represents situations in which (a) the performance of an action conflicts with the performance of another action, or (b) the performance of an action conflicts with a proposition continuing to hold, or (c) two propositions cannot hold simultaneously. To simplify the statement of Axiom 2 we introduce the meta-predicate LEAD. LEAD$(G, \alpha, T_\alpha, prop, \Theta, T_i)$ summarizes the conditions under which, according to the agent's beliefs, an action α directly (Clause 2) or indirectly (Clause 3) leads to *prop* holding. The parameter Θ encodes constraints under which the action α is done. For example, in Variation 4 of the Truckworld scenario, if MT_1 believes it can deliver fuel to MT_2 and that this delivery will enable MT_2 to continue driving to B, then MT_1's delivery-of-fuel action "leads to" TG2's being able to do (CBAG) Shipm2. One constraint on the delivery-of-fuel action is the time period in which it must be completed.

¿From the definition of the LEAD macro, Axiom 2 allows both for cases of α directly bringing about *prop*, and for cases in which α indirectly leads to *prop* through performance of an intermediary action β by another agent or group of agents. In the indirect case, α ensures that the other agent (or group of agents) can do (CBA[G]) β and motivates it (them) to consider doing it (by adopting potential intentions-to or potential intentions-that).

Axiom 2 uses the attitude of potentially intending to do an action, Pot.Int.To, in Clause (3). Potential intentions represent an agent's mental state when it has not yet weighed a possible intention against existing commitments and alternative courses of action [10]. The axiom uses a disjunction of Int.To and Pot.Int.To (in Clause 3b) because we require agents to reconcile a potential new intention with existing intentions before adopting the new intention.[18] When the agent initially considers α, it must weigh doing α against its existing commitments. If it reconciles doing α successfully, then it will have an intention-to do α. The LEAD meta-predicate uses potential intentions-to and potential intentions-that in Clause (3) for analogous reasons.

To accommodate the need for agents to consider the costs of actions and other

---

[17]Clause (3a) repeats the antecedent of Clause 2 because the existential does not distribute over implication.

[18]In addition, turning a potential intention into a full fledged intention requires some preliminary means-ends reasoning [10].

factors (e.g., an agent might be able to do something but only at the cost of losing its job), we use constraints to restrict the range of actions an agent must consider to ones that are "reasonable". In particular, limitations on cost or time may be encoded as constraints. For example, if MT_1 is committed to another, high-priority job immediately following delivery of the rocks to Q, then it may not be able to deliver fuel to MT_2.

Even after adopting an intention-to do α, the agent may not believe *prop* will hold, because it may be unsure of its success in doing α.[19] If the agent drops its intention-to do α or if it fails in doing α, then to keep its intention-that *prop*, it must choose another action that could help, if there are any.

The final clause (Clause 4) of Axiom 2 is needed to avoid requiring an agent to consider repeatedly the same action.[20] From the disjunction, if Clause (4) holds, Clause (3) does not need to. Clause (4) states that the agent has reconciled all possible actions it could take and determined they each conflict in some way with other intentions. Thus, an agent needs to reconcile doing α only once; if it reconciles the potential intention-to do α, discovers a conflict, and decides to keep its current obligations and not to intend-to do α, then it does not need to reconsider α. For example, if MT_1 reconciles the delivery-of-fuel action, determines it conflicts with an obligation to be back at the quarry for another job, and decides it cannot do the delivery-of-fuel, then MT_1 should not consider this action further. In such cases, the agent will have neither an intention-to do α nor a potential intention-to do α, but it may still intend-that *prop*.[21]

This last clause is needed only when the agent has reconciled all the actions it possibly might do to help achieve *prop* and has been unable to adopt an intention to do any of them. In particular, it is not needed if there is another α_i that the agent believes it can do and that "leads to" *prop*, because the existential may be satisfied with any action. For example, if MT_1 can pay someone to take fuel to MT_2 then the payment action satisfies the existential.

In situations that match Clause 4 of this axiom, the intention-that functions passively to keep the agent from doing anything that would conflict with *prop*, rather than actively leading the agent to do something. For example, MT_1 would not adopt an intention to do some action that interfered with another truck delivering fuel to MT_2.

Axiom 2 applies to planning actions as well as to domain actions; planning actions can provide an indirect way to make a proposition true. Thus, this axiom is a mechanism for realizing the planning-instigation function of intentions-that.

---

[19]To have the intention-to, the agent must believe it will be able to perform the action; however, the success of its execution is a different matter.

[20]The last part of this clause is second-order: it quantifies over *prop*, because the axiom needs to cover conflicts doing α may have with an existing intention-that as well as with an existing intention-to. From the implementation perspective, this second-order characteristic does not necessarily present a problem; the formalization is a specification, not input to a theorem prover. One possible implementation strategy is to maintain a database of intentions-that and use it to check explicitly for conflicts.

[21]More specifically, it can keep the intention-that so long as the possibility of doing α was not the sole reason it believed *prop* was possible. The agent must still believe that *prop* is possible.

3.3 Getting from Intentions-that to Intentions-to

Intentions-that play both static and active roles in constraining rational behavior. The static role, realized in intention conflict avoidance axioms [10], serves to prohibit an agent from committing to contradictory states of affairs. The active role, which we consider in this section, serves to generate actions that help bring about the intended proposition. The active role of intentions-that requires means-ends reasoning analogous to that for intentions-to. We use the term Cultivate to refer to the complex activity of doing this reasoning.

There are three different sources of an agent adopting an intention-to in service of an intention-that. First, as represented by Axiom 2, the agent may intend-to do an action that will bring about, or help bring about, the intended proposition. Second, the agent may adopt an intention to do an action that will help another agent when the intended proposition concerns an action by that second agent.[22] For example, MT_1's intention-that TG2 be able to deliver the second shipment of rocks to B may cause MT_1 to radio road-condition information to TG2 (or, if it has sufficient information to know that MT_2 is doing the driving, then just to MT_2) to make the trip shorter and the Shipm2 action overall more efficient (Variation 5). Third, even when an agent believes that *prop* holds, if the time T_{prop} has not passed yet, the intention-that *prop* may generate actions to ensure that *prop* continues to hold. Monitoring actions are instances of such behavior. Variation 6 provides an example: if ST_q has access to information sources unavailable to MT_1 once enroute, then ST_q might monitor road conditions and radio important information to MT_1.

Helpful behavior, the second way in which an agent's intention-that may lead to an intention-to, arises when an agent G_1 intends-that a proposition *prop* holds, *prop* relates in some way to the performance of an action β by another agent (G_2), and G_1 believes that it will be easier for G_2 to do β if G_1 does α. In the Truckworld example of the preceding paragraph, *prop*=CBA(MT_2, drive_to_B(MT_2)...), β = drive_to_B(MT_2), and α=radio_road_info(MT_1). In general, we consider *prop* relevant to G_2's doing β, if *prop* refers to G_2's performance of β or ability to do β. The two situations we model formally are $Do(G_2, \beta...)$ and CBA($G_2, \beta...$) (or CBAG($G_2, \beta...$)).[23]

An agent may monitor a proposition actively or passively. ST_q's monitoring of road conditions exemplifies active monitoring; the agent actively does some action to track whether the proposition holds. In some situations, an agent may be able to rely on someone else to alert it if a proposition changes. For instance, ST_b may assume that MT_2 will let it know if something goes wrong and MT_2 cannot complete the driving subaction of Shipm2. Passive monitoring is daemon-like: notification triggers an adjustment in the SharedPlan. Passive monitoring is less costly than active monitoring (because the agent does not need to do anything unless a condition changes), but cannot always be relied upon. However, a participant in a

---

[22] Axioms that directly tie helpful behavior to a SharedPlan or contracting situation appear in our previous paper [10].

[23] There are additional motivations for helpful behavior, but those are beyond the scope of this paper.

SharedPlan can assume its collaborators will let it know if a proposition relevant to the SharedPlan (e.g., a CBA or an Int.To) ceases to hold. As a result, passive monitoring can be used for these intentions-that.

4 SHAREDPLAN-BASED ARCHITECTURE FOR COLLABORATIVE-AGENT DESIGN

In this section, we present an architecture for collaborative-agent design based on the SharedPlan formalization. This architecture presumes as a base an architecture for an individual, resource-bounded rational agent that has capabilities for scheduling and executing intended actions and for monitoring their effects. A fundamental characteristic of the design of resource-bounded rational agents is that mental actions and domain actions must be treated in a coordinated manner. The agent must be able to interleave thinking (in particular, planning) and acting. At some level of abstraction a single scheduler handles an agenda including both mental and domain actions. Furthermore, the agent architecture must include some basic mechanisms for managing pending and adopted intentions. A variety of designs for such agent architectures have been proposed in current planning architectures and formalizations [23, 5, 25, inter alia].

In this section we address the problem of how to go from the SharedPlan formalization to an implementation that embodies it in an automated rational agent. The specification of collaboration provided by SharedPlans could be used in one of three ways to produce a system: (a) as input for a theorem-proving system; (b) for developing a system tailored to a specific domain in which the system designer decides which components to reason explicitly about and which to hardwire in; or, (c) to develop general algorithms that realize the specifications. We have taken the last approach which we believe to be the most promising. In this section, we describe the most distinctive features of the architecture. We have used this architecture in the implementation of a simple multi-agent system [20] in the Truckworld environment [19] and are currently exploring its use in the implementation of a commercial-exchange system and a collaborative human-computer interface system.

Two processes naturally emerge from the SharedPlan formalization as central to controlling an agent's behavior in accordance with its intentions. The first, ELABORATE_INDIVIDUAL, is responsible for the means-ends reasoning that derives from an intention-to. This process and its relationships to intention-to have been described previously [10]. The second, CULTIVATE, plays an analogous role for intention-that. It both monitors for intention conflicts and initiates an agent's I_ELABORATE_GROUP process. Because SharedPlans are driven off of intentions-that, CULTIVATE plays a central role in the architecture and we describe it before looking at the SharedPlan subsystem and the I_ELABORATE_GROUP processes that CULTIVATE spawns for some intentions-that within the SharedPlan context.

CULTIVATE($Int.Th(G, prop, T_i, T_{prop}, \mathrm{IC}_{prop})$), where $T_i \leq T_{prop}$

```
RECONCILE(prop,intentions-to ∧ intentions-that)
Loop:
   if Bel(G, ¬prop, T_i) then fail
   else if ¬Bel(G, prop, T_i) then
      if know action that may lead to Bel(G, prop, T_{i+1}), then act (Cultivate-Act)
      else set up monitoring triggers (Cultivate-Monitor).
   else if [Bel(G, prop, T_i)]
      set up monitoring triggers (Cultivate-Monitor);
     if prop relevant to action by another agent, then consider helping (Cultivate-Help).
  sleep;
end loop;
```

Figure 7. Pseudocode for CULTIVATE process.

4.1 The cultivate process

CULTIVATE determines the actions that an agent might take as a result of having an intention-that and launches the processes needed for the agent to take (consider taking) them (i.e., creates potential-intentions to do them). It embodies the axioms for intending-that given in our earlier paper. The main job of CULTIVATE for Int.Th($G, prop, ...$) is to identify actions an agent can take to help establish or maintain the proposition *prop*.

Figure 7 shows the main constituent steps of the CULTIVATE process. The *Cultivate-Act* branch applies when the agent [currently] believes that the intended proposition, *prop*, does not hold; the actions an agent takes depend on the particular content of *prop*, and possibly on the domain. The *Cultivate-Help* branch applies for propositions that contain some element referring to an action α by another agent or subgroup, e.g., Do($G_2, \alpha...$); in these cases, even if G_1 believes *prop* holds, it may be able to assist G_2 and in some way make it easier or less costly to do α. Again, the actions depend on *prop* and α. The *Cultivate-Monitor* branches are motivated by situations like the road-condition monitoring examples described earlier.

In the remainder of this section we discuss the Cultivate-Act branch of the algorithm as it applies for the cultivate processes associated with the intentions-that in the SharedPlans definitions. We leave the discussion of the Cultivate-Help and Cultivate-Monitor branches to future work.[24]

[24] As the axioms for helpful behavior in an earlier paper illustrate [10], an agent's decisions about what actions to take to help its collaborators not only refer to domain specific knowledge of action relationships, but also must take into account the cost of actions. The development of cost functions of actions and hence Cultivate-Help are beyond the scope of this paper, as is a discussion of monitoring issues and processes.

The five settings for intentions-that in SharedPlans listed in Figure 5 engender different kinds of CULTIVATE behavior. We discuss each type in turn.

Group activity commitment: If the intention-that is of the AC-IT type, Int.Th(G, Do(GR, $\alpha, T_\alpha, \Theta_\alpha), T_p, T_\alpha$, IC$_\alpha$), then the main action an agent can take is to recruit the other members of the group to work on doing α and, in particular, to begin the formation of a SharedPlan. Hence, the main subprocesses the CULTIVATE process spawns are ones to establish that all other members, G_j, of GR also hold this intention [i.e., ($\forall G_j \in$ GR)Int.Th(G, Do(GR, $\alpha, T_\alpha, \Theta_\alpha), T_p, T_\alpha$, IC$_\alpha$)] and to establish mutual belief among members of GR of everyone's doing so. Because to form a SharedPlan (for doing α) group members must form a full SharedPlan to do the Elaborate_Group action, the CULTIVATE process must also initiate the agent's adoption of an intention-that the group will undertake this planning activity. Thus, the CULTIVATE process for an intention-that related to a domain action spawns an intention-that (and hence a CULTIVATE process) for a mental (planning) action. This second CULTIVATE process spawns an I_ELABORATE_GROUP process. A collaborating group's Elaborate_Group is realized by the sum total of the participating agents' I_ELABORATE_GROUP's spawned in this way.

In our architecture, we treat three instances of AC-IT specially, those associated with the full SharedPlan for selecting a recipe, selecting an agent(s) and elaborating a SharedPlan. These cases correspond to the processes

I_ELABORATE_GROUP	I_SELECT_REC_GR
I_SELECT_AGENT	I_SELECT_SUBGROUP

described in Section 4.2. For each of these cases, CULTIVATE directly initiates the appropriate special purpose process. Treating these AC-IT cases specially enables automated collaborative agents to save on communication and limits the level of recursion in SharedPlans. The advantage of otherwise treating them uniformly with domain actions is that doing so enables a single locus for scheduling and dealing with conflict avoidance.

Core Case: In the Core Case intentions-that (CC-IT), the proposition concerns the ability of another agent(s) to do an action; i.e., it is of the form

$$\text{Int.Th}(G, \text{CBA}(G_k, \beta_k, ...), ...)$$

or

$$\text{Int.Th}(G, \text{CBAG}(\text{GR}_k, \beta_k, ...), ...).$$

In these cases, CULTIVATE first determines whether there is some γ that can lead (directly or indirectly) to CBA (CBAG). If it finds such a γ, it then

1. The first agent that needs a parameter chooses an appropriate entity for the parameter and broadcasts this choice to other agents; this strategy requires some conflict resolution mechanism for the situation in which multiple agents choose simultaneously.
2. Agents decide at the time of forming the initial SharedPlan who will identify each parameter and how others will find out its identity.
3. Agents decide at the time of choosing the recipe who will identify each parameter and how others will find out its identity.

Figure 8. Parameter identification strategies

begins the process for establishing an intention-to do γ.[25] The identification of a γ depends on the domain and the agent's reasons for currently believing that G_k cannot bring about β_k. The example of MT_1 radioing back to TG2 information about road conditions is an instance of the result of this kind of deliberation by a CULTIVATE process.

Parameter identification A variety of strategies can be adopted for parameter identification in SharedPlans,[26] depending on the relationships among group members. Thus we do not give a full specification here, but some possibilities are listed in Figure 8. Two general constraints apply to any strategy. First, the choice of the entity that fills a particular role in an action may depend on the subactions in which the parameter also participates. For example, the time of Shipm1 cannot be finally set until the agents know when ST_q and ST_b can do their parts of the loading and unloading actions. As a result, parameter identification strategies benefit from delaying commitment until as much recipe and agent selection as possible has been completed. This benefit must be balanced against the need to act before the planning process is completed. Second, the agents must have a mechanism for coordinating the choice of parameters for an action with the choice for any subsidiary actions in the recipe (whether single- or multi-agent) that use that parameter. Because handling the general problem of distributed parameter binding is an open issue in Computer Science, our implementations impose a strict ordering and communication regime as a short-term solution.

Constraint satisfaction: The CULTIVATE process associated with an intention-that constraints will be satisfied, (Int.Th$(G_j, \Theta_\alpha, T_p, T_{\Theta_\alpha}, P_\alpha)$), needs to

[25]This process begins with a reconciliation to determine whether the intention is compatible with other commitments.

[26]Individual plans also require parameter identification which is realized in our formalization through an intention-that. However, in this case, the agent acts independently and the CULTIVATE process need only spawn subsidiary elaboration processes for intentions-to do actions that will identify each of the parameters.

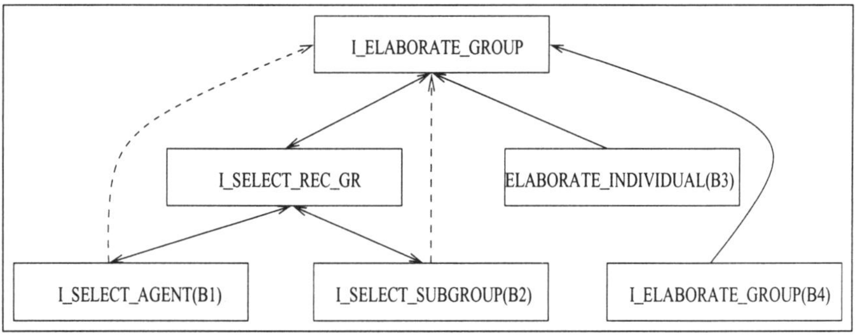

Figure 9. Example of the basic architecture of the SP subsystem in action. In this example, β_1 and β_2 are unresolved (β_k) and β_3 and β_4 are resolved subactions (β_r).

handle the three types of constraints described earlier: execution preconditions, recipe-constraints, and constraints imposed by intention-conflict resolution. For precondition constraints CULTIVATE needs to start processes that coordinate with other agents; if the agent commits to satisfying a precondition, CULTIVATE spawns the corresponding intentions-to. For recipe constraints, CULTIVATE monitors the satisfaction of the constraints and is responsible for coordination with other agents. Handling the general problem of distributed constraint satisfaction is an open issue in Computer Science and beyond the scope of this paper.

Mutual belief satisfaction: The CULTIVATE process for these intentions-that oversees the agent's adoption of the belief and causes communication sufficient for the establishment of mutual belief. If an agent decides not to adopt the belief, then the CULTIVATE process must communicate this to the other agents.

4.2 SP Subsystem

Figure 9 shows the basic architecture of the SP subsystem (adapted from [20, Figure 8]). Each SharedPlan that an agent participates in will yield an instantiation of the processes portrayed in this figure. The major constituent processes in the subsystem are:[27]

- Plan-completion process: I_ELABORATE_GROUP.
- Selection processes: I_SELECT_REC_GR, I_SELECT_AGENT and I_SELECT_SUBGROUP.

[27] We have omitted the CULTIVATE and RECONCILE processes in this figure to avoid cluttering the diagram.

- Intention-management processes: CULTIVATE, ELABORATE_INDIVIDUAL, RECONCILE.

I_ELABORATE_GROUP is responsible for a participating agent's work in helping to create and carry out a group activity. Two major constituents of any Elaborate_Group action are those for selection of a recipe for doing the group action, realized by the I_SELECT_REC_GR processes of the collaborating agents, and for determining which agent(s) will do the constituent actions of the chosen recipe, which is realized by the I_SELECT_AGENT and I_SELECT_SUBGROUP processes of the collaborating agents.

The I_ELABORATE_GROUP, I_SELECT_AGENT, and I_SELECT_SUBGROUP processes each correspond to a requirement in the formalization that the group have a full (i.e., complete) SharedPlan for some complex planning action:

- I_ELABORATE_GROUP corresponds to the requirement in the SP definition (Figure 2) that the group have a full SharedPlan to do an Elaborate_Group; this process is the way in which the agent works toward helping to complete the partial SharedPlan;

- I_SELECT_REC_GR corresponds to the requirement in the PSP definition (Figure 3) that the group have a full SharedPlan to do a Select_Rec_GR; and

- I_SELECT_AGENT and I_SELECT_SUBGROUP correspond to the requirement in the PSP definition that the group have a full SharedPlan to do a Select_Agent or Select_Subgroup as appropriate.

For each of these full SharedPlans the agent must have adopted an AC-IT intention-that which has an associated CULTIVATE process. As described in Section 4.1, the CULTIVATE spawns the corresponding I_ELABORATE_GROUP, I_SELECT_AGENT, and I_SELECT_SUBGROUP. We discuss some details of the other processes in the next section.

4.3 Plan Completion Processes

I_ELABORATE_GROUP is active throughout the time an agent participates in the performance of a group action. If the group succeeds then the I_ELABORATE_GROUP will have taken the agent from its initial intention-that to having the full set of beliefs and intentions needed for it to carry out its part in the group activity. Furthermore, because plans may become more partial (e.g., if some agent is unable to do an action it committed to or a recipe does not work out), I_ELABORATE_GROUP remains active until the group successfully completes the action. In our implementations, the associated CULTIVATE process also tracks success or failure. The I_ELABORATE_GROUP process can be terminated prematurely either if the agent decides to drop out of doing the multi-agent action α (i.e., drops its AC-IT intention-that) or the group elaboration fails for some reason.

I_ELABORATE_GROUP initiates the following processes or monitors them for success or failure:

- Working with the group to determine a way to do I_SELECT_REC_GR, either by choosing a known recipe or by constructing a recipe.
- For each subaction in the recipe working with the group to identify agent(s) to do that action (I_SELECT_AGENT or I_SELECT_SUBGROUP)
- Doing its part to identify parameters of the action and meet constraints on doing the action (e.g., satisfying preconditions)

In addition, I_ELABORATE_GROUP is responsible for establishing and maintaining the mutual beliefs required by the FSP definition. In most cases I_ELABORATE_GROUP will cause some kind of communication among agents. However, in those cases in which the definition of PSP requires only an Int.Th(G, MB...), the process I_ELABORATE_GROUP will spawn cultivate processes for these intentions.

4.4 Selection processes

The recipe- and agent-selection processes I_SELECT_REC_GR, I_SELECT_AGENT, and I_SELECT_SUBGROUP are more complicated than their individual-agent plan counterparts because of the need to coordinate with other agents and reach agreement. The flexibility that characterizes the SharedPlan formalization adds complexity both by not requiring a central manager and by allowing for plans to be partial, including allowing for plans to become more partial because of execution failures or resource limitations that lead agents to change their intentions.

Selecting Recipes

When an agent acts alone, it figures out how to do an action (i.e., finds a recipe) and then carries out the necessary subacts. When agents work together, both the formation or selection of the recipe and the carrying out of the subactions is distributed. The group must reach consensus on how they are going to perform the action (i.e., on the recipe they will use) and on who is going to do the subactions. A major advantage of group action—that an agent does not have to do everything itself—is counterbalanced by the necessity of agents agreeing on their plan and coordinating their activities. Thus, recipe and agent selection in the collaborative setting require interagent negotiation and communication as well as individual agent reconciliation.

The input to an I_SELECT_REC_GR process is the multi-agent action to be done by the group. The I_SELECT_REC_GR process terminates successfully when the group has agreed on a recipe for this action. An agent's basic planning and reasoning processes are important constituents that are used to figure out what the agent knows of how to do the multi-agent action α. In addition, the I_SELECT_REC_GR process requires capabilities for participating in multi-agent decision making.

In the simplest cases, each agent might put forward its favorite recipe; then the agents might vote to choose one of them. (Our implementations so far have taken this approach.) A more complex decision-making scenario is one in which no single agent has a complete recipe. The agents work together to combine pieces of recipes that they individually discover.

Because agents only know that a recipe has worked once they complete performing an action, the I_SELECT_REC_GR process must save enough of its decision making information to allow a new recipe to be selected, should the one selected by it fail. When the I_SELECT_REC_GR terminates it passes back to the spawning I_ELABORATE_GROUP either success and this decision-making record, or a failure and perhaps the reason for failure.

Selecting Agents

The I_SELECT_AGENT and I_SELECT_SUBGROUP processes incorporate decision-making and negotiation strategies that enable an agent to form a consensus with others in the group, and are thus similar to I_SELECT_REC_GR. They are more complex because they also include processes for considering whether the agent itself can do (or participate in doing) an action, forming commitments to other agents' actions, and forming beliefs about other agents' abilities to do actions.

If the action about which the group is deliberating is a single-agent action, then each participating agent's I_SELECT_AGENT must first consider whether the agent can do the action by determining whether it has the capabilities to do it and by reconciling a potential intention to do it. If the agent is selected, its I_SELECT_AGENT will have to spawn the processes needed to form an intention to do the action. These subproceses include those for adopting a commitment to doing the action and for forming a partial individual plan for the action. To do the latter, the agent must have sufficient knowledge to determine a way to do the action. If the agent is not selected, then its I_SELECT_AGENT must spawn processes for adopting intentions that the selected agent can do the action. In both cases it must spawn processes for establishing that the group mutually believe the agent is capable of doing it.

The case is similar if the action about which the group is deliberating is a multi-agent action. The main difference is that the I_SELECT_SUBGROUP must handle the problems of subgroup formation. In particular, it must incorporate processes for deciding whether to participate in a group and processes for initiating a subsidiary SharedPlan.

A further complication arises for I_SELECT_AGENT and I_SELECT_SUBGROUP because rational-agent architectures require that planning and acting be interleaved. As a result, the formalization provides for groups to act on partial recipes. In particular, they may choose agents for a particular action who then might start performing the action before the construction of the full recipe has been completed. An action may regress to the unresolved case if the selected agent or subgroup is unable to perform it successfully. Thus, both I_SELECT_AGENT and I_SELECT_SUBGROUP must be able to restart with information about past attempts. To provide for this possibility in the architecture, we make the I_SELECT_AGENT and I_SELECT_SUBGROUP processes children of both the I_SELECT_REC_GR and I_ELABORATE_GROUP processes. They are spawned by I_SELECT_REC_GR and report success or failure to the I_ELABORATE_GROUP process. Furthermore, if no other agent or group can take over, the agents may need to find another recipe; thus I_SELECT_REC_GR must also be reentrant.

5 CONCLUSIONS

This paper expands the SharedPlan formalization presented earlier [10] by providing a specification of the minimal conditions for an agent to have an intention-that, describing the ways in which having an intention-that can lead to taking action, and describing the basic processes in the architecture of a rational, resource-bounded agent required for that agent to participate in group planning. Two areas in which further research is required are the specification of group decision making processes (e.g., for agent and recipe selection) and the development of intention reconciliation processes that can treat a combination of group-directed and individual-directed intentions.

6 ACKNOWLEDGEMENTS

Support for the first author was provided by Grant No. IRI-9525915 and Grant No. CDA 94-01024 from the National Science Foundation. Support for the second author was provided by NSF Grant No. IRI-9423967. Hanna Pasula did the initial implementation of SharedPlans in the Truckworld domain. We acknowledge also her contribution to the research: many of her questions about the implementation led us to revise the formalization and modify the architecture. We also thank Luke Hunsberger and Charles Ortiz for comments on earlier versions of the paper.

Sarit Kraus (Bar-Ilan University, Israel)
Barbara Grosz (Harvard University, USA)

Type	Notation	Meaning	Section
Modal Operators	Int.To	intend-to	
	Int.Th	intend-that	
	Pot.Int.To	potential intention-to	
	Pot.Int.Th	potential intention-that	
	Bel	belief	
	MB	Mutual belief	
	Do	performance of action	
Meta-Predicates (Plans)	FIP	full individual plan	
	PIP	partial individual plan	
	SP	SharedPlan	
	FSP	full SharedPlan	
	PSP	partial SharedPlan	
Meta-Predicates (Ability)	CBA	can bring about	
	CBAG	can bring about group	
Meta-Predicates (subsidiary)	CONF	actions/propositions conflict	
	PSPC	contracting in PSP	
	CCG	group of agents can contract	

Table 3. Summary of notations (1). If there is no entry in the last column, the concept was described in an earlier paper [10]; a * is used for certain individual planning actions that differ slightly from that paper but are not discussed in this paper.

7 APPENDIX A: FORMAL DEFINITION OF PSP

Tables 3 and 4 together with Figures 10 and 11 comprise a formal definition of PSP.

REFERENCES

[1] Cecile T. Balkanski. Modelling act-type relations in collaborative activity. Technical Report 23-90, Harvard University, 1990.

[2] Michael E. Bratman. *Intention, Plans, and Practical Reason*. Harvard University Press, Cambridge, MA, 1987.

[3] Michael E. Bratman. What is intention? In P.R. Cohen, J. Morgan, and M.E. Pollack, editors, *Intentions in Communication*, chapter 2, pages 15–31. MIT Press, Cambridge, MA, 1990.

[4] Michael E. Bratman. Shared cooperative activity. *The Philosophical Review*, 101(2):327–341, 1992.

[5] Michael E. Bratman, David J. Israel, and Martha E. Pollack. Plans and resource-bounded practical reasoning. *Computational Intelligence*, 4(4):349–355, 1988.

[6] T.L. Dean and M. P. Wellman. *Planning and Control*. Morgan Kaufman, Publishers, California, 1991.

[7] Barbara Grosz and Sarit Kraus. Collaborative plans for group activities. In Ruzena Bajcsy, editor, *Proceedings of the 1993 International Joint Conference on Artificial Intelligence (IJCAI-93)*, pages 367–373, San Mateo, CA, 1993. Morgan Kaufmann Publishers, Inc.

Act-types for Planning Actions	Select_Rec	agent selects (extends) recipe	
	Select_Rec_GR	group of agents selects (extends) a recipe	
	Select_Agent	group of agents selects a member for a subaction	2.3
	Select_Subgroup	group of agents selects a subgroup for a subaction	2.3
	Elaborate_Individual	agent extends partial plan	
	Cultivate	cultivate an intention-that	3.3
	Elaborate_Group	group of agents extends a partial SharedPlan	
	I_Select_Rec_GR	individual planning action constituting Select_Rec_GR	*
	I_Select_Agent	individual planning action constituting Select_Agent	*
	I_Select_Subgroup	individual planning action constituting Select_Subgroup	*
	I_Elaborate_Group	individual planning action constituting Elaborate_Group	2.1
	Reconcile	reconcile new intention with old ones	
Predicates (subsidiary)	single.agent	single-agent action	
	multi.agent	group action	
	id.params	parameters had been identified	3.1
	selected.rec	selected recipe	*
Functions	cost	cost of action	
	econ	relativize cost (for benefit comparison)	
	inherit	inherit constraints	2.3
	prec	preconditions	2.3
	con	constraints of action	2.3
Processes	CULTIVATE	process which realized Cultivate	4.1
	I_ELABORATE_GROUP	process which realized I_Elaborate_Group	4.2
	I_SELECT_AGENT	process which realized I_Select_Agent	4.2
	I_SELECT_SUBGROUP	process which realized I_Select_Subgroup	4.2
	I_SELECT_REC_GR	process which realized I_Select_Rec_GR	4.2
	RECONCILE	process which realized Reconcile	4.2

Table 4. Summary of notations (2). If there is no entry in the last column, the concept was described in an earlier paper [10]; a * is used for certain individual planning actions that differ slightly from that paper but are not discussed in this paper.

[8] Barbara Grosz and Candace Sidner. Plans for discourse. In P. Cohen, J. Morgan, and M. Pollack, editors, *Intentions in Communication*, pages 417–444. Bradford Books/MIT Press, Cambridge, MA, 1990.

[9] Barbara J. Grosz. Collaborative systems: 1994 AAAI presidential address. *AI Magazine*, 2(17), Summer 1996.

[10] Barbara J. Grosz and Sarit Kraus. Collaborative plans for complex group action. *Artificial Intelligence*, 86(2):269–357, 1996.

[11] P. Haddawy. *Representing Plans under Uncertainty: A Logic of Time, Chance and Action*. PhD thesis, University of Illinois, 1991. University of Illinois Tech. Report UIUCDCS-R-91-1719.

[12] Steve Hanks, Dat Nguyen, and Chris Thomas. A beginner's guide to the truckworld simulator. Technical Report 93-06-09, Department of Computer Science and Engineering, University of Washington, 1993.

$\text{PSP}(P_\alpha, \text{GR}, \alpha, T_p, T_\alpha, \Theta_\alpha, IC_\alpha)$
(0) $\text{MB}(\text{GR}, (\forall G_j \in \text{GR})\ \text{Int.Th}(G_j, \text{Do}(\text{GR}, \alpha, T_\alpha, \Theta_\alpha), T_p, T_\alpha, \text{IC}_\alpha), T_p)\ \wedge$
$(0a)$ $\text{MB}(\text{GR}(\forall G_j \in \text{GR})\text{Int.Th}(G_j, \text{id.params}(\text{GR}, \alpha, T_\alpha), T_p, T_\alpha, P_\alpha), T_p)\ \wedge$
$(0b)$ $\text{MB}(\text{GR}(\forall G_j \in \text{GR})\text{Int.Th}(G_j, \Theta_\alpha, T_p, T_{\Theta_\alpha}, P_\alpha), T_p)$
$(\exists R_\alpha, \text{B}, \text{B}_{res}, \text{B}_{unres}, \text{C})[$
$(1a)$ $R_\alpha = (\text{B}, \text{C})\ \wedge\ \text{B} = \text{B}_{res} \cup \text{B}_{unres}\ \wedge$
$(1b)$ $[\text{MB}(\text{GR}, R_\alpha \in Recipes(\alpha) \wedge \text{selected.rec}(\text{GR}, R_\alpha, P_\alpha), T_p) \otimes$
$(1c)$ $[\text{MB}(\text{GR}, (\exists R^f_\alpha, \text{B}^f, \text{C}^f)R^f_\alpha = (\text{B}^f, \text{C}^f) \wedge R^f_\alpha \in Recipes(\alpha) \wedge \text{B} \subset \text{B}^f \wedge \text{C} \subset \text{C}^f, T_p) \wedge$
$(1d)$ $(\exists T_{select.rec.g}, P_{select.rec.g}, R_{select.rec.g})$
$(1d1)$ $\text{FSP}(P_{select.rec.g}, \text{GR}, \text{Select\_Rec\_GR}(\text{GR}, \alpha, R_\alpha, T_{select.rec.g}), T_p, T_{select.rec.g},$
$R_{select.rec.g}, \Theta_{select.rec.g}, P_\alpha)]]\ \bigwedge$
$(\forall \beta_r \in \text{B}_{res})[$
$(2a)$ $[[\text{single.agent}(\beta_r) \wedge (\exists G_k \in \text{GR})$
$(2a1a)$ $[\text{Int.To}(G_k, \beta_r, T_p, T_{\beta_r}, \text{inherit}(\Theta_\alpha, \beta_r) \wedge \bigwedge_{\rho_j \in \text{C}} \rho_j \wedge \text{prec}(\beta_r), P_\alpha) \wedge$
$(2a1b)$ $\text{MB}(\text{GR}, \text{Int.To}(G_k, \beta_r, T_p, T_{\beta_r}, \text{inherit}(\Theta_\alpha, \beta_r) \wedge \bigwedge_{\rho_j \in \text{C}} \rho_j \wedge \text{prec}(\beta_r), P_\alpha), T_p) \wedge$
$(2a2)$ $(\forall G_i \in \text{GR})\ \text{Int.Th}(G_i, \text{MB}(\text{GR}, (\exists R_{\beta_r})\text{CBA}(G_k, \beta_r, R_{\beta_r}, T_{\beta_r},$
$(2a2a)$ $\text{inherit}(\Theta_\alpha, \beta_r) \wedge \bigwedge_{\rho_j \in \text{C}} \rho_j \wedge \text{prec}(\beta_r)), T_{\beta_r}), T_p, T_{\beta_r}, P_\alpha) \wedge$
$(2a3)$ $\text{MB}(\text{GR}, (\forall G_j \in \text{GR}, G_j \neq G_k)\text{Int.Th}(G_j, (\exists R_{\beta_r})\text{CBA}(G_k, \beta_r, R_{\beta_r}, T_{\beta_r},$
$(2a3a)$ $\text{inherit}(\Theta_\alpha, \beta_r) \wedge \bigwedge_{\rho_j \in \text{C}} \rho_j \wedge \text{prec}(\beta_r)), T_p, T_{\beta_r}, P_\alpha), T_p)]] \bigotimes$
$(2b)$ $[\text{multi.agent}(\beta_r) \wedge (\exists \text{GR}_k \subset GR)$
$(2b1)$ $[(\exists P_{\beta_r})\text{SP}(P_{\beta_r}, \text{GR}_k, \beta_r, T_p, T_{\beta_r}, \text{inherit}(\Theta_\alpha, \beta_r) \wedge \bigwedge_{\rho_j \in \text{C}} \rho_j \wedge \text{prec}(\beta_r), P_\alpha) \wedge$
$(2b2)$ $\text{MB}(\text{GR}, (\exists P_{\beta_r})\text{SP}(P_{\beta_r}, \text{GR}_k, \beta_r, T_p, T_{\beta_r},$
$(2b2a)$ $\text{inherit}(\Theta_\alpha, \beta_r) \wedge \bigwedge_{\rho_j \in \text{C}} \rho_j \wedge \text{prec}(\beta_r), P_\alpha), T_p) \wedge$
$(2b3)$ $(\forall G_i \in \text{GR})\text{Int.Th}(G_i, \text{MB}(\text{GR}, (\exists R_{\beta_r})\text{CBAG}(\text{GR}_k, \beta_r, R_{\beta_r}, T_{\beta_r},$
$(2b3a)$ $\text{inherit}(\Theta_\alpha, \beta_r) \wedge \bigwedge_{\rho_j \in \text{C}} \rho_j \wedge \text{prec}(\beta_r)), T_{\beta_r}), T_p, T_{\beta_r}, P_\alpha) \wedge$
$(2b4)$ $\text{MB}(\text{GR}, (\forall G_j \in \text{GR} \setminus \text{GR}_k)\text{Int.Th}(G_j, (\exists R_{\beta_r})\text{CBAG}(\text{GR}_k, \beta_r, R_{\beta_r}, T_{\beta_r},$
$(2b4a)$ $\text{inherit}(\Theta_\alpha, \beta_r) \wedge \bigwedge_{\rho_j \in \text{C}} \rho_j \wedge \text{prec}(\beta_r)), T_p, T_{\beta_r}, P_\alpha), T_p)]]] \bigotimes$
(3) $[\text{PSPC}(\text{GR}, \beta_r, T_{\beta_r}, \text{inherit}(\Theta_\alpha, \beta_r) \wedge \bigwedge_{\rho_j \in \text{C}} \rho_j \wedge \text{prec}(\beta_r))]]\ \bigwedge$

(4) $(\forall \beta_k \in \text{B}_{unres})[$
$(4a1)$ $[\text{single.agent}(\beta_k)\ \wedge$
$(4a1a)$ $\text{MB}(\text{GR}, (\exists G_k \in \text{GR}, R_{\beta_k}, T_{\beta_k})\text{CBA}(G_k, \beta_k, R_{\beta_k}, T_{\beta_k},$
$(4a1a1)$ $\text{inherit}(\Theta_\alpha, \beta_k) \wedge \bigwedge_{\rho_j \in \text{C}} \rho_j \wedge \text{prec}(\beta_k)) \vee$
$(4a1a2)$ $(\exists T_{\beta_k})\text{CCG}(\text{GR}, \beta_k, T_{\beta_k}, \text{inherit}(\Theta_\alpha, \beta_k) \wedge \bigwedge_{\rho_j \in \text{C}} \rho_j \wedge \text{prec}(\beta_k)), T_p) \wedge$
$(4a1b)$ $(\exists T_{select.agent}, P_{select.agent}, R_{select.agent})\ \text{FSP}(P_{select.agent}, \text{GR}$
$(4a1b1)$ $\text{Select\_Agent}(\text{GR}, \beta_k, T_{select.agent}), T_p, T_{select.agent},$
$(4a1b2)$ $\text{inherit}(\Theta_\alpha, \beta_k) \wedge \bigwedge_{\rho_j \in \text{C}} \rho_j \wedge \text{prec}(\beta_k), P_\alpha)] \otimes$
$(4a2)$ $[\text{multi.agent}(\beta_k) \wedge$
$(4a2a)$ $\text{MB}(\text{GR}, (\exists \text{GR}_k \subseteq \text{GR}, R_{\beta_k}, T_{\beta_k})CBAG(\text{GR}_k, \beta_k, R_{\beta_k}, T_{\beta_k},$
$(4a2a1)$ $\text{inherit}(\Theta_\alpha, \beta_k) \wedge \bigwedge_{\rho_j \in \text{C}} \rho_j \wedge \text{prec}(\beta_k)) \vee$
$(4a2a1)$ $(\exists T_{\beta_k})\text{CCG}(\text{GR}, \beta_k, T_{\beta_k}, \text{inherit}(\Theta_\alpha, \beta_k) \wedge \bigwedge_{\rho_j \in \text{C}} \rho_j \wedge \text{prec}(\beta_k)), T_p) \wedge$
$(4a2b)$ $(\exists T_{select.subgroup}, P_{select.subgroup}, R_{select.subgroup})\ \text{FSP}(P_{select.subgroup}, \text{GR},$
$(4a2b1)$ $\text{Select\_Subgroup}(\text{GR}, \beta_k, T_{select.subgroup}), T_p, T_{select.subgroup}, \Theta_{ssg}, P_\alpha)]]]$

Figure 10. Formal Definition of PSP

$$
\begin{aligned}
&MENT: \\
&t.rec.g \Leftrightarrow \\
&{}_{\alpha}^{f}, \mathrm{B}^f, \mathrm{C}^f)[\\
&\quad R^f = (\mathrm{B}^f, \mathrm{C}^f) \wedge \mathrm{MB}(\mathrm{GR}, R_\alpha^f \in \mathit{Recipes}(\alpha) \wedge \mathrm{B} \subseteq \mathrm{B}^f \wedge \mathrm{C} \subseteq C^f, T_{select.rec.g}) \wedge \\
&\quad (\forall \delta_v \in \mathrm{B}^f)\mathrm{MB}(\mathrm{GR}, (\exists T_{\delta_v}) \\
&\qquad [[\mathrm{single.agent}(\delta_v) \wedge \\
&\qquad\quad (\exists G_{\delta_v} \in \mathrm{GR}, R_{\delta_v})\mathrm{CBA}(G_{\delta_v}, \delta_v, T_{\delta_v}, R_{\delta_v}, \\
&\qquad\quad \mathrm{inherit}(\Theta_\alpha, \delta_v) \wedge \textstyle\bigwedge_{\kappa_e \in \mathrm{C}} \kappa_e \wedge \mathrm{prec}(\delta_v))] \otimes \\
&\qquad [\mathrm{multi.agent}(\delta_v) \wedge \\
&\qquad\quad (\exists \mathrm{GR}_{\delta_v} \subseteq \mathrm{GR}, R_{\delta_v})\mathrm{CBAG}(\mathrm{GR}_{\delta_v}, \delta_v, T_{\delta_v}, R_{\delta_v}, \\
&\qquad \mathrm{inherit}(\Theta_\alpha, \delta_v) \wedge \textstyle\bigwedge_{\kappa_e \in \mathrm{C}} \kappa_e \wedge \mathrm{prec}(\delta_v)) \vee \\
&\qquad\qquad\quad \mathrm{CCG}(\mathrm{GR}, \delta_v, T_{\delta_v}, \\
&\mathrm{nherit}(\Theta_\alpha, \delta_v) \wedge \textstyle\bigwedge_{\kappa_e \in \mathrm{C}} \kappa_e \wedge \mathrm{prec}(\delta_v))]], T_{select.rec.g})] \textstyle\bigwedge \Phi_{select.rec.g}
\end{aligned}
$$

Figure 11. Formal Definition of PSP (cont)

[13] Nick R. Jennings. Controlling cooperative problem solving in industrial multi-agent systems using joint intentions. *Artificial Intelligence Journal*, 75(2):1–46, 1995.

[14] L. Kaelbling. An architecture for intelligence reactive systems. In M. Georgeff and A. Lansky, editors, *Reasoning about Actions and Plans*. Morgan-Kaufmann, Los Altos, California, 1987.

[15] H. Levesque, P. Cohen, and J. Nunes. On acting together. In *Proceedings of the Annual Conference of the American Association for Artificial Intelligence (AAAI-90)*, pages 94–99, 1990.

[16] Karen Lochbaum. *Using Collaborative Plans to Model the Intentional Structure of Discourse*. PhD thesis, Harvard University, 1994. Available as Tech Report TR-25-94.

[17] Karen Lochbaum, Barbara Grosz, and Candace Sidner. Models of plans to support communication: An initial report. In *Proceedings of AAAI-90*, pages 485–490, Boston, MA, 1990.

[18] Karen E. Lochbaum. The use of knowledge preconditions in language processing. In Chris S. Mellish, editor, *Proceedings of the International Joint Conference on Artificial Intelligence (IJCAI-95)*, volume 2, pages 1260–1266, San Mateo, CA, 1995. Morgan Kaufmann Publishers, Inc.

[19] D. Nguyen, S. Hanks, and C. Thomas. The TRUCKWORLD manual. Technical Report TR 93-09-08, Dept. of Computer Science and Engineering, Univ. of Washington, 1993.

[20] Hanna Pasula. Design of a collaborative planning system. Harvard University, Senior Honors Thesis, 1996.

[21] Martha E. Pollack. Plans as complex mental attitudes. In P.N. Cohen, J.L. Morgan, and M.E. Pollack, editors, *Intentions in Communication*. Bradford Books, MIT Press, 1990.

[22] C. Rich and C. L. Sidner. Adding a collaborative agent to direct-manipulation interfaces. Technical Report 96-11, Mitsubishi Electric Research Laboratories, Cambridge, MA, May 1996.

[23] S. J. Russell and E. H. Wefald. Principles of meta-reasoning. In *Proceedings of the First International Conference of Knowledge Representation and Reasoning*, pages 400–411, 1989.

[24] John R. Searle. Collective intentions and actions. In *Intentions in Communication*, chapter 19. The MIT Press, 1990.

[25] Y. Shoham. Agent oriented programing. *Artificial Intelligence*, 1(60):51–92, 1993.

[26] S. Zilberstein and S.J. Russell. Optimal composition of real-time systems. *Artificial Intelligence*, 82 (1-2):181–213, 1996.

S. R. THOMAS

A SURVEY OF AGENT-ORIENTED PROGRAMMING

1 INTRODUCTION

Many researchers have considered the problem of representing an agent in terms of its mental state. Originally, these researchers believed that agents must have informational attitudes, such as beliefs or knowledge, in order to store information to be used in reasoning, and motivational attitudes, such as desires, in order for the agents to have a reason to act. Bratman [2] showed that these attitudes are not sufficient, and added the attitude of intention to the framework. This gave rise to a good deal of work on BDI-architectures, in which the mental states of agents consisted of the attitudes beliefs, desires, and intentions.

Some of the work on BDI-architectures was largely theoretical, concerned with finding logical frameworks which allowed a natural representation of the three attitudes and the relationships between them. Since Hintikka's possible-worlds semantics for epistemic modal logics (logics of knowledge and belief) were widely known and reasonably intuitive, many proposals extended these semantics to allow the representation of desires and intentions. Since desires and intentions typically refer to the future, time had to be incorporated into these proposals, either by assigning each possible world a time, or by giving each possible world temporal extent.

The basic ideas underlying agent-oriented programming extend Hewitt's Actors formalism, in which modules, each with its own internal state, communicate with each other and have methods for dealing with incoming messages. In agent-oriented programming,

- the modules are called agents,
- each agent has an internal state, which consists of attitudes such as those described above; this is now called the mental state, and
- each message sent or received by an agent performs a speech act, as described in the philosophical literature [1, 7, 12].

M. Wooldridge and A. Rao (eds.), Foundations of Rational Agency, 263–274.

2 THE ORIGINAL PROPOSAL

In Shoham's original paper concerning agent-oriented programming [13], he outlines the ideas described above and further specifies certain aspects of the AOP approach. In particular,

- Agents communicate by exchanging messages, each of which is explicitly labeled according to the type of speech act it performs.
- Each agent's internal state consists only of the "mental state" - e.g., the beliefs, desires and intentions of the agent. The AOP approach does not dictate which attitudes ought to be represented as part of the mental state; the choice is up to the designer of a particular system, and may depend on characteristics of the domain of application.
- The behavior of an agent depends only on its mental state and the inputs it receives. These inputs may include both messages from other agents and sensor input.
- An agent's program will consist of at least an initial mental state for the agent and a set of rules that collectively specify a transition function from the mental state and inputs to the new mental state.
- A computation consists of a collection of agents communicating with each other and/or acting in the world.

3 REALIZATIONS OF THE PROPOSAL

3.1 Agent-0

As part of his original paper on AOP [13], Shoham describes a very simple AOP language called Agent-0. Agent-0 is meant to be an extremely simple language, used only to help demonstrate the potential power of the AOP approach.

The mental state of an Agent-0 agent did not follow the usual belief-desire-intention model; rather, the Agent-0 mental state consists of beliefs, abilities, obligations, and decisions. It was thought that desires are too ill-formed a notion to play the motivational role they are typically given in agent systems, and so they were omitted from the mental state. Since Agent-0 was meant to be a very simple language, the subtle and difficult attitude of intention was omitted as well.

Shoham added the idea of abilities to the mental state; abilities are important in planning. For Agent-0, each agent has a fixed set of actions that the agent is capable of performing, if conditions are right. That is, an agent has an ability exactly if the agent can perform the relevant action whenever some specified set of conditions obtain in the world. For example, one agent might have the ability to open a door whenever the door in unlocked. This tells us nothing about what the agent might be able to do when the door is in fact locked. (Note: Syntactically, Agent-0 omitted actions in favor of facts; a primitive action is represented by the fact that it brings about. For example, instead of having a raise-arm action, an agent is capable of achieving the arm-is-raised fact.)

An obligation in Agent-0 is a social commitment. There are two agent arguments for each commitment: the agent who is making the commitment and the agent to whom the commitment is made. If these two agents are in fact the same, then the obligation may be called a decision. An obligation is essentially a promise that the obligated agent will take a specified action at a specified time. It is through making obligations and decisions that an agent controls its future actions; in the implementations of Agent-0 done by Shoham's research group at Stanford at the time of development of Agent-0, the actions that had been committed to were automatically performed at the appropriate times without further reflection by the agent, as long as the actions were actually performable.

The Agent-0 agent communication language allows messages of three types. "Inform" messages convey information, and an agent who receives an inform message incorporates the content information into its set of beliefs. (This is another example of the simplicity of Agent-0: there is no provision for filtering or evaluating information received from other agents; all is automatically believed.) "Request" messages ask another agent to create a new obligation to the requesting agent, that some particular action will be carried out at a particular time. An "unrequest" message cancels an earlier request. The syntax of these messages is that of a list whose first element is the speech act type, whose second element is a time point, the time at which communication is to take place, whose third element is the agent to receive the message, and whose fourth element is the content of the message. The message content will be a fact (remember that primitive actions are represented by facts) or another communication action. For example, the following are all legal messages

```
(INFORM t a (t1 (on block1 block2)))

(REQUEST t1 a (REQUEST t2 b (t3 (on block1 block2))))

(REQUEST t1 a (INFORM t2 b (t3 (on block1 block2))))
```

An Agent-0 program consists of

- a time grain, which specifies how often the agent will go through its cycle of processing messages and updating its mental state, as well as how often the agent will perform actions; the agent executes one primitive (indivisible) action per time grain
- an initial mental state, which consists only of beliefs,
- a set of commitment rules, which specifies the transition function from mental state and incoming message(s) to new mental state.

The commitment rules are really the heart of the Agent-0 program. A commitment rule has three parts: a condition on the mental state which may or may not be met (such as, the agent currently believes that the door is unlocked), a condition on the just-received messages (for example, a message has just been received from Agent72), and a group of new obligations to be added to the mental state if and only if both conditions are met. Commitment rules can only add obligations to the mental state; they cannot remove obligations or affect the beliefs of the agent. Obligations are removed from the mental state in one of two ways: either the time arrives at which the obligation is to be carried out, in which case the obligation is removed whether the action is successfully completed or not, or the agent to whom the obligation is made cancels the obligation by sending an "unrequest" message, thus freeing the obligated agent from its commitment. An agent's beliefs can be changed by the receipt of an "inform" message, which causes the information from the message to be incorporated into the agent's beliefs. Presumably, sensor input can also cause changes to an agent's beliefs, although this was not made explicit.

As an example, examine the following commitment rule (taken from [13]):

```
(COMMIT (?a REQUEST ?action)
        (B (now (myfriend ?a)))
        (?a ?action))
```

In this rule, strings which begin with the character '?' name variables, and the symbol "now" evaluates to the current time. The message condition is "(?a REQUEST ?action)," which specifies that a message must have been received from some agent ?a requesting the performance of action ?action. The specification of ?action will include the time at which the action is to be carried out. The mental condition is "(B (now (myfriend ?a)))," that is, I currently believe that ?a is currently my friend. The final part of the commitment rule is the list of new commitments to be added to the mental state. In this case there is only one, a commitment to agent ?a to perform ?action.

At least two interpreters for Agent-0 were produced by Shoham's research group, and test programs written and executed. Since Agent-0 is such a simple AOP language, it was not expected that highly complex programs would be easy to write. However, this phase of development was naturally very useful in evaluating and refining Agent-0.

Limitations of Agent-0 included the following:

- An agent's beliefs could only be changed by receipt of an "inform" message from some agent, and such messages were always believed.
- Agents had no planning capability, and in fact a request to achieve some high-level goal could not even be legally communicated, since the only proper content of a request message is a primitive, nondecomposable action.

The next AOP language to be developed, PLACA, addressed these two limitations.

3.2 *PLACA*

PLACA (which stands for PLAnning Communicating Agents) was developed by the author while a graduate student in Shoham's research group. The primary motivation for the development of PLACA was the second limitation described above, which (unlike the first limitation) is both serious and fundamental to the design of Agent-0, though not to the AOP approach. The fact that agents cannot plan and cannot communicate about high-level goals seriously compromises the goal of providing a way to program heterogeneous systems of agents while allowing maximum autonomy. First, if one agent can only request of a second agent that the second agent perform one of its primitive actions, this means that the first agent must know about all the relevant primitive actions of the second agent. Is that second agent capable of the climb-stairs action? What about the press-elevator-button action? Agents often will not have this level of information about the capabilities of other agents in the system. In addition, the responsibility for planning now rests squarely with the first agent; no matter which agent ends up carrying out the required action, it is the first agent who must develop a fully-specified plan for achieving its high-level goal. This may seem reasonable at first, but consider the effect of this situation on replanning in the face of unanticipated failure. If one agent has dispatched another via the elevator to fetch a book from another floor, and it turns out that the elevator is out of service, the second agent cannot generate a new plan for fetching the book, even if the second agent is capable of planning! The reason is that the second agent does not know what end its actions are meant to bring about. Fetching the book may be a secondary effect of its actions, which are really meant to test (for example) whether the agent's battery has enough power to allow it to carry the book, or whether the book's owner is willing to lend out books to an artificial agent. (Note: if the second agent is equipped with goal-recognition capabilities, it may be able to recognize the most likely overall goal of its actions and replan accordingly; however this seems a rather roundabout way of providing for persistence in the face of failure!)

PLACA, then, assumes that agents are capable of generating plans and provides for the communication of higher-level goals. This allows agents to remain ignorant of the exact actions available to other agents, and to delegate both planning and execution of some goals to other agents. In addition, the delegated-to agent can now be expected to be persistent in pursuing the goals it adopts even in the face of failure, since it know what those goals are.

The mental state of a PLACA agent retains beliefs as in the Agent-0 mental state. It omits obligations and decisions, instead including both intentions and three forms of plans. Recall that Agent-0 omitted actions in favor of facts; each action was represented by the fact it achieved. In PLACA, both actions and facts are allowed. An agent can have an intention to achieve some goal or fact. Similarly, an agent can have a plan to achieve a fact, or a plan to do or not to do some primitive action at a particular time. An intention provides the agent with a goal to be achieved and with the problem of deciding how to achieve this goal, consistent with all the agent's other intentions. A plan is a sequence of actions (only linear plans are allowed) which are anticipated to achieve some intention of the agent's; having a plan entails not only knowing that the plan exists and should achieve the intention, but also actually meaning to carry out all the actions in the plan. PLACA agents also have capabilities, but they are different from Agent-0 capabilities. There are two forms of capabilities in PLACA: abilities to perform actions and abilities to achieve goals. The first kind of ability, CanDo, is roughly like the Agent-0 capability; the CanAch capability is new to PLACA.

Armed now with an expanded mental state which allows the representation of non-primitive intentions, we must similarly expand the programming language from Agent-0 to allow for the communication and processing of such intentions. A PLACA program is similar to an Agent-0 program in that it consists of an initial mental state and a set of mental-change rules which together specify a transition function from mental state and inputs to a new mental state. However, the rules in PLACA are more general. In particular, rules now consist of four pieces: a mental condition and message condition, similar to Agent-0's; a list of mental changes, which can add or drop beliefs and/or intentions to/from the mental state; and a list of messages to be sent immediately. This last piece is merely a convenience; the rule could instead specify that the agent should adopt an intention to send a message.

Notice that only beliefs and intentions are affected by the mental-change rules. Abilities to perform actions are not affected, because like Agent-0 abilities, they are assumed not to change. Plans are not affected, because they are maintained by the agent's planner, and should not be affected by incoming messages. Thus deliberation about which intentions to adopt is part of the agent program; pragmatic planning about how to achieve intentions is carried about separately.

To clarify, let's look at a few example rules. In the following, strings which begin with the character '?' name variables, and "Fred" refers to the agent whose program includes these rules.

```
((TO Fred, FROM boss, REQUEST, ?x),
 (),
 ((ADOPT (INTEND ?x))),
 ((TO boss, FROM Fred, INFORM, (*now* (intend ?x))))
)

((TO Fred, FROM ?agent, INFORM, ?s),
 (BELIEVE (*now* (unreliable ?agent))),
 (),
 ()
)

((TO Fred, FROM ?agent, INFORM, (?t (on ?thing1 ?thing2))),
 (AND (BELIEVE (*now* (block-expert ?agent)))
      (BELIEVE (?t (block ?thing1)))
```

```
      (BELIEVE (?t (block ?thing2)))),
 ((ADOPT (BELIEVE (?t (on ?thing1 ?thing2))))),
 ()
)

((TO Fred, FROM ?agent, REQUEST, (?t (did ?action))),
 (AND (BELIEVE (*now* (friendly ?agent)))
      (NOT (PLAN-NOT-DO ((- ?t 1) ?action)))),
 ((ADOPT (INTEND (?t (did ?action))))),
         ((TO ?agent,
           FROM Fred,
           INFORM, (*now* (intend (?t (did ?action)))))
         )
)
```

The first rule says that if our agent Fred receives a message from the boss containing a request to achieve ?x (notice ?x specifies not only what is to be achieved but when), then Fred will adopt a new intention to do ?x and inform the boss of this intention. Notice that the mental-condition is (), which is always fulfilled. The second rule says that if Fred should receive a message containing information from an agent whom Fred believes to be currently unreliable, then Fred will ignore the information contained in the message. The third rule, in contrast, says that if Fred is informed that one block is on another, by an agent whom Fred believes to be a block-expert, then Fred will adopt this information as a belief. Finally, the fourth rule says that if Fred receives a message containing a request that Fred perform ?action at time (- ?t 1) (that is, that Fred make it the case that at time ?t Fred has just done the action), and Fred believes the requesting agent is currently friendly and furthermore Fred hasn't already decided not to do ?action, then Fred will adopt an intention to do ?action, and will so inform the agent.

As can be seen from the above rules, the syntax of messages is slightly different in PLACA. A message has four pieces: the receiver, which follows the word "TO", the sender, which follows the word "FROM", the message type, and finally the message content. The message content is a fact for all message types; this is to preserve the idea that agents need not know which actions are available to other agents. Thus one agent does not request that another agent perform an action, but that the second agent bring about some fact. The fact will always have an associated time, but the message itself will not. Implicitly, the time of the message is now - the messages in the message-condition part of the rule have just been received, and the messages in the fourth part of the rule are to be sent immediately.

To summarize, then, PLACA is an extension of Agent-0 in which

- Agents are assumed to be able to plan.
- Agents can communicate about high-level goals.
- Agents need not know about the detailed capabilities of other agents.

- Agents are responsible for any replanning required to achieve their goals.
- Agents' beliefs can be changed by incoming messages, but are not necessarily changed by every "inform" message.

Limitations of PLACA include:

- Despite being richer than Agent-0, PLACA is still a small and simple language, used mainly for testing ideas about agent-oriented programming rather than for the development of any realistic systems.
- Both PLACA and Agent-0 have rather impoverished agent communication languages, with only three types of speech acts allowed.
- PLACA, as implemented, did not incorporate a state of the art planner, but rather allowed only linear plans.
- PLACA was intended for domains where close cooperation between agents is not required. In other words, PLACA is more appropriate for domains in which agents "subcontract out" some of their own work than for domains in which agents have joint intentions and may have to work closely together.

3.3 Agent-K

Agent-K is another extension of Agent-0, created by Winton Davies [4].

Agent-K extends Agent-0 by adopting KQML [11] as the agent communication language, replacing Shoham's inform, request, and unrequest speech act types with a larger range of possible speech acts.

KQML, like the communication languages of Agent-0 and PLACA, is based on speech act theory and labels each method explicitly with the type of speech act being performed. KQML's set of legal speech acts is much larger than the inform/request/unrequest types available in earlier AOP languages. The syntax of a KQML message is that the speech act type is given first, followed by keyword-value pairs that specify such things as the sender and receiver of the message, the message content, etc.

Like PLACA, Agent-K removes the difficulty in Agent-0 that agents believe everything of which they are informed.

The use of KQML as the agent communication language means that Agent-K agents can, ideally at least, communicate across networks with any other type of information system that understands and uses KQML. In fact, Agent-K uses Prolog to represent the mental state of agents, which means that these agents will not be able to understand the content fields of messages from systems that use Lisp-based syntax. The developers of Agent-K mention the future possibility of using KIF as a substitute language for message contents.

In Agent-K, the syntax of messages is altered from that used in earlier AOP languages. A message takes the form kqml(time, [performative, keyword(value), keyword(value), ...]). The time is the time of the message. The performative specifies the speech act type performed by the message, and the list of keyword-value pairs give the other relevant information, including the sender and receiver of the message and the message content.

Agent-K adapts the Agent-0 commitment rule, with the changes made necessary by the new message syntax. An example Agent-K rule (from [4]) is

```
COMMIT(['ask-one', content(b([t, fact]))],
       [clock(Now), b([t, fact])],
       S, kqml(Now, [reply,
                     receiver(S),
                     sender(c3po),
                     content(b([t, fact])),
                     language(prolog)]
       )
)
```

This rule specifies the response of an agent to an 'ask-one' message querying about the agents state of belief in fact at time t. If the agent does in fact currently believe that fact is true at time t, then the agent should respond with the message specified in the third and fourth lines of the rule.

4 RELATED WORK

The agent-oriented programming languages discussed here are languages for programming individual agents, not for specifying communications protocols. Such protocols as contract nets [5], partial global plans [6], or negotiation protocols [3, 17] could be used in conjunction with these languages, with the protocol specifying how agent interactions should occur and the AOP language used to program the individual agents, as constrained by the requirements of the protocol.

Others involved in work on agent-oriented programming have considered the problem of developing conventions at the societal level. For example, Shoham and Tennenholtz discuss in [14] the problem of developing useful social laws. These laws specify conditions under which certain actions are prohibited. One advantage of having such social laws is that it may simplify the problem of reasoning about what actions other agents may take, by allowing all agents to assume that all other agents will at least obey all social laws. (The possibility of building "outlaw agents" exists, of course, but is under the control of the system designers.) The related problem of making a society of agents develop, or "co-learn," social laws is an interesting one.

McCarthy has proposed Elephant [9], a programming language based on speech acts. Correctness of these programs depends on whether these speech acts are performed correctly[1]. Elephant programs may have two kinds of specifications: illo-

[1] For a proposal on a semantics for determining whether speech acts are "correct," see [15].

cutionary specifications, which relate output to inputs, and perlocutionary specifications, which concern the effects of programs on the world. Programmers will usually want to specify how the two kinds of specifications relate to one another. McCarthy speculates that keeping a history of interactions may eliminate the need for explicit data structures such as our mental state. Work on agent-oriented programming has been influenced by the ideas underlying Elephant.

Shoham discussed the similarities of agent-oriented programming and object-oriented programming (and Hewitt's Actors formalism [8]) in [13]. To paraphrase, AOP is a specialization of OOP, where we have "agents" in place of objects, and the internal states of these agents are constrained to be "mental states" as described above. In AOP, messages are categorized into types corresponding to types of speech acts. While I have been assuming that the agents in question are obviously distinct entities, such as physically-instantiated robots or processes running on different machines, there is no reason not to use agent-oriented programming languages for general-purpose programming. When AOP are used in this way, the similarities between AOP and OOP are more noticeable. One major difference between AOP languages and the OOP languages in use today, as Shoham points out in [13], is that the concept of inheritance within AOP has not been explored in any depth. It might make sense to define classes of agents, where agents of a particular class all include the mental-change rules specified for the class in their individual agent programs. These rules shared by all agents of a class might be used, for example, to ensure that all these agents adhere to some protocol by responding to messages in some minimal acceptable way; an individual agent might then have more specific rules for responding to a particular message as well. It is easy to see how a hierarchy of classes might be built, much like the class hierarchies found in OOP programs.

In [16] Tan and Weihmayer discuss a framework for cooperative problem solving which integrates Agent Oriented Programming with planning. Specifically, the AOP language AGENT-0 was integrated with the planner PRODIGY [10]. Their thesis is that by separating communication concerns from the agents' local problem-solving abilities, we allow increased agent heterogeneity, since only the communication language needs to be mutually understandable. PLACA differs from their system in several ways; most importantly, for them planning, which occurs directly as a result of firing a mental-change rule (which are called "commitment rules" in AGENT-0), is not interruptible; it acts as a single primitive action. Since many rules may fire during one cycle, many of these planning processes may be triggered. On a single-processor machine, this means that the cycle must be quite long if all planning processes are to be finished within one tick of the clock.

5 WHAT THE FUTURE HOLDS

It is logical that future work in the Agent-Oriented Programming paradigm should continue in the direction indicated by Agent-K; namely, developing agents for use over networks, using the best-established agent communications languages available. It would be enlightening to develop an application in which agents serve as database servers, mediating between organized collections of data on the one hand, and users (including other agents) who would like access to this data on the other.

One important problem to be solved if agent-oriented programming is to become widely used is the problem of maintaining the mental state. There is no requirement that the mental state be implemented as a database of logical sentences, but this is certainly a natural approach. This approach brings with it all the problems of maintaining consistency and of propagating changes throughout the database that have bedeviled many other systems. It may therefore be productive to investigate alternative means of implementing agents, in which logical sentences are not explicitly represented. I have not discussed in this paper the logic work done by both Shoham and Thomas in the development of their respective AOP languages, but both provide logical descriptions of the mental states of agents and use these logical frameworks to specify certain properties that the mental states must have - for example, that an agent can introspect about its beliefs and obligations. If logic is no longer to be used in the implementation of the mental state, then the new representation (whatever it is) should similarly describe and enforce desirable properties of the mental state, including particularly reasonable constraints among the different components such as beliefs and intentions.

It will be important to apply the techniques of AOP to nontrivial problems, as Davies plans to do. Both Agent-0 and PLACA were applied only to small problems, because both languages were developed in an investigatory mode rather than with the idea of providing a real-world programming language. Agent-K has been developed with an eye toward applying it to larger, more realistic problems in distributed data-gathering and discovery, and so it should push the frontier of work in the AOP paradigm in a desirable way.

S. Rebecca Thomas (Marist College, USA)

REFERENCES

[1] J. L. Austin. *How to Do Things with Words*. Harvard University Press, 1955.

[2] M. E. Bratman. *Intention, Plans, and Practical Reason*. Harvard University Press, 1987.

[3] Susan E. Conry, Robert A. Meyer, and Victor R. Lesser. Multistage negotiation in distributed planning. COINS 86-67, University of Massachusetts, 1986.

[4] Winton H. E. Davies and Peter Edwards. Agent-K: An integration of AOP and KQML.

[5] Randall Davis and Reid G. Smith. Negotiation as a metaphor for distributed problem solving. *Artificial Intelligence*, 20:63–109, 1983.

[6] Edmund H. Durfee and Victor R. Lesser. Using partial global plans to coordinate distributed problem solvers. In *Proceedings of the Tenth International Joint Conference on Artificial Intelligence*, pages 875–883, 1987.

[7] P. Grice. *Studies in the Ways of Words*. Harvard University Press, 1989.

[8] Carl Hewitt. Viewing control structures as patterns of passing messages. *Artificial Intelligence*, 8:323–364, 1977.

[9] John McCarthy. Elephant 2000: A programming language based on speech acts, 1990. Unpublished manuscript. Incomplete preliminary draft of March 8, 1990.

[10] Steven Minton, Craig A. Knoblock, Daniel R. Kuokka, Yolanda Gil, Robert L. Joseph, and Jaime G. Carbonell. Prodigy 2.0: The manual and tutorial. Technical Report CMU-CS-89-146, Carnegie Mellon University, 1989.

[11] R.S. Patil, R.E. Fikes, P.F. Patel-Schneider, D. MacKay, T. Finin, T. Gruber, and R. Neches. The DARPA knowledge sharing effort: Progress report. In *Proceedings of KR'92 - The Annual International Conference on Knowledge Representation*, 1992.

[12] J. R. Searle. *Speech Acts: An Essay in the Philosophy of Language*. Cambridge University Press, 1969.

[13] Y. Shoham. Agent-oriented programming. *Artificial Intelligence*, 60(1):51–92, 1993.

[14] Yoav Shoham and Moshe Tennenholtz. On the synthesis of useful social laws for artificial agent societies (preliminary report). In *Proceedings of the Tenth National Conference on Artificial Intelligence*, pages 276–281, 1992.

[15] Munindar P. Singh. Towards a formal theory of communication for multiagent systems. In *Proceedings of the International Joint Conference on Artificial Intelligence*, pages 69–74, 1991.

[16] Ming Tan and Robert Weihmayer. Integrating agent-oriented programming and planning for cooperative problem solving. In *Notes from the AAAI-92 Workshop on Cooperation among Heterogeneous Intelligent Systems*, 1992.

[17] Gilad Zlotkin and Jeffrey S. Rosenschein. Negotiation and conflict resolution in non-cooperative domains. In *Proceedings of the Eighth National Conference on Artificial Intelligence*, pages 100–105, 1990.

Y. LESPÉRANCE, H. J. LEVESQUE, AND R. REITER

A SITUATION CALCULUS APPROACH TO MODELING AND PROGRAMMING AGENTS

1 INTRODUCTION

The notion of computational *agents* has become very fashionable lately [24, 32]. Building such agents seems to be a good way of congenially providing services to users in networked computer systems. Typical applications are information retrieval over the internet, automation of common user activities, smart user interfaces, integration of heterogenous software tools, intelligent robotics, business and industrial process modeling, etc. The term "agent" is used in many different ways, so let us try to clarify what we mean by it. We take an agent to be any active entity whose behavior is usefully described through mental notions such as knowledge, goals, abilities, commitments, etc. (This is pretty much the standard usage in artificial intelligence, in contrast to the common view of agents as scripts that can execute on remote machines). Moreover, we will focus on the approach to building applications that involves designing a system as a collection of interacting agents.

Agent programming [30] can be viewed as a generalization of object-oriented programming. But the notion of an agent is much more complex than that of an object. Because of this, it is crucial that tools for modeling and designing agents be based on solid theoretical foundations. For some time now, we have been working on a logical theory of agency and on programming tools based on it. The theoretical framework includes a formalization of action that incorporates a solution to the frame problem, thus relieving the designer from having to specify what aspects of the world don't change when an action is performed. The framework also includes a model of what agents know and how they can acquire information by doing knowledge-producing actions, such as sensing the environment with vision or sonar, or interacting with users or other agents. And finally, we have an account of complex actions and processes that inherits the solution to the frame problem for simple actions. The framework also has attractive computational properties.

The set of complex action expressions defined can be viewed as a programming language for agents. It can be used to model the behavior of a set of agents and/or to actually implement them. Given a declarative specification of the agents' primitive actions, the designer/modeler can specify complex behaviors for the agents procedurally in the programming language. The behavior specification can be in

M. Wooldridge and A. Rao (eds.), Foundations of Rational Agency, 275–299.

terms of very high-level actions and can refer to conditions in effect in the environment — the interpreter automatically maintains a world model based on the specifications. The approach focuses on high-level programming rather than planning. But the programs can be nondeterministic and search for appropriate actions. When an implementation of the primitive actions is provided, the programs can be executed in a real environment; otherwise, a simulated execution is still possible.

Most of our work so far on the theory and implementation of agents has been concerned with single agents. Here, we extend our framework to deal with multi-agent systems. The treatment proposed is somewhat preliminary and we identify various problems that need to be solved before we have a completely satisfactory theoretical account and implementation.

The approach will be presented with the help of an example — a multi-agent system that helps users schedule meetings. Each user has a "schedule manager" agent that knows something about his schedule. When someone wants to organize a meeting, he creates a new "meeting organizer" agent who contacts the participants' schedule managers and tries to set up the meeting. The example is a very simplified version of such a system. To be truly useful, a schedule manager agent should know the user's preferences about meeting times, when and how it should interact with the user in response to requests from meeting organizer agents, perhaps the hierarchical structure of the office, etc. Meeting organizer agents should have robust scheduling and negotiating strategies. Our formalization of the application includes a simple generic agent communication module that can be used for other applications. Each agent has a set of messages waiting for it and abstract communication acts are defined (e.g., INFORM, REQUEST, QUERYWHETHER, etc.).

In the next section, we outline our theory of simple actions. Then, we discuss how knowledge and knowledge-producing actions can be modeled. Next, we present our account of complex actions, and explain how it can be viewed as an agent programming language. Section 6 develops a set of simple tools for agent communication and section 7 completes our specification of the meeting scheduling application. In the following section, we discuss various architectural issues that arise in implementing our framework, describe the status of the implementation, and sketch what experimental applications have been implemented. We conclude by summarizing the main features of our approach and discussing the problems that remain.

2 THE SITUATION CALCULUS AND THE FRAME PROBLEM

The situation calculus [17] is a first-order language (with some second-order features) for representing dynamically changing worlds. All changes to the world are the result of named *actions*. A possible world history, which is simply a sequence of actions, is represented by a first order term called a *situation*. The constant S_0 is used to denote the initial situation, namely that situation in which no actions have yet occurred. There is a distinguished binary function symbol *do* and the term $do(\alpha, s)$ denotes the situation resulting from action α being performed in situation s. Actions may be parameterized. For example, $\text{PUT}(agt, x, y)$ might stand for the action of agent agt putting object x on object y, in which case $do(\text{PUT}(agt, x, y), s)$ denotes that situation resulting from agt placing x on y when the world is in situation s. Notice that in the situation calculus, actions are denoted by function symbols, and situations (world histories) are also first order terms. For example,

$$do(\text{PUTDOWN}(\text{AGT}, \text{A}), do(\text{WALK}(\text{AGT}, \text{P}), do(\text{PICKUP}(\text{AGT}, \text{A}), S_0)))$$

is a situation denoting the world history consisting of the sequence of actions

$$[\text{PICKUP}(\text{AGT}, \text{A}), \text{WALK}(\text{AGT}, \text{P}), \text{PUTDOWN}(\text{AGT}, \text{A})].$$

Notice that the sequence of actions in a history, in the order in which they occur, is obtained from a situation term by reading off its action instances from right to left.

Relations whose truth values vary from situation to situation, called *relational fluents*, are denoted by predicate symbols taking a situation term as their last argument. For example, $\text{HOLDING}(agt, x, s)$ might mean that agt is holding object x in situation s. Functions whose denotations vary from situation to situation are called *functional fluents*. They are denoted by function symbols with an extra argument taking a situation term, as in $\text{POS}(agt, s)$, i.e., the position of agt in situation s.

An action is specified by first stating the conditions under which it can be performed by means of a *precondition axiom* of the following form:

$$Poss(\alpha(\vec{x}), s) \Leftrightarrow \pi_\alpha(\vec{x}, s)$$

Here, $\pi_\alpha(\vec{x}, s)$ is a formula specifying the preconditions for action $\alpha(\vec{x})$. For example, the precondition axiom for the action ADDTOSCHED might be:

$$\begin{aligned}(1)\quad & Poss(\text{ADDTOSCHED}(agt, user, period, activity, organizer), s) \Leftrightarrow \\ & agt = \text{SCHEDULEMANAGER}(user) \wedge \\ & \neg\exists activity', organizer' \\ & \quad \text{SCHEDULE}(user, period, activity', organizer', s)\end{aligned}$$

This says that it is possible for agent agt to add an activity to $user$'s schedule in situation s iff agt is $user$'s schedule manager and there is nothing on $user$'s schedule for that period in s.

Secondly, one must specify how the action affects the state of the world with *effect axioms*. For example,

$$\begin{aligned}& Poss(\text{ADDTOSCHED}(agt, user, period, activity, organizer), s) \Rightarrow \\ & \text{SCHEDULE}(user, period, activity, organizer, \\ & \quad do(\text{ADDTOSCHED}(agt, user, period, activity, organizer), s))\end{aligned}$$

Effect axioms provide the "causal laws" for the domain of application.

The above axioms are not sufficient if one wants to reason about change. It is usually necessary to add *frame axioms* that specify when fluents remain unchanged by actions; for example,

$$\begin{aligned}&Poss(\text{RAISESALARY}(company, user_1, amount), s) \wedge \\ &\text{SCHEDULE}(user_2, per, activ, org, s) \Rightarrow \\ &\text{SCHEDULE}(user_2, per, activ, org, \\ &\quad do(\text{RAISESALARY}(company, user, amount), s))\end{aligned}$$

The frame problem arises because the number of these frame axioms is very large, in general, of the order of $2 \times \mathcal{A} \times \mathcal{F}$, where $\mathcal{A}$ is the number of actions and $\mathcal{F}$ the number of fluents.

Our approach incorporates the solution to the frame problem described in [20]. First, for each fluent F, one can collect all effects axioms involving F to produce two *general effect axioms* of the following form:

$$Poss(a, s) \wedge \gamma_F^+(\vec{x}, a, s) \Rightarrow F(\vec{x}, do(a, s))$$

$$Poss(a, s) \wedge \gamma_F^-(\vec{x}, a, s) \Rightarrow \neg F(\vec{x}, do(a, s))$$

Here $\gamma_F^+(\vec{x}, a, s)$ is a formula describing under what conditions doing the action a in situation s leads the fluent $F(\vec{x})$ to become true in the successor situation $do(a, s)$ and similarly $\gamma_F^-(\vec{x}, a, s)$ is a formula describing the conditions under which performing action a in situation s results in the fluent $F(\vec{x})$ becoming false in situation $do(a, s)$. For example, the following might be the general effect axioms for the fluent SCHEDULE:

$$\begin{aligned}&Poss(a, s) \wedge \exists agt\, a = \text{ADDTOSCHED}(agt, user, per, activ, org) \\ &\Rightarrow \text{SCHEDULE}(user, per, activ, org, do(a, s))\end{aligned} \tag{2}$$

$$\begin{aligned}&Poss(a, s) \wedge \exists agt\, a = \text{RMVFROMSCHED}(agt, user, per) \\ &\Rightarrow \neg\text{SCHEDULE}(user, per, activ, org, do(a, s))\end{aligned} \tag{3}$$

The solution to the frame problem rests on a *completeness assumption*. This assumption is that the general effect axioms characterize all the conditions under which action a can lead to a fluent $F(\vec{x})$'s becoming true (respectively, false) in the successor situation. Therefore, if action a is possible and $F(\vec{x})$'s truth value changes from *false* to *true* as a result of doing a, then $\gamma_F^+(\vec{x}, a, s)$ must be *true* and similarly for a change from *true* to *false*. Additionally, *unique name axioms* are added for actions and situations. From the general effect axioms and the completeness assumption, one can derive a *successor state axiom* of the following form for the fluent F:

$$\begin{aligned}&Poss(a, s) \Rightarrow [F(\vec{x}, do(a, s)) \Leftrightarrow \\ &\quad \gamma_F^+(\vec{x}, a, s) \vee (F(\vec{x}, s) \wedge \neg\gamma_F^-(\vec{x}, a, s))]\end{aligned}$$

By quantifying over actions, this single axiom provides a parsimonious solution to the frame problem. Similar successor state axioms can be written for functional fluents.[1]

Applying this to our example, from the general effect axioms 2 and 3 we obtain the following successor state axiom for SCHEDULE:

$$(4)\quad \begin{aligned}&\mathit{Poss}(a,s) \Rightarrow\\ &[\mathrm{SCHEDULE}(user, per, activ, org, do(a,s)) \Leftrightarrow\\ &\exists agt\; a = \mathrm{ADDTOSCHED}(agt, user, per, activ, org)\\ &\;\vee \mathrm{SCHEDULE}(user, per, activ, org, s) \wedge\\ &\;\;\neg\exists agt\; a = \mathrm{RMVFROMSCHED}(agt, user, per)]\end{aligned}$$

i.e., an activity is on $user$'s schedule following the performance of action a in situation s iff either the action is some agent adding the activity to $user$'s schedule, or the activity was already on $user$'s schedule in s and the action is not some agent removing the activity from $user$'s schedule. Now note for example that if $\mathrm{SCHEDULE}(u,p,i,o,S_0)$, then by the unique names axioms for actions, it also follows that $\mathrm{SCHEDULE}(u,p,i,o,do(\mathrm{RAISESALARY}(c,u',d),S_0))$.

In multi-agent domains, it is often useful to refer to the agent of an action a. We use the term $agent(a)$ for this. We require that for each primitive action one provide an axiom specifying who its agent is, for example:

$$agent(\mathrm{ADDTOSCHED}(agt, user, per, activ, org)) = agt$$

In general, a particular domain of application will be specified by the union of the following sets of axioms:

- Axioms describing the initial situation, S_0.
- Action precondition axioms, one for each primitive action.
- Successor state axioms, one for each fluent.
- Unique names axioms for the primitive actions.
- Axioms specifying the *agent* of each primitive action.
- Some foundational, domain independent axioms.

The latter foundational axioms include unique names axioms for situations, and an induction axiom. They also introduce the relation $<$ over situations. $s < s'$ holds iff s' is the result of some sequence of actions being performed in s, where each action in the sequence is possible in the situation in which it is performed; $s \leq s'$ stands for $s < s' \vee s = s'$. Since the foundational axioms play no special role

[1] In the above, we have assumed that there were no state constraints which might contribute ramifications, i.e., indirect effects of actions. In [15], the approach presented is extended to deal with state constraints by compiling their effects into the successor state axioms.

in this paper, we omit them. For details, and for some of their metamathematical properties, see Lin and Reiter [15] and Reiter [21].

For domain theories of this kind, there are very clean characterizations of various reasoning tasks, for instance planning [7]:

> **Classical Planning:** Given a domain theory *Axioms* as above, and a goal formula $\phi(s)$ with a single free-variable s, the planning task is to find a sequence of actions $\vec{a}$ such that:
>
> $$\mathit{Axioms} \models S_0 \leq do(\vec{a}, S_0) \wedge \phi(do(\vec{a}, S_0))$$
>
> where $do([a_1, \ldots, a_n], s)$ is an abbreviation for
>
> $$do(a_n, do(a_{n-1}, \ldots, do(a_1, s) \ldots)).$$

In other words, the task is to find a sequence of actions that is executable (each action is executed in a context where its precondition is satisfied) and that achieves the goal (the goal formula ϕ holds in the final situation that results from performing the actions in sequence). If you have a particular planning algorithm, you can show that it is sound by proving that it only returns answers that satisfy the specification given above.

3 KNOWLEDGE AND KNOWLEDGE PRODUCING ACTIONS

Knowledge can be represented in the situation calculus by adapting the possible world model of modal logic (as first done by Moore [18]). The idea is to model someone's uncertainty (lack of knowledge) about what is true using the set of situations he/she considers possible. We introduce a fluent *K*, where $K(agt, s', s)$ means that in situation s, the agent agt thinks the world could be in situation s' (in modal logic terms, *K* is the knowledge accessibility relation). Then we introduce the abbreviation:

$$\mathbf{Know}(agt, \phi, s) \stackrel{\text{def}}{=} \forall s'(K(agt, s', s) \Rightarrow \phi(s')).$$

Thus, agt knows in s that ϕ holds iff ϕ holds in all the situations s' that agt considers possible in s.[2]

With this in place, we can then consider knowledge-producing actions (as they occur in perception or communication). Such actions affect the mental state of the agent rather than the state of the external world. For example, consider the action of an agent sensing what messages he has with the following effect axiom:

$$\begin{aligned}&\mathit{Poss}(\textsc{senseMsgs}(agt), s) \Rightarrow\\ &\mathbf{KWhether}(agt, \textsc{MsgRcvd}(agt, sender, msgId, msg, now),\\ &\quad do(\textsc{senseMsgs}(agt), s))\end{aligned}$$

[2]In this, ϕ stands for a situation calculus formula with all situation arguments suppressed; $\phi(s')$ will denote the formula obtained by restoring situation variable s' to all fluents appearing in ϕ. For clarity, we sometimes use the special constant *now* to represent the situation bound by the enclosing **Know**; so $\mathbf{Know}(agt, \phi(now), s)$ stands for $\forall s'(K(agt, s', s) \Rightarrow \phi(s'))$.

This says that after performing the action SENSEMSGS, the agent agt knows exactly which messages it has received and not yet processed, who sent them, and what their message IDs are (**KWhether**(agt, ϕ, s) is an abbreviation for the formula **Know**$(agt, \phi, s) \vee$ **Know**$(agt, \neg\phi, s)$).

Scherl and Levesque [27] have shown how one can generalize the solution to the frame problem of the previous section to deal with knowledge and knowledge-producing actions. But they only consider domains where there is a single agent. For multi-agents settings, their solution can used with minimal changes provided we assume that all actions are *public*, i.e., that agents are aware of every action that happens. For instance, if we make this assumption and the only knowledge producing action in the domain is SENSEMSGS, then we can use the following successor state axiom for the knowledge fluent *K*:

$$
\begin{array}{l}
Poss(a, s) \Rightarrow \\
[K(knower, s'', do(a, s)) \Leftrightarrow \\
\;\exists s'(K(knower, s', s) \wedge s'' = do(a, s') \wedge Poss(a, s') \wedge \\
\qquad [a = \text{SENSEMSGS}(knower) \Rightarrow \\
\qquad\;\; \forall sndr, mId, m(\text{MSGRCVD}(knower, sndr, mId, m, s') \Leftrightarrow \\
\qquad\qquad \text{MSGRCVD}(knower, sndr, mId, m, s))])]
\end{array}
$$

Let's look at what this says. There are two cases. If the action a is not a knowledge producing action performed by the agent under consideration $knower$ (i.e., $a \neq \text{SENSEMSGS}(knower)$), then the axiom says that in the resulting situation $do(a, s)$, $knower$ considers possible any situation s'' that is the result of a being performed in a situation s' that $knower$ used to consider possible before the action. Thus, $knower$ only acquires the knowledge that the action a has been performed. If on the other hand, the action a is a knowledge producing action performed by $knower$ (i.e., $a = \text{SENSEMSGS}(knower)$), then we get that in the resulting situation $do(a, s)$, $knower$ considers possible any situation s'' that is the result of a being performed in a situation s' that $knower$ used to consider possible before the action, and where the fluent $\text{MSGRCVD}(knower, \ldots)$ holds exactly for the messages for which it holds in the "real" situation s. Thus after doing SENSEMSGS, an agent knows that it has performed this action and knows exactly which messages it has received and not yet processed. This can be extended to an arbitrary number of knowledge-producing actions in a straightforward way.

However, the assumption that all actions are public is too strong for many multi-agent domains; agents need not be aware of the actions of other agents (exogenous actions). Another workable approach is to be very "conservative" and have agents allow for the occurrence of an arbitrary number of exogenous actions at every step. For our meeting scheduling application, this yields the following successor state axiom:

$$
\begin{array}{l}
Poss(a, s) \Rightarrow \\
[K(knower, s'', do(a, s)) \Leftrightarrow \\
\exists s'(K(knower, s', s) \wedge \\
\quad (agent(a) \neq knower \Rightarrow ExoOnly(knower, s', s'')) \wedge \\
\quad (agent(a) = knower \Rightarrow \exists s^*(ExoOnly(knower, s', s^*) \wedge \\
\quad\quad s'' = do(a, s^*) \wedge Poss(a, s^*) \wedge \\
\quad\quad [a = \text{SENSEMSGS}(knower) \Rightarrow \\
\quad\quad\quad \forall sndr, mId, m(\text{MSGRCVD}(knower, sndr, mId, m, s^*) \Leftrightarrow \\
\quad\quad\quad\quad \text{MSGRCVD}(knower, sndr, mId, m, s))]))))] \\
\text{where } ExoOnly(agt, s, s') \stackrel{\text{def}}{=} s \leq s' \wedge \\
\quad\quad\quad\quad\quad\quad \forall s^* \forall a(s < do(a, s^*) \leq s' \Rightarrow agent(a) \neq agt)
\end{array}
$$

Let us explain how this works. When an action a is performed by some agent other than the $knower$, the specification says that in the resulting situation, $knower$ considers possible any situation that is the result of any number of exogenous actions occurring in a situation that used to be considered possible; this means that $knower$ allows for the occurrence of an arbitrary number of exogenous actions, and thus loses any knowledge it may have had about fluents that could be affected by exogenous actions. When a is a non-knowledge-producing action (e.g., $\text{ADDTOSCHED}(knower, u, p, a, o)$) performed by $knower$, the specification states that in the resulting situation, $knower$ considers possible any situation that is the result of its doing a preceded by any number of exogenous actions occurring in a situation that used to be considered possible; thus, $knower$ acquires the knowledge that it has just performed a, but loses any knowledge it may have had about fluents that could be affected by exogenous actions and are not reset by a. Finally, when a is a knowledge-producing action (i.e., SENSEMSGS) performed by $knower$, we get the same as above, plus the fact that the agent acquires knowledge of the values of the fluents associated with the sensing action, in this case what messages it has received and not yet processed.

In general, allowing for the occurrence of an arbitrary number of exogenous actions at every step as we do here would probably leave agents with too little knowledge. But our meeting scheduling domain is neatly partitioned: a user's schedule can only be updated by that user's schedule manager agent. Thus, schedule manager agents always know what their user's schedule (as recorded) is. In other circumstances, it may be appropriate to assume that actions are all public, as discussed earlier. In other cases, it seems preferable for agents to assume that no exogenous actions occur and to revise their beliefs when an inconsistency is discovered; a formalization of this approach is being investigated.

4 COMPLEX ACTIONS AND GOLOG

A very general and flexible approach to designing agents involves using a planner. When the agent gets a goal, the planner is invoked to generate a plan that achieves the goal, and then the plan is executed. A problem with this approach is that plan synthesis is often computationally infeasible in complex domains, especially when the agent does not have complete knowledge and there are exogenous actions. An alternative approach that is showing promise is that of *high-level program execution* [14]. The idea, roughly, is that instead of searching for a sequence of actions that would take the agent from an initial state to some goal state, the task is to find a sequence of actions that constitutes a legal execution of some high-level non-deterministic program. As in planning, to find such a sequence it is necessary to reason about the preconditions and effects of the actions within the body of the program. However, if the program happens to be almost deterministic, very little searching is required; as more and more non-determinism is included, the search task begins to resemble traditional planning. Thus, in formulating a high-level program, the user gets to control the search effort required. The hope is that in many domains, what an agent needs to do can be conveniently expressed using a suitably rich high-level programming language.[3]

Our proposal for such a language is Golog[14], a logic-programming language whose primitive actions are those of a background domain theory of the form described earlier. It includes the following constructs:

α,	primitive action
$\phi?$,	wait for a condition[4]
$(\sigma_1; \sigma_2)$,	sequence
$(\sigma_1 \mid \sigma_2)$,	nondeterministic choice between actions
$\pi x.\sigma$,	nondeterministic choice of arguments
σ^*,	nondeterministic iteration
if ϕ **then** σ_1 **else** σ_2,	conditional
while ϕ **do** σ,	loop
proc $\beta(\vec{x})\ \sigma$,	procedure definition[5]

Here's a simple example to illustrate some of the more unusual features of the language:

proc REMOVEABLOCK
    $\pi b\,[\text{ONTABLE}(b)?; \text{PICKUP}(b); \text{PUTAWAY}(b)]$
endProc;
REMOVEABLOCK^*;
$\neg\exists block\ \text{ONTABLE}(block)?$

Here we first define a procedure to remove a block from the table using the non-deterministic operator π. $\pi x\,[\sigma(x)]$ means nondeterministically pick an individual

[3]This is not to imply that the planning approach or belief-desire-intention models of agents are never useful, quite the opposite. Later on, we will see for instance, that modeling goals would be quite useful in dealing with communication. But the fact remains that these approaches are computationally problematic.

[4]Because there are no exogenous actions or concurrent processes in Golog, waiting for ϕ amounts to testing that ϕ holds in the current situation.

[5]For space reasons, we ignore these here.

x, and for that x, perform the program $\sigma(x)$. The wait action ONTABLE(b)? succeeds only if the individual chosen, b, is a block that is on the table. The main part of the program uses the nondeterministic iteration operator; it simply says to execute REMOVEABLOCK zero or more times until the table is clear.

In its most basic form, the high-level program execution task is a special case of the planning task discussed earlier:

> **Program Execution:** Given a domain theory *Axioms* as above, and a program σ, the execution task is to find a sequence of actions $\vec{a}$ such that:
>
> $$Axioms \models Do(\sigma, S_0, do(\vec{a}, S_0))$$
>
> where $Do(\sigma, s, s')$ is an abbreviation for a formula of the situation calculus which says that program σ when executed starting in situation s has s' as a legal terminating situation.

In [14], a simple inductive definition of *Do* was presented, containing rules such as:

$$Do([\sigma_1; \sigma_2], s, s') \stackrel{\text{def}}{=} \exists s''.\ Do(\sigma_1, s, s'') \wedge Do(\sigma_2, s'', s')$$
$$Do([\sigma_1 \mid \sigma_2], s, s') \stackrel{\text{def}}{=} Do(\sigma_1, s, s') \vee Do(\sigma_2, s, s')$$

one for each construct in the language. This kind of semantics is sometimes called *evaluation semantics* [8] since it is based on the complete evaluation of the program.

It is difficult to extend this kind of semantics to deal with concurrent actions. Since these are required in multi-agent domains, a more refined kind of semantics was developed in [2]. This kind of semantics called *computational semantics* [8], is based on "single steps" of computation, or *transitions*[6]. A step here is either a primitive action or testing whether a condition holds in the current situation. Two special predicates are introduced, *Final* and *Trans*, where $Final(\sigma, s)$ is intended to say that program σ may legally terminate in situation s, and where $Trans(\sigma, s, \sigma', s')$ is intended to say that program σ in situation s may legally execute one step, ending in situation s' with program σ' remaining.

Final and *Trans* are characterized by a set of equivalence axioms, each depending on the structure of the first argument. These quantify over programs and so, unlike in [14], it is necessary to encode Golog programs as first-order terms, including introducing constants denoting variables, and so on. As shown in [4], this is laborious but quite straightforward[7]. We omit all such details here and simply use programs within formulas as if they were already first-order terms.

The equivalence axioms for *Final* are as follows (universally closing on s):[8]

---

[6] Both types of semantics belong to the family of structural operational semantics introduced in [19].
[7] Observe that *Final* and *Trans* cannot occur in tests, hence self-reference is disallowed.
[8] It is convenient to include a special "empty" program *nil*.

$$
\begin{aligned}
&\mathit{Final}(\mathit{nil}, s) \Leftrightarrow \mathit{TRUE}\\
&\mathit{Final}(\alpha, s) \Leftrightarrow \mathit{FALSE}\\
&\mathit{Final}(\phi?, s) \Leftrightarrow \mathit{FALSE}\\
&\mathit{Final}([\sigma_1; \sigma_2], s) \Leftrightarrow \mathit{Final}(\sigma_1, s) \wedge \mathit{Final}(\sigma_2, s)\\
&\mathit{Final}([\sigma_1 \mid \sigma_2], s) \Leftrightarrow \mathit{Final}(\sigma_1, s) \vee \mathit{Final}(\sigma_2, s)\\
&\mathit{Final}(\pi x.\sigma, s) \Leftrightarrow \exists x.\mathit{Final}(\sigma, s)\\
&\mathit{Final}(\sigma^*, s) \Leftrightarrow \mathit{TRUE}\\
&\mathit{Final}(\textbf{if } \phi \textbf{ then } \sigma_1 \textbf{ else } \sigma_2, s) \Leftrightarrow\\
&\qquad \phi(s) \wedge \mathit{Final}(\sigma_1, s) \ \vee\ \neg\phi(s) \wedge \mathit{Final}(\sigma_2, s)\\
&\mathit{Final}(\textbf{while } \phi \textbf{ do } \sigma, s) \Leftrightarrow \phi(s) \wedge \mathit{Final}(\sigma, s) \ \vee\ \neg\phi(s)
\end{aligned}
$$

The equivalence axioms for *Trans* are as follows (universally closing on s, δ, s'):

$$
\begin{aligned}
&\mathit{Trans}(\mathit{nil}, s, \delta, s') \Leftrightarrow \mathit{FALSE}\\
&\mathit{Trans}(\alpha, s, \delta, s') \Leftrightarrow \mathit{Poss}(\alpha, s) \wedge \delta = \mathit{nil} \wedge s' = \mathit{do}(\alpha, s)\\
&\mathit{Trans}(\phi?, s, \delta, s') \Leftrightarrow \phi(s) \wedge \delta = \mathit{nil} \wedge s' = s\\
&\mathit{Trans}([\sigma_1; \sigma_2], s, \delta, s') \Leftrightarrow\\
&\qquad \mathit{Final}(\sigma_1, s) \wedge \mathit{Trans}(\sigma_2, s, \delta, s') \vee \exists \delta'.\delta = (\delta'; \sigma_2) \wedge \mathit{Trans}(\sigma_1, s, \delta', s')\\
&\mathit{Trans}([\sigma_1 \mid \sigma_2], s, \delta, s') \Leftrightarrow \mathit{Trans}(\sigma_1, s, \delta, s') \vee \mathit{Trans}(\sigma_2, s, \delta, s')\\
&\mathit{Trans}(\pi x.\sigma, s, \delta, s') \Leftrightarrow \exists x.\mathit{Trans}(\sigma, s, \delta, s')\\
&\mathit{Trans}(\sigma^*, s, \delta, s') \Leftrightarrow \exists \delta'.\delta = (\delta'; \sigma^*) \wedge \mathit{Trans}(\sigma, s, \delta', s')\\
&\mathit{Trans}(\textbf{if } \phi \textbf{ then } \sigma_1 \textbf{ else } \sigma_2, s, \delta, s') \Leftrightarrow\\
&\qquad \phi(s) \wedge \mathit{Trans}(\sigma_1, s, \delta, s') \quad \vee \neg\phi(s) \wedge \mathit{Trans}(\sigma_2, s, \delta, s')\\
&\mathit{Trans}(\textbf{while } \phi \textbf{ do } \sigma, s, \delta, s') \Leftrightarrow\\
&\qquad \phi(s) \wedge \exists \delta'.\ \delta = (\delta'; \textbf{while } \phi \textbf{ do } \sigma) \wedge \mathit{Trans}(\sigma, s, \delta', s')
\end{aligned}
$$

With *Final* and *Trans* in place, *Do* may be defined as:

$$\mathit{Do}(\sigma, s, s') \stackrel{\text{def}}{=} \exists \delta.\mathit{Trans}^*(\sigma, s, \delta, s') \wedge \mathit{Final}(\delta, s')$$

where Trans^* is the transitive closure of *Trans*, defined as the (second-order) situation calculus formula:

$$\mathit{Trans}^*(\sigma, s, \sigma', s') \stackrel{\text{def}}{=} \forall T[\ldots \Rightarrow T(\sigma, s, \sigma', s')]$$

where the ellipsis stands for:

$$
\begin{aligned}
&\forall s.\ T(\sigma, s, \sigma, s) \quad \wedge\\
&\forall s, \delta', s', \delta'', s''.\ T(\sigma, s, \delta', s') \wedge \mathit{Trans}(\delta', s', \delta'', s'') \ \Rightarrow\ T(\sigma, s, \delta'', s'').
\end{aligned}
$$

In other words, $\mathit{Do}(\sigma, s, s')$ holds iff it is possible to repeatedly single-step the program σ, obtaining a program δ and a situation s' such that δ can legally terminate in s'. In [4], it is shown that this definition of *Do* is equivalent to that in [14].

On the surface, Golog looks a lot like a standard procedural programming language. It is indeed a programming language, but one whose execution, like planning, depends on reasoning about actions. An interpreter for Golog essentially searches for a sequence of primitive actions that can be proven to lead to a final situation of the program. Thus, a crucial part of a Golog program is the *declarative* part: the precondition axioms, the successor state axioms, and the axioms characterizing the initial situation. A Golog program together with the definition of *Do* and some foundational axioms about the situation calculus *is* a formal logical theory about the possible behaviors of an agent in a given environment.

The declarative part of a Golog program is used by the Golog interpreter in two ways. The successor state axioms and the axioms specifying the initial situation are used to evaluate the conditions that appear in the program (wait actions and **if/while** conditions) as the program is interpreted. The action preconditions axioms are used (with the other axioms) to check whether the next primitive action is possible in the situation reached so far. Golog programs are often nondeterministic and a failed precondition or test action causes the interpreter to backtrack and try a different path through the program. For example, given the program $(a; \phi?) \mid (b; c)$, the Golog interpreter might determine that a is possible in the initial situation S_0, but upon noticing that ϕ is false in $do(a, S_0)$, backtrack and return the final situation $do(c, do(b, S_0))$ after confirming that b is possible initially and that c is possible in $do(b, S_0)$.

Thus in a way, the Golog interpreter is automatically maintaining a model of the world state for the programmer using the axioms. If a program is going to maintain a model of its environment, it seems that having it done automatically from declarative specifications is much more convenient and less error prone than having to program such model updating from scratch. The Golog programmer can work at a much higher level of abstraction.

And to reiterate the main idea, Golog aims for a middle ground between runtime planning and explicit programming down to the last detail. It supports search for appropriate actions through nondeterminism as well as explicit programming. Thus for example, the program

$$\textbf{while } \exists b\, \textsc{OnTable}(b) \textbf{ do } \pi b.\, \textsc{remove}(b) \textbf{ endWhile}$$

leaves it to the Golog interpreter to find a legal sequence of actions that clears the table.

5 CONCURRENT ACTIONS AND CONGOLOG

To implement multiple agents in a single program, we need concurrent processes. In [2, 3], an extended version of Golog that incorporates a rich account of concurrency is developed. This extended language is called ConGolog. Let us now review the syntax and semantics of ConGolog (this section is a quasi-verbatim reproduction of part of [2]). The ConGolog account of concurrency is said to be 'rich' because it handles:

- concurrent processes with possibly different priorities,

- high-level interrupts,
- arbitrary exogenous actions.

As is commonly done in other areas of computer science, concurrent processes are modeled as interleavings of the primitive actions in the component processes. A concurrent execution of two processes is one where the primitive actions in both processes occur, interleaved in some fashion. So in fact, there is never more than one primitive action happening at the same time. As discussed in [3, 22], to model actions that intuitively could occur simultaneously, *e.g.* actions of extended duration, one can use instantaneous start and stop (*i.e.* clipping) actions, where once again interleaving is appropriate.

An important concept in understanding concurrent execution is that of a process becoming *blocked*. If a deterministic process σ is executing, and reaches a point where it is about to do a primitive action a in a situation s but where $Poss(a, s)$ is false (or a wait action $\phi?$, where $\phi(s)$ is false), then the overall execution need not fail as in Golog. In ConGolog, the current interleaving can continue successfully provided that a process other than σ executes next. The net effect is that σ is suspended or blocked, and execution must continue elsewhere.[9]

The ConGolog language is exactly like Golog except with the following additional constructs:

$(\sigma_1 \parallel \sigma_2)$,	concurrent execution
$(\sigma_1 \rangle\!\rangle\, \sigma_2)$,	concurrency with different priorities
$\sigma^{\parallel}$,	concurrent iteration
$<\phi \rightarrow \sigma>$,	interrupt.

$(\sigma_1 \parallel \sigma_2)$ denotes the concurrent execution of the actions σ_1 and σ_2. $(\sigma_1 \rangle\!\rangle\, \sigma_2)$ denotes the concurrent execution of the actions σ_1 and σ_2 with σ_1 having higher priority than σ_2. This restricts the possible interleavings of the two processes: σ_2 executes only when σ_1 is either done or blocked. The next construct, $\sigma^{\parallel}$, is like nondeterministic iteration, but where the instances of σ are executed concurrently rather than in sequence. Finally, $<\phi \rightarrow \sigma>$ is an interrupt. It has two parts: a trigger condition ϕ and a body, σ. The idea is that the body σ will execute some number of times. If ϕ never becomes true, σ will not execute at all. If the interrupt gets control from higher priority processes when ϕ is true, then σ will execute. Once it has completed its execution, the interrupt is ready to be triggered again. This means that a high priority interrupt can take complete control of the execution. For example, $<\mathit{TRUE} \rightarrow \mathit{ringBell}>$ at the highest priority would ring a bell and do nothing else. With interrupts, one can easily write agent programs that can stop whatever task they are doing to handle various concerns as they arise. They are, dare we say, more reactive.

[9]Just as actions in Golog are external (*e.g.* there is no internal variable assignment), in ConGolog, blocking and unblocking also happen externally, via *Poss* and wait actions. Internal synchronization primitives are easily added.

Let us now explain how *Final* and *Trans* are extended to handle these constructs. (Interrupts are handled separately below.) For *Final*, the extension is straightforward:

$$\begin{aligned}
&Final([\sigma_1 \parallel \sigma_2], s) \Leftrightarrow Final(\sigma_1, s) \wedge Final(\sigma_2, s)\\
&Final([\sigma_1 \rangle\!\rangle\, \sigma_2], s) \Leftrightarrow Final(\sigma_1, s) \wedge Final(\sigma_2, s)\\
&Final(\sigma^{\parallel}, s) \Leftrightarrow TRUE
\end{aligned}$$

Observe that the last clause says that it is legal to execute the σ in $\sigma^{\parallel}$ zero times. For *Trans*, we have the following:

$$\begin{aligned}
&Trans([\sigma_1 \parallel \sigma_2], s, \delta, s') \Leftrightarrow\\
&\quad \exists\delta'.\delta = (\delta' \parallel \sigma_2) \wedge Trans(\sigma_1, s, \delta', s') \vee \delta = (\sigma_1 \parallel \delta') \wedge Trans(\sigma_2, s, \delta', s')\\
&Trans([\sigma_1 \rangle\!\rangle\, \sigma_2], s, \delta, s') \Leftrightarrow\\
&\quad \exists\delta'.\delta = (\delta' \rangle\!\rangle\, \sigma_2) \wedge Trans(\sigma_1, s, \delta', s') \quad \vee\\
&\qquad \delta = (\sigma_1 \rangle\!\rangle\, \delta') \wedge Trans(\sigma_2, s, \delta', s') \wedge \neg\exists\delta'', s''.Trans(\sigma_1, s, \delta'', s'')\\
&Trans(\sigma^{\parallel}, s, \delta, s') \Leftrightarrow \exists\delta'.\delta = (\delta' \parallel \sigma^{\parallel}) \wedge Trans(\sigma, s, \delta', s')
\end{aligned}$$

In other words, you single step $(\sigma_1 \parallel \sigma_2)$ by single stepping either σ_1 or σ_2 and leaving the other process unchanged. The $(\sigma_1 \rangle\!\rangle\, \sigma_2)$ construct is identical, except that you are only allowed to single step σ_2 if there is no legal step for σ_1.[10] This ensures that σ_1 will execute as long as it is possible for it to do so. Finally, you single step $\sigma^{\parallel}$ by single stepping σ, and what is left is the remainder of σ as well as $\sigma^{\parallel}$ itself. This allows an unbounded number of instances of σ to be running.

Exogenous actions are primitive actions that may occur without being part of a user-specified program. It is assumed that in the background theory, the user declares using a predicate *Exo* which actions can occur exogenously. A special program is defined for exogenous events:

$$\delta_{EXO} \stackrel{\text{def}}{=} (\pi\, a.Exo(a)?; a)^*$$

Executing this program involves performing zero, one, or more nondeterministically chosen exogenous events. Then, the user's program is made to run concurrently with δ_{EXO}:

$$\delta_{EXO} \parallel \delta$$

This allows exogenous actions whose preconditions are satisfied to occur during the execution of the user's program.

Finally, regarding interrupts, it turns out that these can be explained using other constructs of ConGolog:

$$<\phi \rightarrow \sigma> \stackrel{\text{def}}{=} \textbf{while } Interrupts\_running \textbf{ do} \\ \textbf{if } \phi \textbf{ then } \sigma \textbf{ else } FALSE?$$

[10]It is true, though not immediately obvious, that *Trans** remains properly defined even with these axioms containing negative occurrences of *Trans*. See [4] for details.

To see how this works, first assume that the special fluent $Interrupts\_running$ is always true. When an interrupt $<\phi \rightarrow \sigma>$ gets control, it repeatedly executes σ until ϕ becomes false, at which point it blocks, releasing control to anyone else able to execute. Note that according to the above definition of *Trans*, no transition occurs between the test condition in a while-loop or an if-then-else and the body. In effect, if ϕ becomes false, the process blocks right at the beginning of the loop, until some other action makes ϕ true and resumes the loop. To actually terminate the loop, one uses a special primitive action $stop\_interrupts$, whose only effect is to make $Interrupts\_running$ false. Thus, to execute a program σ containing interrupts, one would actually execute the program

$$\{start\_interrupts\,;\,(\sigma\ \rangle\!\rangle\ stop\_interrupts)\}$$

which has the effect of stopping all blocked interrupt loops in σ at the lowest priority, *i.e.* when there are no more actions in σ that can be executed.

6 COMMUNICATION IN CONGOLOG

Multi-agent applications usually require some kind of inter-agent communication facility. A popular choice is the KQML communication language [5] and its associated tools. However according to Cohen and Levesque [1], the KQML definition has many deficiencies, in particular the lack of a formal semantics. One of our objectives is to show that ConGolog is suitable for various implementation tasks, so here we define our own simple communication toolkit. The specification can be viewed as a generic package that can be included into specific applications. We first specify a set of basic message passing actions; later, some abstract communication actions are defined in terms of the primitives. In a given situation, each agent is taken to have a set of messages it has received and not yet processed; this is modeled using the predicate fluent $\text{MsgRcvd}(agt, sender, msgId, msg, s)$, meaning that in situation s, agt has received a message msg with message ID $msgId$ from $sender$ (which it has yet to process). Note that this is more general than a simple queue; the agent need not process the messages in the order in which they arrive. We assume that message IDs are generated using a global message counter represented by the functional fluent $\text{msgCtr}(s)$ (it is straightforward to generalize this to use agent-relative IDs). There are three primitive action types that operate on these fluents:

- $\text{sendMsg}(agt, recipient, msg)$: agt sends message msg to $recipient$; the current value of the message counter is used as message ID and the message is added to the set of messages $recipient$ has received and not yet processed; the value of the message counter is also incremented;

- $\text{senseMsgs}(agt)$: agt senses what messages he has received and not yet processed; and

- $\text{RMVMSG}(agt, msgId)$: agt removes the message with ID $msgId$ from his set of messages received and not yet processed.

The preconditions of these actions are as follows: $\text{RMVMSG}(agt, msgId)$ is possible in s iff agt has received a message with the given ID and not yet processed it in s:

$$Poss(\text{RMVMSG}(agt, msgId), s) \Leftrightarrow \\ \exists sender, msg\, \text{MSGRCVD}(agt, sender, msgId, msg, s).$$

$\text{SENDMSG}(agt, rcpt, msg)$ and $\text{SENSEMSG}(agt)$ are always possible (we leave out the formal statements).

The effects of these actions are as described above, which yields the following successor state axioms for the MSGRCVD and MSGCTR fluents:

$$\begin{array}{l} Poss(a, s) \Rightarrow \\ [\text{MSGRCVD}(agt, sndr, mId, m, do(a, s)) \Leftrightarrow \\ \;\exists m(a = \text{SENDMSG}(sndr, agt, m) \wedge \text{MSGCTR}(s) = mId \\ \;\vee \text{MSGRCVD}(agt, sndr, mId, m, s) \wedge a \neq \text{RMVMSG}(agt, mId)]. \end{array}$$

$$\begin{array}{l} Poss(a, s) \Rightarrow \\ [\text{MSGCTR}(do(a, s)) = n \Leftrightarrow \\ \;\exists agt, agt', m(a = \text{SENDMSG}(agt, agt', m) \wedge \text{MSGCTR}(s) = n - 1 \\ \;\vee \text{MSGCTR}(s) = n \wedge \neg\exists agt, agt', m\, a = \text{SENDMSG}(agt, agt', m)]. \end{array}$$

Given these primitives, we can now define some useful abstract communication actions:

proc $\text{INFORM}(agt, agt', \phi)$
  $\mathbf{Know}(agt, \phi)?; \text{SENDMSG}(agt, agt', \ulcorner \text{INFORM}(\phi) \urcorner)$
end

proc $\text{INFORMWHETHER}(agt, agt', \phi)$
  $\text{INFORM}(agt, agt', \phi) \mid \text{INFORM}(agt, agt', \neg\phi)$
  $\mid \text{INFORM}(agt, agt', \neg\mathbf{KWhether}(agt, \phi))$
end

proc $\text{REQUEST}(agt, agt', \sigma)$
  $\text{SENDMSG}(agt, agt', \ulcorner \text{REQUEST}(\sigma) \urcorner)$
end

proc $\text{QUERYWHETHER}(agt, agt', \phi)$
  $\text{REQUEST}(agt, agt', \text{INFORMWHETHER}(agt', agt, \phi))$
end

Note that the above definitions use quotation.

We can show that the INFORM abstract communication act behaves as one would expect. Let us take $\text{SINCERE}(sender, rcpt, s)$ as meaning that up to situation s, every INFORM message sent to $rcpt$ by $sender$ was truthfully sent:

$$\text{SINCERE}(sender, rcpt, s) \stackrel{\text{def}}{=}$$
$$\forall s'[do(\text{SENDMSG}(sender, rcpt, \ulcorner \text{INFORM}(\phi)\urcorner), s') \leq s \Rightarrow$$
$$\mathbf{Know}(sender, \phi, s')].$$

Then, we can show that after an agent sends an INFORM(ϕ) message to someone and the recipient senses his messages, the recipient will know that at some prior time the sender knew that ϕ, provided that the recipient knows that he had no messages initially and that the sender has been sincere with him over that period:

Proposition

$$\mathbf{Know}(rcpt, \neg\exists sndr, mId, m\ \text{MSGRCVD}(rcpt, sndr, mId, m, now), S_0) \wedge$$
$$\mathbf{Know}(rcpt, \text{SINCERE}(sndr, rcpt, now),$$
$$do(\text{SENSEMSGS}(rcpt), do(\text{SENDMSG}(sndr, rcpt, \ulcorner \text{INFORM}(\phi)\urcorner), S_0))) \Rightarrow$$
$$\mathbf{Know}(rcpt, \exists s'[s' \leq now \wedge \mathbf{Know}(sndr, \phi, s')],$$
$$do(\text{SENSEMSGS}(rcpt), do(\text{SENDMSG}(sndr, rcpt, \ulcorner \text{INFORM}(\phi)\urcorner), S_0)))$$

It is not possible to prove useful general results about REQUEST, because we have not provided a formalization of goals and intentions. Such a formalization is developed in [28, 29]. In the next section, we show that the simple communication tools specified above are sufficient for developing interesting applications. We are in the process of refining the specification and extending it to handle other types of communicative acts. Eventually, we would like to have a comprehensive communication package that handles most applications.

7 MEETING SCHEDULING AGENTS IN CONGOLOG

To define our simple meeting scheduling system, we first have to complete our specification of the primitive actions that manipulate users' schedule databases. The precondition axiom for ADDTOSCHED was given earlier (1).

For RMVFROMSCHED, we take it to be possible for an agent to remove the activity scheduled for a user at a period iff the agent is the user's schedule manager and there is currently something on the user's schedule for that period:

$$Poss(\text{RMVFROMSCHED}(agt, user, period), s) \Leftrightarrow$$
$$agt = \text{SCHEDULEMANAGER}(user) \wedge$$
$$\exists activ, org\ \text{SCHEDULE}(user, period, activ, org, s)$$

The effects of these actions on the SCHEDULE fluent are captured in the successor state axiom given earlier (4).

We are now ready to use ConGolog to define the behavior of our agents. We start with the "meeting organizer" agents. These will be running the procedure in figure 1. The procedure uses two abbreviations. First, it uses an iteration construct

proc ORGANIZEMEETING($agt, organizer, Participant, period$)
  for $p : Participant(p)$ **do**
    REQUEST(agt, SCHEDULEMANAGER(p), ADDTOSCHEDULE(
      SCHEDULEMANAGER(p), p, $period$, MEETING, $organizer$));
    QUERYWHETHER(agt, SCHEDULEMANAGER(p),
      AGREEDTOMEET($p, agt, period, organizer$))
  endFor;
  while $\neg$**KWhether**($agt, \forall p[Participant(p) \Rightarrow$
      AGREEDTOMEET($p, agt, period, organizer$)]) **do**
    SENSEMSG(agt);
    for $mId : \exists sn, m$ MSGRCVD(agt, sn, mId, m) **do** RMVMSG(agt, mId) **endFor**
  endWhile;
  INFORMWHETHER($agt, organizer$,
    $\forall p[Participant(p) \Rightarrow$ AGREEDTOMEET($p, agt, period, organizer$)]);
  if $\neg\forall p[Participant(p) \Rightarrow$ AGREEDTOMEET($p, agt, period, organizer$)] **then**
    % release participants from commitment
    for $p : Participant(p)$ **do**
      REQUEST(agt, SCHEDULEMANAGER(p),
        RMVFROMSCHED(SCHEDULEMANAGER(p), p, $period$))
    endFor
  endIf
endProc

Figure 1. Procedure run by the "meeting organizer" agents.

for $x : \phi(x)$ **do** $\delta(x)$ **endFor** that performs $\delta(x)$ for all x's such that $\phi(x)$ holds (at the beginning of the loop).[11] Secondly, it uses the abbreviation:

AGREEDTOMEET($participant, requester, period, organizer, s$) $\stackrel{\text{def}}{=} \exists s'($
$do($ADDTOSCHEDULE(SMP, $participant, period$, MEETING, $organizer$), $s') \leq s$
$\wedge$ **Know**(SMP,
  $\exists mId$ MSGRCVD(SMP, $requester, mId$, ⌜REQUEST(SMP,
    ADDTOSCHEDULE(SMP, $participant, period$, MEETING, $organizer$))⌝,
   now),
  s')),

where SMP $\stackrel{\text{def}}{=}$ SCHEDULEMANAGER($participant$).

Thus, we take the participant to have agreed to a meeting request iff at some prior situation, the participant's schedule manager added the meeting to his/her schedule

[11] **for** $x : \phi(x)$ **do** $\delta(x)$ **endFor** is defined as:

[**proc** $P(Q)$ /* where P is a new predicate variable */
  if $\exists y\, Q(y)$ **then** $\pi\, y, R\, [Q(y) \wedge \forall z(R(z) \Leftrightarrow Q(z) \wedge z \neq y)?; \delta(y); P(R)]$ **endIf**
endProc;
$\pi\, Q\, [\forall z(Q(z) \Leftrightarrow \phi(z))?; P(Q)]]$

```
proc MANAGESCHEDULE(agt, user)
  ⟨ requester, msgId, period :
      Know(agt, MSGRCVD(agt, requester, msgId,
          ⌜REQUEST(RMVFROMSCHEDULE(agt, user, period))⌝)) → [
      if agt = SCHEDULEMANAGER(user) ∧
          ∃activ SCHEDULE(user, period, activ, OWNER(requester))
          then RMVFROMSCHEDULE(agt, user, period) endIf;
      RMVMSG(agt, msgId)] ⟩
  ⟫
  ⟨ requester, msgId, period, activ, organizer :
      Know(agt, MSGRCVD(agt, requester, msgId, ⌜ REQUEST(
          ADDTOSCHEDULE(agt, user, period, activ, organizer))⌝)) → [
      if PERMITTEDTOADDTOSCHED(agt, organizer) ∧
          Poss(ADDTOSCHEDULE(agt, user, period, activ, organizer))
          then ADDTOSCHEDULE(agt, user, period, activ, organizer) endIf;
      RMVMSG(agt, msgId)] ⟩
  ||
  ⟨ queryer, msgId, p :
      Know(agt, MSGRCVD(agt, queryer, msgId, ⌜ QUERYWHETHER(p)⌝)) → [
      INFORMWHETHER(agt, queryer, p);
      RMVMSG(agt, msgIg)] ⟩
  ||
  ⟨ ∃informer, msgId, p
      Know(agt, MSGRCVD(agt, informer, msgId, ⌜ INFORM(p)⌝)) → [
      % if message is INFORM(p), nothing to do
      RMVMSG(agt, msgId)] ⟩
  ⟫
  ⟨ True → [      % if no new messages
     SENSEMSGS(agt)] ⟩
endProc
```

Figure 2. Procedure run by the "schedule manager" agents.

while knowing that it had received the request.[12]

Meanwhile, users' "schedule manager" agents run the procedure in figure 2[13]. Because these agents are "event-driven", we program them as a set of interrupts running concurrently. Interrupts handling "more urgent" events are assigned a higher priority. For instance in the example, requests to remove an activity from the schedule are handled at the highest priority in order to minimize the chance of scheduling conflicts. Note that the procedure given does not handle cases where a user wants to be released from a commitment.

[12]This is quite a simplistic way of modeling agreement to a request. We should for instance, talk about the most recent instance of the request.

[13]Here we use the abbreviation $\langle\, \vec{x} : \phi \rightarrow \sigma \,\rangle \stackrel{\text{def}}{=} \langle\, \exists \vec{x}\, \phi \rightarrow \pi \vec{x}.[\phi?; \sigma] \,\rangle$.

To run a meeting scheduling system, one could, for example, give the following program to the ConGolog interpreter:

$$\begin{array}{l}\text{MANAGESCHEDULE}(\text{SM}_1, \text{USER}_1) \parallel \\ \text{MANAGESCHEDULE}(\text{SM}_2, \text{USER}_2) \parallel \\ \text{MANAGESCHEDULE}(\text{SM}_3, \text{USER}_3) \parallel \\ \text{ORGANIZEMEETING}(\text{MO}_1, \text{USER}_1, \{\text{USER}_1, \text{USER}_3\}, \text{NOON}) \parallel \\ \text{ORGANIZEMEETING}(\text{MO}_2, \text{USER}_2, \{\text{USER}_2, \text{USER}_3\}, \text{NOON}).\end{array}$$

Here, the meeting organizers will both try to obtain USER_3's agreement for a meeting at noon; there will thus be two types of execution sequences, depending on who obtains this agreement.

8 IMPLEMENTATION AND EXPERIMENTATION

Prototype interpreters have been implemented in Prolog for both Golog [14] and ConGolog [3]. The implementations require that the program's precondition axioms, successor state axioms, and axioms about the initial situation be expressible as Prolog clauses. This is a limitation of the implementations, not the theory.

For programs that are embedded in real environments and run for long periods or perform sensing, it is often advantageous to interleave the interpretation of the program with its execution. In the current implementation, whenever the interpreter reaches a sensing action, it commits to the primitive actions generated so far and executes them, performs the sensing action, and generates exogenous actions to appropriately update the values of the sensed fluent. One can also add directives to programs to force the interpreter to commit when it gets to that point in the program. As well, whenever the interpreter commits and executes part of the program, it rolls its database forward to reflect the execution of the actions, and the situation reached behaves like a new initial situation [16].[14]

Note however, that committing to a sequence of action as soon as a sensing action is reached could lead to problems when the program to be executed is nondeterministic. Perhaps we are on a branch that does not lead to a final situation. To avoid this, we need to lookahead over sensing actions and generate a kind of conditional plan that is guaranteed to lead to a final situation no matter how the sensing turns out. A prototype interpreter that does this kind of lookahead has been implemented. The account of planning in the presence of sensing developed in [13] clarifies these issues. A general account of when an agent *knows how* to execute a program (i.e., of the knowledge preconditions of actions) has also been developed [10]. There are still some discrepancies between the Golog implementation and our theory of agency in the way knowledge, sensing, exogenous events,

[14] To evaluate whether a condition holds in a given situation, Golog regresses the condition to the initial situation and then uses the axioms about the initial situation to evaluate the regressed condition. This becomes less efficient as the number of action grows. After a while it becomes necessary to roll the database forward.

and the relation between planning and execution are treated. We are working to bridge this gap.

Experiments in the use of Golog and ConGolog to develop various applications have been conducted. Our longest running application project is in the area of robotics. High-level controllers have been programmed in Golog and ConGolog to get a robot to perform mail delivery in an office environment [9, 12]. These have been used to drive RWI-B21 robots at the University of Toronto and the University of Bonn and RWI-B12 and Nomad200 robots at York University.

Our first multi-agent application involved a personal banking assistant system [11, 26]. Users can perform transactions (in a simulated financial environment) using the system and have it monitor their financial situation for particular conditions and take action when they arise, either by notifying them or by performing transactions on their behalf. The system was implemented as a collection of Golog agents that communicate using TCP/IP.

A version of the meeting scheduling application described in this paper has also been implemented. The agents are realized as a set of processes in a single ConGolog program. Associated Tcl/Tk processes are used to implement the agents' user interfaces. The behavior of the implemented meeting organizer agent is more sophisticated than that of the simple meeting organizer in the previous section. When a meeting request is rejected by one of the participants' agent, the meeting organizer agent will collect information about the schedule of all users involved in the conflict and plan a set of rescheduling actions that would resolve it; if a plan is found, the meeting organizer will then request the agents involved to actually perform the rescheduling actions.

Another project under way involves developing ConGolog-based tools for modeling business and organizational processes [33]. In contrast to the operational view of conventional modeling tools, ConGolog takes a logical view of processes. This should prove advantageous when it comes to modeling system behavior under incompletely known conditions and proving properties about the system. So far we have used ConGolog to model a simple mail order business, as well as a section of a nuclear power plant; in the latter we model potential faults that can occur and how the operators deal with them.

9 DISCUSSION

One project that is closely related to ours is work on the AGENT-0 programming language [30]. But it is hard to do a systematic comparison between ConGolog and AGENT-0 as there are numerous differences. The latter includes a model of commitments and capabilities, and has simple communication acts built-in; its agents all have a generic rule-based architecture; there is also a global clock and all beliefs are about time-stamped propositions. However, there is no automatic maintenance of the agents' beliefs based on a specification of primitive actions as in ConGolog and only a few types of complex actions are handled; there also seems to be less emphasis on having a complete formal specification of the system.

Another agent language based on a logic is Concurrent MetateM [6]. Here, each agent's behavior is specified in a subset of temporal logic. The specifications are executed using iterative model generation techniques. A limitation of the approach is that neither the interactions between agents nor their mental states are modeled within the logic. In [31], Wooldridge proposes a richer logical language where an agent's knowledge and choices could be specified; he also sketches how model generation techniques could be used to synthesize automata satisfying the specifications. This follows the situated automata view of Rosenschein and Kaelbling [25], which allows knowledge to be attributed to agents without any commitment to a symbolic architecture.

We believe that much of the brittleness of current AI systems derives from a failure to provide an adequate theoretical account of the task to be accomplished. Accordingly, we are trying to develop agent design tools that are based on a solid theoretical foundation. We think the framework we have developed so far represent a significant step towards this objective. But clearly, more work is required on both implementation issues and the underlying theory. As mentioned earlier, we are examining various ways of supporting deliberation in the presence of sensing and exogenous actions, as well as the interleaving of deliberation with execution. We are also looking at mechanisms to facilitate the use of Golog nondeterminism for planning. Also under investigation are issues such as handling uncertainty and belief revision, as well as agent-relative (indexical) representations for robotics. A version of Golog that supports temporal constraints has also been developed [23].

In terms of its support for multi-agent interaction, the current framework is rather limited. When agents interact with others without having complete knowledge of the situation, it is advantageous for them to view other agents as having goals, intentions, commitments, and abilities, and as making rational choices. This allows them to anticipate and influence the behavior of other agents, and cooperate with them. It also supports an abstract view of communication acts as actions that affect other agents' mental states as opposed to mere message passing. We have started extending our framework to model goals, intentions, ability, and rational choice [28, 29, 10], and considering possible implementation mechanisms. Communication raises numerous issues: How far should agent design tools go in handling intricacies that arise in human communication (e.g., deception, irony)? What's a good set of communication primitives and how do we implement them? With respect to coordination, what sort of infrastructure should we provide? Can we come up with a set of general purpose coordination policies? We hope to examine all these questions. Of course in the end, the usefulness of our approach will have to be evaluated empirically.

10 ACKNOWLEDGMENTS

This paper is a much revised version of "Foundations of a Logical Approach to Agent Programming" which appeared in *Intelligent Agents Volume II — Proceedings of the 1995 Workshop on Agent Theories, Architectures, and Languages (ATAL-95)*. The work described involved contributions by many people over a number of years, in particular, Giuseppe De Giacomo, Fangzhen Lin, Daniel Marcu, Richard Scherl, and David Tremaine. This research received financial support from the Information Technology Research Center (Ontario, Canada), the Institute for Robotics and Intelligent Systems (Canada), and the Natural Science and Engineering Research Council (Canada). Many of our team's papers are available at:

`http://www.cs.toronto.edu/~cogrobo/.`

Yves Lespérance (York University, Canada)
Hector Levesque (University of Toronto, Canada)
Raymond Reiter (University of Toronto, Canada)

REFERENCES

[1] Philip R. Cohen and Hector J. Levesque. Communicative actions for artificial agents. In Victor Lesser and Les Gasser, editors, *Proceedings of the First International Conference on Multiagent Systems*, San Francisco, CA, June 1995. AAAI Press/MIT Press.

[2] Giuseppe De Giacomo, Yves Lespérance, and Hector J. Levesque. Reasoning about concurrent execution, prioritized interrupts, and exogenous actions in the situation calculus. In *Proceedings of the Fifteenth International Joint Conference on Artificial Intelligence*, pp. 1221–1226, Nagoya, August, 1997.

[3] Giuseppe De Giacomo, Yves Lespérance, and Hector J. Levesque. *ConGolog*, a Concurrent Programming Language based on the Situation Calculus: Language and Implementation. Submitted, 1998.

[4] Giuseppe De Giacomo, Yves Lespérance, and Hector J. Levesque. *ConGolog*, a Concurrent Programming Language based on the Situation Calculus: Foundations. Submitted, 1998.

[5] ARPA Knowledge Sharing Initiative External Interfaces Working Group. Specification of the KQML agent-communication language. Working Paper, June 1993.

[6] M. Fisher. A survey of Concurrent METATEM — the language and its applications. In D. M. Gabbay and H. J. Ohlbach, editors, *Temporal Logic — Proceedings of the First International Conference (LNAI Volume 827)*, pages 480–505. Springer-Verlag, July 1994.

[7] C.C. Green. Theorem proving by resolution as a basis for question-answering systems. In B. Meltzer and D. Michie, editors, *Machine Intelligence*, volume 4, pages 183–205. American Elsevier, New York, 1969.

[8] M. Hennessy. *The Semantics of Programming Languages*. John Wiley & Sons, 1990.

[9] Yves Lespérance, Hector J. Levesque, Fangzhen Lin, Daniel Marcu, Raymond Reiter, and Richard B. Scherl. A logical approach to high-level robot programming – a progress report. In Benjamin Kuipers, editor, *Control of the Physical World by Intelligent Agents, Papers from the 1994 AAAI Fall Symposium*, pages 109–119, New Orleans, LA, November 1994.

[10] Yves Lespérance, Hector J. Levesque, Fangzhen Lin, and Richard B. Scherl. Ability and knowing how in the situation calculus. Unpublished manuscript, 1997.

[11] Yves Lespérance, Hector J. Levesque, and Shane J. Ruman. An experiment in using Golog to build a personal banking assistant. In Lawrence Cavedon, Anand Rao, and Wayne Wobcke, editors, *Intelligent Agent Systems: Theoretical and Practical Issues (Based on a Workshop Held at PRICAI '96 Cairns, Australia, August 1996)*, volume 1209 of *LNAI*, pages 27–43. Springer-Verlag, 1997.

[12] Yves Lespérance, Kenneth Tam, and Michael Jenkin. Reactivity in a Logic-Based Robot Programming Framework. In *Cognitive Robotics — Papers from the 1998 AAAI Fall Symposium*, pp. 98–105, Orlando, FL, October, 1998, Technical Report FS-98-02, AAAI Press.

[13] Hector J. Levesque. What is planning in the presence of sensing? In *Proceedings of the Thirteenth National Conference on Artificial Intelligence*, pages 1139–1146, Portland, OR, August 1996.

[14] Hector J. Levesque, Raymond Reiter, Yves Lespérance, Fangzhen Lin, and Richard B. Scherl. GOLOG: A logic programming language for dynamic domains. *Journal of Logic Programming*, **31**, 59–84, 1997.

[15] Fangzhen Lin and Raymond Reiter. State constraints revisited. *Journal of Logic and Computation*, 4(5):655–678, 1994.

[16] Fangzhen Lin and Raymond Reiter. How to progress a database. *Artificial Intelligence*, **92**, 131–167, 1997.

[17] John McCarthy and Patrick Hayes. Some philosophical problems from the standpoint of artificial intelligence. In B. Meltzer and D. Michie, editors, *Machine Intelligence*, volume 4, pages 463–502. Edinburgh University Press, Edinburgh, UK, 1979.

[18] Robert C. Moore. A formal theory of knowledge and action. In J. R. Hobbs and Robert C. Moore, editors, *Formal Theories of the Common Sense World*, pages 319–358. Ablex Publishing, Norwood, NJ, 1985.

[19] G. Plotkin. A structural approach to operational semantics. Technical Report DAIMI-FN-19, Computer Science Dept. Aarhus Univ. Denmark, 1981.

[20] Raymond Reiter. The frame problem in the situation calculus: A simple solution (sometimes) and a completeness result for goal regression. In Vladimir Lifschitz, editor, *Artificial Intelligence and Mathematical Theory of Computation: Papers in Honor of John McCarthy*, pages 359–380. Academic Press, San Diego, CA, 1991.

[21] Raymond Reiter. Proving properties of states in the situation calculus. *Artificial Intelligence*, pages 337–351, December 1993.

[22] Raymond Reiter. Natural actions, concurrency and continuous time in the situation calculus. In *Proc. of the 5th Int. Conf. on Principles of Knowledge Representation and Reasoning (KR'96)*, pages 2–13, 1996.

[23] Raymond Reiter. Sequential, temporal GOLOG. In A.G. Cohn and L.K. Schubert, editors, *Principles of Knowledge Representation and Reasoning: Proceedings of the Sixth International Conference (KR'98)*, pages 547–556, Trento, Italy, Morgan Kaufmann, 1998.

[24] D. Riecken (editor). Communications of the ACM **37** (7), special issue on intelligent agents, July 1994.

[25] Stanley J. Rosenschein and Leslie P. Kaelbling. A situated view of representation and control. *Artificial Intelligence*, 73:149–173, 1995.

[26] Shane J. Ruman. GOLOG as an agent-programming language: Experiments in developing banking applications. Master's thesis, Department of Computer Science, University of Toronto, 1996.

[27] Richard B. Scherl and Hector J. Levesque. The frame problem and knowledge-producing actions. In *Proceedings of the Eleventh National Conference on Artificial Intelligence*, pages 689–695, Washington, DC, July 1993. AAAI Press/The MIT Press.

[28] Steven Shapiro, Yves Lespérance, and Hector J. Levesque. Goals and rational action in the situation calculus — a preliminary report. In *Working Notes of the AAAI Fall Symposium on Rational Agency: Concepts, Theories, Models, and Applications*, pages 117–122, Cambridge, MA, November 1995.

[29] Steven Shapiro, Yves Lespérance, and Hector J. Levesque. Specifying Communicative Multi-Agent Systems with ConGolog. In *Working Notes of the AAAI Fall 1997 Symposium on Communicative Action in Humans and Machines*, Cambridge, MA, November, 1997, AAAI Press.

[30] Yoav Shoham. Agent-oriented programming. *Artificial Intelligence*, 60(1):51–92, 1993.

[31] Michael J. Wooldridge. Time, knowledge, and choice. In M. Wooldridge, J. P. Müller, and M. Tambe, editors, *Intelligent Agents Volume II — Proceedings of the 1995 Workshop on Agent Theories, Architectures, and Languages (ATAL-95)*, Lecture Notes in Artificial Intelligence. Springer-Verlag, 1996.

[32] Michael J. Wooldridge and Nicholas R. Jennings. Intelligent agents: Theory and practice. *Knowledge Engineering Review*, 10(2), 1995.

[33] Eric K.S. Yu, John Mylopoulos, and Yves Lespérance. AI models for business process reengineering. *IEEE Expert*, 11:16–23, August 1996.

APPLIED LOGIC SERIES

1. D. Walton: *Fallacies Arising from Ambiguity.* 1996 ISBN 0-7923-4100-7
2. H. Wansing (ed.): *Proof Theory of Modal Logic.* 1996 ISBN 0-7923-4120-1
3. F. Baader and K.U. Schulz (eds.): *Frontiers of Combining Systems.* First International Workshop, Munich, March 1996. 1996 ISBN 0-7923-4271-2
4. M. Marx and Y. Venema: *Multi-Dimensional Modal Logic.* 1996 ISBN 0-7923-4345-X
5. S. Akama (ed.): *Logic, Language and Computation.* 1997 ISBN 0-7923-4376-X
6. J. Goubault-Larrecq and I. Mackie: *Proof Theory and Automated Deduction.* 1997 ISBN 0-7923-4593-2
7. M. de Rijke (ed.): *Advances in Intensional Logic.* 1997 ISBN 0-7923-4711-0
8. W. Bibel and P.H. Schmitt (eds.): *Automated Deduction - A Basis for Applications.* Volume I. Foundations - Calculi and Methods. 1998 ISBN 0-7923-5129-0
9. W. Bibel and P.H. Schmitt (eds.): *Automated Deduction - A Basis for Applications.* Volume II. Systems and Implementation Techniques. 1998 ISBN 0-7923-5130-4
10. W. Bibel and P.H. Schmitt (eds.): *Automated Deduction - A Basis for Applications.* Volume III. Applications. 1998 ISBN 0-7923-5131-2
    (Set vols. I-III: ISBN 0-7923-5132-0)
11. S.O. Hansson: *A Textbook of Belief Dynamics.* Theory Change and Database Updating. 1999 including *Solutions to exercises.* 1999. Hb: ISBN 0-7923-5324-2; Set: (Pb): ISBN 0-7923-5329-3
12. R. Pareschi and B. Fronhöfer (eds.): *Dynamic Worlds.* From the Frame Problem to Knowledge Management. 1999 ISBN 07923-5535-0
13. D.M. Gabbay and H. Wansing (eds.): *What is Negation?.* 1999 ISBN 0-7923-5569-5
14. M. Wooldridge and A. Rao (eds.): *Foundations of Rational Agency.* 1999 ISBN 0-7923-5601-2

KLUWER ACADEMIC PUBLISHERS – DORDRECHT / BOSTON / LONDON